Analysis of Surveyor 3 material and photographs returned by Apollo 12

Scientific and Technical Information Office   1972
NATIONAL AERONAUTICS AND SPACE ADMINISTRATION
Washington, D.C.

Preface

Surveyor 3 was one of five automated spacecraft that successfully soft-landed and operated on the lunar surface, acquired a vast amount of new scientific and engineering data, and provided a firm foundation for subsequent manned landings on the Moon.

When we designed and launched these Surveyors, there was no plan for them to be visited by astronauts in subsequent manned missions. Some of us, however, had the quiet hope that, at some later date, astronauts would walk up to a landed Surveyor, examine and photograph it and the surrounding terrain, and remove and return to Earth selected components for engineering and scientific studies.

Such an opportunity was provided by the Apollo 12 mission. Thirty-one months after Surveyor 3 landed, the crew of Apollo 12 photographed the spacecraft and its landing site, and removed and brought back a number of selected components. These parts, which included the television camera, were analyzed to determine the effects on the hardware of the long exposure to the lunar environment.

The returned material and photographs have been studied and evaluated by 40 teams of engineering and scientific investigators over a period of more than 1 year. A few tasks are still in process and several proposals for additional studies have been received.

This report represents a compilation of the main engineering and scientific results to date.

Engineering studies of the television camera show that the complex electromechanical components, optics, and solid-state electronics were remarkably resistant to the severe lunar surface environment over 32 lunar day/night cycles with their extremes of temperature and long exposure to solar and cosmic radiation. These results indicate that the state of technology, even as it existed some years ago, is capable of producing reliable hardware that makes feasible long-life lunar and planetary installations.

Scientific studies of the returned Surveyor parts provide new data in many fields and provide further confirmation that specially designed recoverable experiments should have great value in the study of the space environment.

BENJAMIN MILWITZKY
Assistant Director, Engineering
(Special Projects)
Apollo Program

MAY 1971

Contents

		PAGE
I	INTRODUCTION W. F. Carroll, R. Davis, M. Goldfine, S. Jacobs, L. D. Jaffe, L. Leger, B. Milwitzky, and N. L. Nickle	1
II	SUMMARY AND CONCLUSIONS N. L. Nickle and W. F. Carroll	9
III	RETURNED SURVEYOR 3 HARDWARE: ENGINEERING RESULTS W. F. Carroll, P. M. Blair, Jr., E. I. Hawthorne, S. Jacobs, and L. Leger	15
IV	SPACECRAFT CHANGES	23

PART A. Lunar Dust and Radiation Darkening of Surveyor 3 Surfaces   23
W. F. Carroll and P. M. Blair, Jr.

PART B. Characterization of Dust on Clear Filter From Returned Surveyor 3 Television Camera   29
D. M. Robertson, E. L. Gafford, H. Tenny, and R. S. Strebin, Jr.

PART C. Debris on the Surveyor 3 Mirror   46
M. H. Carr and S. J. Proudfoot

PART D. Dynamic Considerations of Dust on the Television Camera Mirror   51
N. L. Nickle

PART E. Changes in Optical Properties of the Surveyor 3 Camera   60
J. Rennilson, H. Holt, and K. Moll

PART F. Particle Impact and Optical Property Analysis of the Surfaces of Surveyor 3 Materials   76
D. L. Anderson, B. E. Cunningham, R. G. Dahms, and R. G. Morgan

PART G. Examination of Surveyor 3 Parts With the Scanning Electron Microscope and Electron Microprobe   89
A. A. Chodos, J. R. Devaney, and K. C. Evans

PART H. Sputter-Ion Source Mass Spectrometer Analysis of Samples Cut From the Surveyor 3 Camera   91
F. G. Satkiewicz and F. F. Marmo

PART I. Blowing of Lunar Soil by Apollo 12: Surveyor 3 Evidence   94
L. D. Jaffe

			PAGE
	Part J.	Low-Temperature Oxygen-Plasma Effects on Surveyor Plasmo-Clay Coating R. B. Gillette	96
	Part K.	Examination of the Surveyor 3 Surface Sampler Scoop R. F. Scott and K. A. Zuckerman	100
	Part L.	Movement of the Surveyor 3 Spacecraft R. F. Scott, T.-D. Lu, and K. A. Zuckerman	114
	Part M.	Analysis of Surveyor 3 Television Cable F. C. Gross and J. J. Park	119
V	ORGANIC CONTAMINATION ANALYSIS		127
		High-Resolution Mass Spectrometric Analysis of Surface Organics on Selected Areas of Surveyor 3 B. R. Simoneit and A. L. Burlingame	127
VI	MICROMETEORITE IMPACT ANALYSES		143
	Part A.	Primary and Secondary Micrometeoroid Impact Rate on the Lunar Surface: A Direct Measurement D. Brownlee, W. Bucher, and P. Hodge	143
	Part B.	Preliminary Examination of Surveyor 3 Components for Impact Phenomena D. S. Hallgren, A. T. Laudate, R. P. Schwarz, W. D. Radigan, and C. L. Hemenway	151
	Part C.	Examination of Sample of Surveyor 3 Strut for Meteoroid Impacts L. Zernow	153
	Part D.	Surface Studies on Surveyor 3 Tubing Sections E. A. Buvinger	154
	Part E.	Results of Examination of the Returned Surveyor 3 Samples for Particulate Impacts B. G. Cour-Palais, R. E. Flaherty, R. W. High, D. J. Kessler, D. S. McKay, and H. A. Zook	158
	Part F.	Microcrater Investigations on Surveyor 3 Material E. Schneider, G. Neukum, A. Mehl, and H. Fechtig	167
	Part G.	Lunar Surface: Changes in 31 Months and Micrometeoroid Flux L. D. Jaffe	173
VII	RADIOACTIVITY AND RADIATION DAMAGE ANALYSES		177
	Part A.	Examination of Returned Surveyor 3 Camera Visor for Alpha Radioactivity T. E. Economou and A. L. Turkevich	177
	Part B.	Preliminary Results on Tritium in Surveyor 3 Material E. L. Fireman	180

B. G. Cour-Palais, R. E. Flaherty, R. W. High, D. J.

			PAGE
	PART C.	High-Voltage Transmission Microscopy of Surveyor 3 Camera Shrouds R. M. Fisher, W. R. Duff, L. E. Thomas, and S. V. Radcliffe	184
	PART D.	Solar and Galactic Cosmic-Ray Exposure of Surveyor 3 as Determined From Cosmogenic Radionuclide Measurements L. A. Rancitelli, R. W. Perkins, N. A. Wogman, and W. D. Felix	196
VIII	SOLAR WIND RARE GAS ANALYSIS		201
	Trapped Solar Wind Helium and Neon in Surveyor 3 Material F. Bühler, P. Eberhardt, J. Geiss, and J. Schwarzmüller		201
IX	PARTICLE TRACK ANALYSES		
	PART A.	Solar Particle Tracks in Glass From Surveyor 3 G. Crozaz and R. M. Walker	209
	PART B.	Very Heavy Solar Cosmic Rays: Energy Spectrum and Implications for Lunar Erosion R. L. Fleischer, H. R. Hart, Jr., and G. M. Comstock	213
	PART C.	Enhanced Emission of Iron Nuclei in Solar Flares P. B. Price, I. D. Hutcheon, R. Cowsik, and D. J. Barber	217
	PART D.	Solar Flares, the Lunar Surface, and Gas-Rich Meteorites D. J. Barber, R. Cowsik, I. D. Hutcheon, P. B. Price, and R. S. Rajan	220
X	SOIL PROPERTY ANALYSES		227
	PART A.	Bearing Strength of the Lunar Soil L. D. Jaffe	227
	PART B.	Cracking of the Lunar Soil L. D. Jaffe	233
	PART C.	Whiskers on the Moon D. Brownlee, W. Bucher, and P. Hodge	236
XI	MICROBE SURVIVAL ANALYSES		239
	PART A.	Surveyor 3: Bacterium Isolated From Lunar-Retrieved Television Camera F. J. Mitchell and W. L. Ellis	239
	PART B.	Microbiological Sampling of Returned Surveyor 3 Electrical Cabling M. D. Knittel, M. S. Favero, and R. H. Green	248
	APPENDIX A.	SPACECRAFT ORIENTATION AND EXPOSURE TO ENVIRONMENT N. L. Nickle	253

		PAGE
APPENDIX B.	SURVEYOR 3 MATERIAL ANALYSIS PLAN	261
	N. L. Nickle and W. F. Carroll	
APPENDIX C.	SURVEYOR TELEVISION CAMERA — SELECTED MATERIALS AND ELECTRONIC COMPONENTS	278
	W. F. Carroll	
APPENDIX D.	CATALOG OF SURVEYOR 3 PHOTOGRAPHS FROM APOLLO 12	284
APPENDIX E.	INDEX OF CONTRIBUTING AUTHORS	294

I. Introduction

W. F. Carroll, R. Davis, M. Goldfine, S. Jacobs, L. D. Jaffe,
L. Leger, B. Milwitzky, and N. L. Nickle

In November 1969, the Apollo 12 astronauts visited the Surveyor 3 spacecraft, which had landed on the lunar surface 31 months earlier. During the visit, the astronauts examined and photographed the spacecraft and removed selected parts and enclosed soil for return to Earth. The parts, soil, crew observations and photographs have been evaluated to obtain information concerning the spacecraft hardware that could be of value to engineering design and to obtain scientific information that could provide a better understanding of lunar and space environments. This evaluation has been undertaken by individuals and groups in various organizations in the United States and abroad. A summary of the engineering and scientific results is presented in chapter II of this document.

The primary examination of the hardware relative to engineering performance was conducted by Hughes Aircraft Co. (HAC) under contracts from the Jet Propulsion Laboratory (JPL) and the Manned Spacecraft Center (MSC). The evaluation team included many key people who had been associated with the initial design, test, and operation of the Surveyor spacecraft series. The results of this effort are summarized in chapter III.

The scientific investigations were conducted by 40 teams of specialists in the fields of surface changes and characteristics, organic chemistry, micrometeorite impacts, naturally induced radioactivity, radiation damage, solar wind rare gases, particle tracks, soil characteristics, and microbe analysis. Results of most of these investigations are contained in chapters IV through XI of this document. The findings are presented as individual articles written by the investigators. Because the articles were written independently of one another, some differences in interpretation may exist among them. Some of the investigations are not yet complete and will be reported in appropriate technical journals.

Rationale and Objectives

The reasons for biasing an Apollo mission to land near a Surveyor spacecraft on the Moon and for expending extravehicular activity (EVA) time to examine, photograph, and collect material from a Surveyor and its immediate vicinity, and for returning this material, can be summarized as

(1) To improve the technology for designing, fabricating, and testing future spacecraft and lunar and planetary stations.

(2) To increase the understanding of lunar surface processes and rates by determining the changes that occurred on the lunar surface and in Surveyor 3 during 31 months in the lunar environment.

(3) To check the validity of the techniques used for interpretation of remote observations and analyses of lunar and planetary surfaces.

From observations made by the astronauts, from photographs of the Surveyor and rephotographs of lunar areas televised by Surveyor, and from examinations of returned material, it was expected [1] that information could be obtained concerning:

(1) Effects on spacecraft surfaces of micrometeoroid impact, physical changes due to solar and cosmic radiation, and effects of thermal cycling.

[1] Memorandum, B. Milwitzky (NASA) to Director, Apollo Lunar Exploration Office, NASA Headquarters, Jan. 10, 1969.

(2) Extent of vacuum welding of movable spacecraft elements.

(3) Effects of prolonged exposure on as many types of spacecraft material and components as possible.

(4) Spacecraft movement due to thermal cycling and to seismic disturbances.

(5) Dust deposits on the spacecraft.

(6) Evidence of creep of fine surface material.

(7) New craters, blocks, or other changes in surface features.

(8) Changes in footpad imprints, surface sampler trenches, vernier-engine blast areas, and other disturbances of the lunar surface made by Surveyor during the intervening time.

(9) Changes in the optical characteristics of darker material, which appeared wherever the lunar surface was disturbed by Surveyor.

(10) Correlation between film and remotely controlled television data with regard to lunar photometry, colorimetry, and polarimetry.

(11) Comparison of the bearing strength and other mechanical properties of lunar fines returned from the vicinity of the Surveyor with those properties obtained by remote-control techniques during the Surveyor mission.

(12) Assessment of the original analyses and interpretations made by the Surveyor Science Team by means of the returned lunar rocks and soil viewed by Surveyor.

Surveyor 3 and the Planning of Apollo 12

Surveyor 3 landed on the Moon on April 20, 1967. The landing site was in the southwest part of Oceanus Procellarum, about 370 km south of the crater Copernicus at selenographic coordinates, in the ACIC coordinate system, 2.99° S, 23.34° W, or, in the AMS coordinate system, 3°12'0.36" S, 23°22'54.2" W. The spacecraft came to rest in a subdued, rounded crater about 200 m in diameter and was inclined about 12° to the horizontal on the eastern slope of the crater. Details of the landed spacecraft's orientation are given in appendix A. The results of the Surveyor 3 mission are contained in references 1 and 2.

The decision to target Apollo 12 to land next to a Surveyor was based on two primary considerations: (1) the desire to use a landed spacecraft as a target to demonstrate a point-landing capability, (2) the engineering and scientific information to be gained from the return of Surveyor components and photographs of its landing site (ref. 3). Surveyor 3 was chosen as the specific target for Apollo 12 because it was located in one of the prime sites previously established for the Apollo missions.

After the decision to land Apollo 12 near Surveyor 3, NASA developed a plan for astronaut activities at the Surveyor 3 site. The planning was a low-level effort, as the chance of returning Surveyor material was considered slight. Inputs on specific tasks that would provide the most valuable engineering information were prepared by HAC (which had designed and built Surveyor), by JPL, and by MSC. Inputs as to tasks that would provide the most valuable scientific information were obtained primarily through JPL, which, at NASA's request, contacted a number of scientists, especially those familiar with Surveyor 3. Many valuable suggestions were received from individual scientists and engineers. These suggestions were first reviewed and screened by the organizations mentioned.[2,3] They were examined and screened again at MSC for compatibility with astronaut and other Apollo constraints and with the mission schedule and time line.

The tasks finally selected were:

(1) Obtain stereo photographs giving general views of the lunar surface close to Surveyor and of specific pre-selected lunar objects televised by Surveyor (dust fillets around rock, layered flat rock).

(2) Obtain stereo photographs of lunar surface disturbances produced by Surveyor soil mechanics surface sampler and footpads.

(3) Kick up fresh material near previously disturbed and undisturbed areas. Photograph together, to reveal effects of exposure on the albedo of disturbed lunar soil.

(4) Inspect and photograph Surveyor from all sides.

(5) Inspect and photograph polished aluminum and gold on vernier engine, glass tops of

---

[2] Letter, L. D. Jaffe (JPL) to Director, Apollo Lunar Exploration Office, NASA Headquarters, Aug. 7, 1969.

[3] Enclosure 2 to Letter, E. I. Hawthorne (HAC) to G. M. Low, NASA MSC, Aug. 22, 1969.

electronic compartments, glass-covered solar-cell array, and painted tops of footpads.

(6) Wipe metal mirror of television camera and glass mirror of electronic compartment. Inspect and photograph them before and after.

(7) Return television camera (if feasible).

(8) Return scoop of soil mechanics surface sampler (at astronaut option).

(9) Return unpainted aluminum structural tubing.

(10) Return, sealed in vacuum, tubing with inorganic white paint.

(11) Return, in sterile fashion, cable with aluminized Mylar foil wrapping.

(12) Return glass from top of electronic compartment (if feasible).

(13) Return soil from vicinity of Surveyor (at astronaut option).

(14) Return (as part of field geology experiment) specimens representing material televised by Surveyor: sharp rocks around a specified nearby crater ("Blocky Crater"), presumed ray material from crater Lansberg, and a layered rock.

The purpose of each of these tasks is discussed in reference 3 (also see footnotes 2 and 3).

Many other desirable tasks were omitted as not practical or as hazardous. For example, return of solar panel cells would have been desirable, but the solar panel was too high to reach with any degree of safety.

A detailed mission plan was prepared incorporating the selected tasks. The Apollo 12 astronauts were briefed and trained, using a full-scale model of Surveyor 3 set up in its lunar configuration.

Mission Operations and Returned Material

The Apollo 12 Lunar Module (LM) landed on the Moon on November 19, 1969. During its descent, the LM passed from east to west across the northern rim of the crater within which Surveyor 3 rested. LM touchdown occurred on the northwest rim of this crater, 155 m from Surveyor 3. (See fig. 1.) Thus, the objective of demonstrating the point-landing capability of Apollo was attained. Post-flight evidence indicated that lunar material blown by LM exhaust during landing impinged on the Surveyor. (See chs. IV and VI of this document.) The landing coordinates of the LM were 3°11'51" S, 23°23'7.5" W, in the AMS lunar coordinate system (ref. 4).

During their second EVA, astronauts Charles Conrad and Alan Bean reached Surveyor 3 on November 20, 1969, at 06:27 GMT. They spent about 25 minutes at Surveyor and an additional 10 minutes at a nearby small crater ("Blocky Crater"), which had previously been televised by Surveyor. They took 56 black-and-white photographs of the Surveyor and its vicinity in accordance with the mission plan. Many of these were taken as stereopairs, by photographing, taking one step to the side, and rephotographing. A catalog of Surveyor-related photographs from Apollo 12 is included as appendix D of this document.

The astronauts inspected the Surveyor spacecraft, paying particular attention to items specified in the mission plan, and conducted the pre-planned swipe of the television camera mirror and electronic compartment top. Their observations are recorded in the mission commentary transcript and summarized in reference 5. The observations have been amplified in formal debriefing sessions and subsequent informal discussions at MSC and JPL.

The astronauts removed the following material from Surveyor 3 with a pair of shearing cutters:

(1) The complete television camera with its associated optical and mechanical components, electronics, pieces of cabling, and support struts. (See fig. 2.) A more detailed inventory of the television camera components is presented in appendix C.

(2) The scoop from the soil mechanics surface sampler (fig. 3), together with more than 6.5 g of lunar soil which it contained.

(3) A 19.7-cm section of unpainted aluminum tube from the strut supporting the radar altimeter and doppler velocity sensor (RADVS) (fig. 4).

(4) A 10-cm section of aluminum tube from one of the camera support struts. This tube was coated with inorganic white paint.

(5) About 13 cm of television cable, with its wrappings of aluminized plastic film.

Figure 5 shows the location of the components removed from the spacecraft. Glass from a com-

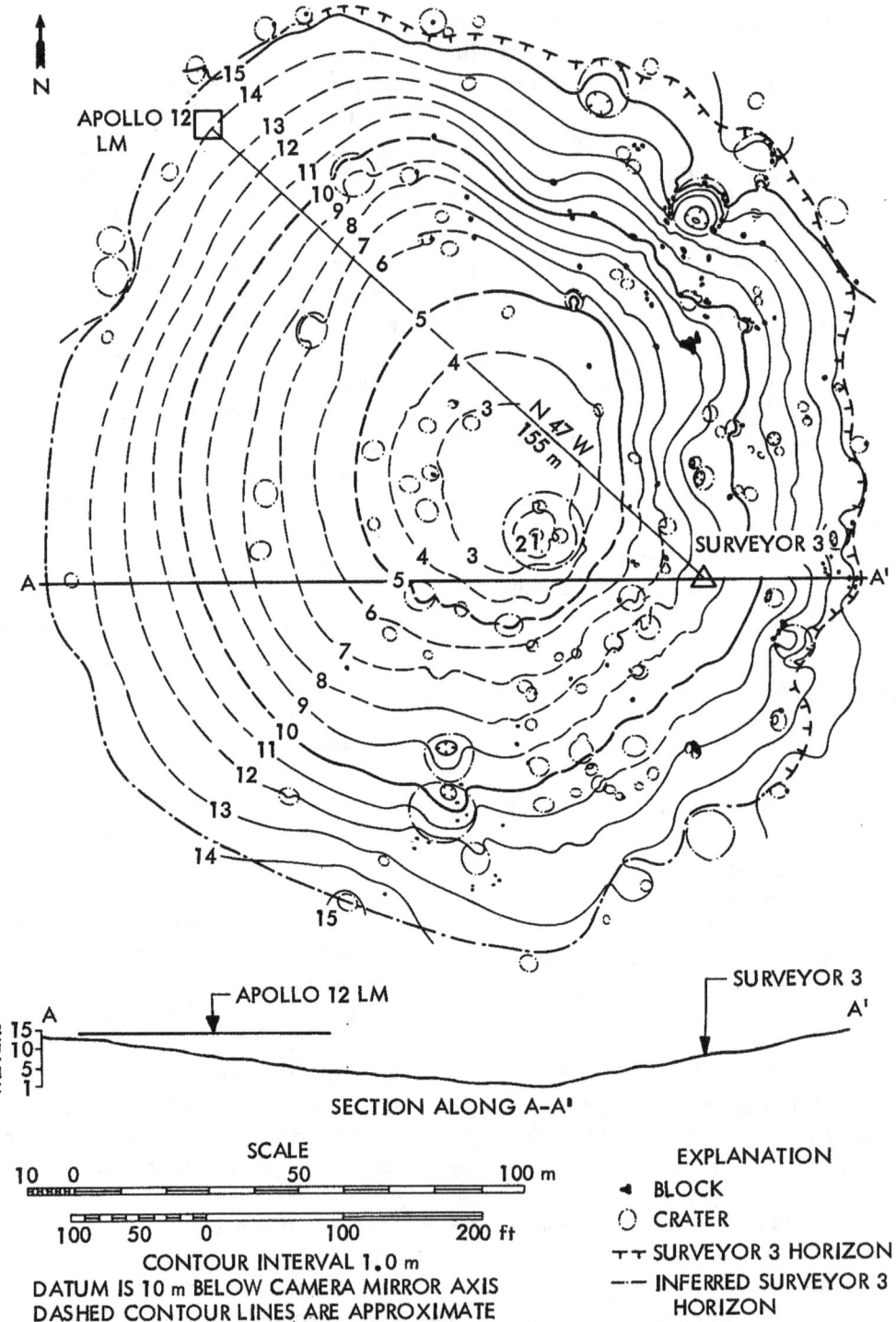

FIGURE 1.—"Surveyor Crater" showing the relative position of Surveyor 3 and Apollo 12 in plan and cross-section view. The LM was situated 155 m away from, N 47° W of, and at a ground level of 4.3 m higher than the television camera on Surveyor 3 (see ref. 7). Base map from Batson (see ref. 8).

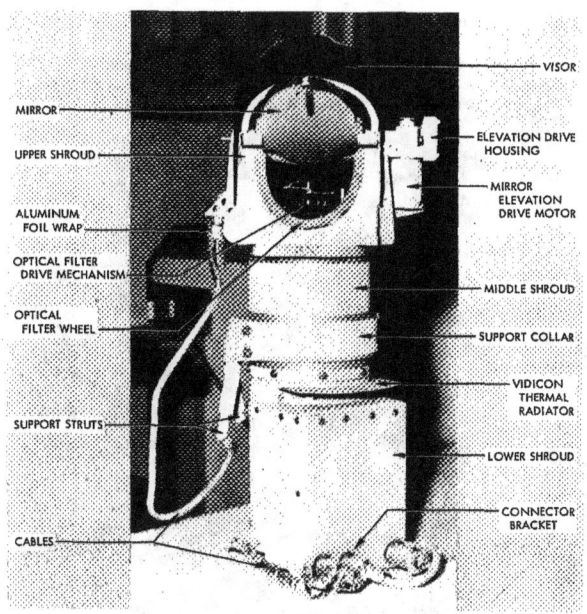

FIGURE 2.—Surveyor 3 television camera as it was unbagged at the LRL after its return from the Moon. Dents in the visor occurred during transport from the lunar surface.

partment top was not removed; the astronauts could not break it free of its support (app. D, frame AS12-48-7137). The hardware taken included samples representative of many spacecraft engineering subsystems, with a wide variety of electrical and electronic components, optics, functional mechanisms, lubricants, and temperature control devices and coatings.

The astronauts also collected a number of rocks from the lunar surface close to Surveyor and at nearby "Blocky Crater" as part of the field geology experiment. The analysis of these specimens is not given in this document.

Handling of Material

Handling of recovered parts on the Moon was planned to minimize contamination to the extent considered practical. As the camera, scoop, and unpainted tube were cut from the spacecraft and handled by astronauts Bean and Conrad, they were placed in pockets in the Surveyor tote bag (back pack). The bag was constructed from beta-cloth, a woven glass fabric coated with FEP Teflon, identical to the material of the astronauts' suits. In accordance with the plan, the astronauts

FIGURE 3.—Closeup of scoop of Surveyor 3 surface sampler. Photograph was taken in the LRL after the scoop was returned from the Moon.

let the painted tube and the cable fall, with a minimum of handling, directly into an Apollo sealed environmental sample container (SESC; see fig. 6). They sealed the container, and placed it, in turn, in the tote bag.

The parts were carried back to the LM; they remained in the tote bag during transit and during the multiple transfers to the Command Module in orbit, to the Mobile Quarantine Facility onboard the recovery ship, and to the Crew Reception Area (CRA) in the Lunar Receiving Laboratory (LRL). This handling is known to have at least caused abrasion of the exposed outer surfaces of the returned materials with partial removal of adhering lunar fines, and contamination of exposed surfaces with beta-cloth fibers and organic and biologic species.

While in quarantine in the CRA, the returned material was removed from the tote bag, the camera and scoop were photographed on a table top, and all parts were individually heat sealed in two polyethylene bags (fig. 7). The bagged parts were placed in bonded storage, where they remained until quarantine was lifted on January

FIGURE 4.—Section of unpainted aluminum tube from Surveyor 3, mounted on jig in LRL after its return from the Moon.

Figure 5.—Photograph of Surveyor 3 on the Moon, with astronaut Charles Conrad, Jr. Locations of the removed Surveyor parts are indicated. Apollo sample tongs appear immediately below surface sampler scoop, in circle. The Apollo 12 LM is in the background, on northwest rim of "Surveyor Crater."

7, 1970. All parts were then transferred to the astronaut debriefing room, where a temporary laboratory had been prepared.

The bagged parts were inspected and photographed and the parts and recovery discussed with the astronauts. The camera was taken to the low-level radiation counting laboratory in the LRL, where it remained overnight. Most parts were then unbagged, examined, and documentary photographs were taken of the surfaces. The camera and unpainted aluminum tube were unbagged on a laminar flow bench and mounted on special jigs. The scoop was not opened on the laminar flow bench for fear of losing lunar fines contained in and on the scoop. The SESC was not removed from its bag.

The camera support collar was taken off to permit mounting the camera for additional examination and photography and to facilitate biological sampling. The cable connectors and bracket from the camera front and the lower shroud of the camera were removed to gain access for internal biological assay. A quantity of dark particulate material was found inside the support collar recess. Most of the material was collected for subsequent analysis; a small amount (less than 0.5 mg) was separately collected for preliminary emission spectrographic analysis (ref. 6). Biological samples were collected from various sites. (See ch. XI, pt. A, of this document.)

The camera's lower shroud was replaced and the camera remained on a special mount in the laminar flow bench until January 15. The camera

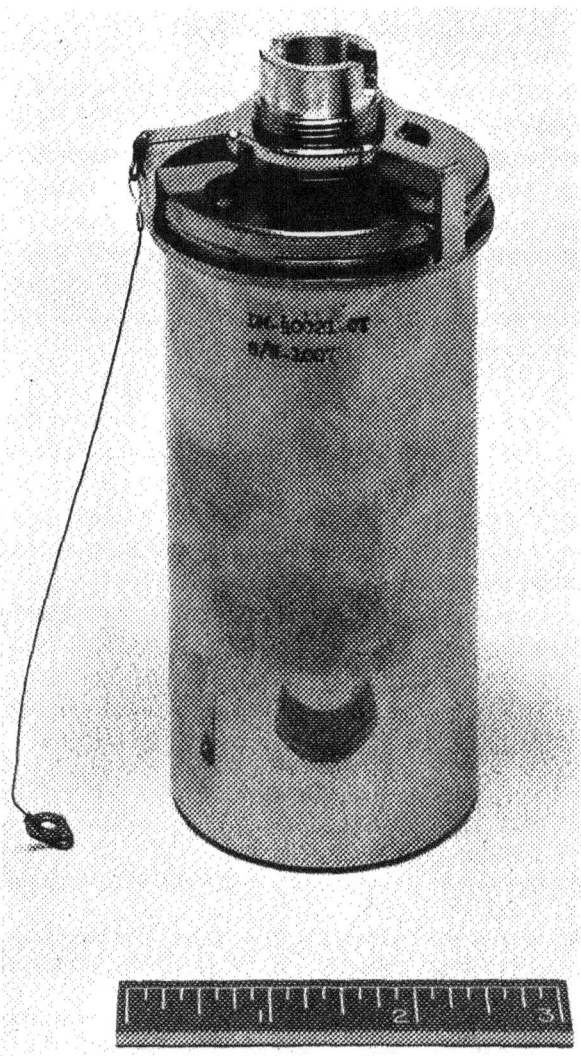

FIGURE 6.—Apollo sealed environmental sample container (SESC), containing Surveyor 3 cabling and painted tube, just before container was opened at JPL. The SESC is 15½ cm high and is 6.0 cm wide at its base.

and unpainted aluminum tube received preliminary examination for micrometeoroid impacts. (See ch. VI, pt. E.)

The unpainted aluminum tube was sectioned into six pieces, which were then individually packaged to protect the outer surface from additional damage.

The camera and removed parts were wrapped in FEP Teflon; the scoop was rewrapped in the plastic bag in which it had been stored during quarantine. All parts, except three of the six

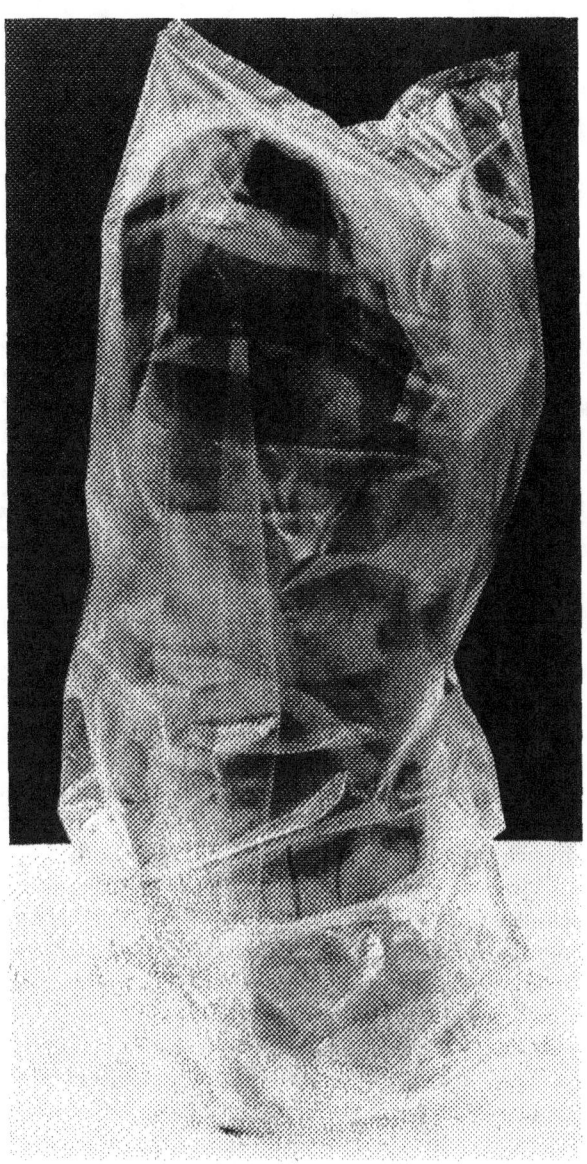

FIGURE 7.—Returned Surveyor 3 television camera, sealed in polyethylene bag. Photograph was taken in the LRL after the camera was returned from the Moon.

pieces of the polished tubing which were to remain at MSC, were packed in foam-lined shipping containers and flown to HAC, Culver City, Calif., on January 16, 1970.

HAC provided a limited-access clean room for their many engineering tests. The room contained two class 100 laminar flow benches, which were used in all operations in which a dust-free environment was desirable. All parts were placed in containers or covered with Teflon sheeting when not in actual use and stored in a floor vault for security.

Subsequently, parts called out in the Material Analysis Plan (see app. B) were transferred to JPL for distribution to engineering investigators outside of HAC, and to the science investigators in the United States and elsewhere.

The sealed SESC, containing the cable and painted tube, was opened at JPL in a sterile glove box under high purity argon and red light. It was found that the SESC had leaked, admitting air or air and oxygen; apparently a good seal was not obtained on the Moon.[4] The cable and painted tube were sectioned in the glove box; parts not tested there were resealed for further distribution. (See ch. XI, pt. B, of this document.)

Analysis Plan

The analysis of the returned parts, soil, and photographs was conducted under a comprehensive analysis plan. The plan was designed to insure retrieval of a maximum amount of information, while the integrity of the material was maintained as far as possible along the sequence of scientific and engineering investigations.

Most of the engineering investigations were conducted by HAC, using some of the equipment and personnel employed in the construction of the spacecraft prior to the Surveyor missions. (See ch. III of this document.)

Science and engineering investigators outside of HAC were individually invited to submit proposals that were brief, but which included a statement of objectives, the amount and type of material of interest, the type of tests to be performed, and the expected degree of alteration to the material. The proposals were reviewed for their scientific merit by a JPL Review Committee,[5] which recommended to NASA the type and amount of material to be allocated. Another group, the Surveyor Parts Steering Group (SPSG),[6] was later authorized to allocate material to those investigators planning tests not previously included in the analysis plan.

The analysis plan included 40 teams of investigators in nine categories; during a period of 16 months, approximately 275 tasks were performed, some on no more than one-half of a given part if the tests were destructive or had some effect on the material. This policy preserved material for possible future testing, as information and new ideas became available. The complete analysis plan, which includes both completed tests and those still in progress, can be found in appendix B.

Status

Most investigations originally included in the analysis plan have been completed. Some analyses are still in process, and a few investigators are awaiting results of other analyses before proceeding. Results of these analyses are expected to be published in the open literature.

One condition imposed upon each investigator was that he document the treatment that each part received while in his possession. This information has been compiled at JPL and can be made available for specific parts upon request.

References

1. *Surveyor III—A Preliminary Report*, NASA SP-146, Washington, D.C., 1967.
2. *Surveyor Program Results*, NASA SP-184, Washington, D.C., 1969.
3. *Mission Requirements, H-1 Type Mission Lunar Landing*, Revision A SPD9-R-051, Manned Spacecraft Center, Houston, Tex., 1969.
4. MISSION EVALUATION TEAM: *Apollo 12 Mission Report*, MSC-01855, Houston, Tex., 1970.
5. *Apollo 12 Preliminary Science Report*, NASA SP-235, Washington, D.C., 1970.
6. JOHNSON, P. H.; BENSON, R. E.; COUR-PALAIS, B. G.; GIDDINS, L. E., JR.; JACOBS, S.; MARTIN, J. R.; MITCHELL, F. J.; AND RICHARDSON, K. A.: "Preliminary Results from Surveyor 3 Analysis." *Apollo 12 Preliminary Science Report*, NASA SP-235, Washington, D.C., 1970, pp. 217–223.
7. JAFFE, L. D.: "Blowing of Lunar Soil by Apollo 12: Surveyor 3 Evidence." *Science*, vol. 171, 1971, pp. 798–799.
8. BATSON, R. M.: "Landing Site Maps." *Surveyor Program Results*, NASA SP-184, Washington, D.C., 1969, pp. 45–56.

---

[4] M. A. Adams and M. Knittel (JPL), private communication.

[5] Membership in the JPL Review Committee consisted of L. Jaffe (Chairman), W. Carroll, D. Nash, and C. Snyder.

[6] Membership in the SPSG consisted of N. Nickle (Chairman) and W. Carroll of JPL; B. Doe of USGS, Denver (formerly of NASA Headquarters); D. Senich of NASA Headquarters; S. Jacobs of MSC; and G. Wasserburg of Caltech. F. Fanale of JPL served as an alternate in G. Wasserburg's absence.

II. Summary and Conclusions

N. L. Nickle and W. F. Carroll

The successful return of the Surveyor 3 hardware, lunar soil, and photographs taken by the Apollo 12 astronauts permitted 36 studies to be made by more than 80 investigators.

Chapter III contains the significant engineering results obtained from these studies. Chapters IV through XI contain the results of the scientific investigations. Because the papers were written individually by members of the investigating teams and therefore are presented in a different format than are chapters I through III, some redundancy or differences in interpretation may occur.

This chapter is a summary of the engineering and scientific results derived from the investigations.

Engineering Results

Results of the engineering investigations were essentially "nonspectacular"; the primary value lies in the fact that no failures or serious adverse environment effects on the hardware were uncovered that, to some degree, had not been anticipated. The absence of detected major effects and the resulting implications for future space vehicles are significant. However, the absence of effects should not be construed to indicate that the problems associated with material and component selections, test, design, assembly, and systems test can be ignored.

Spacecraft Changes

Measured reflectance data have been analyzed in order to separate and understand the effects of lunar dust and radiation damage. The radiation-induced discoloration on various surfaces was found to be proportional to the degree of solar illumination, and is in reasonable agreement with laboratory simulations. The discoloration was found to be subject to photo-induced oxygen bleaching. This bleaching was responsible for a considerable change in color during the several months of exposure since return to Earth. Organic contamination is not a significant factor in the observed discoloration of the external surfaces.

Almost all exposed surfaces on the camera were partially covered with a fine layer of lunar dust. Substantial variations existed in the quantity and apparent particle size of dust on the various surfaces. The dust distribution indicates that the fines were disturbed and implanted upon the spacecraft primarily by the initial Surveyor landing and by the approach and landing of the Apollo 12 Lunar Module (LM). The presence of dust, even in very small quantities, can have a significant effect on temperature control and optical performance of hardware on the lunar surface.

Lunar dust adhering to the camera's optical filters consists of less than 1- to 40-μm-wide particles of calcic plagioclase, clinopyroxene, tridymite, and glass. Most particulates are complex mixtures of more than one crystalline phase and not micrometer-sized pieces of single-phase minerals. The assumed parent material of this dust is a fine-grained breccia or a soil from such a rock type.

Dust on the camera's mirror consists of particles large enough to see with the unaided eye (contaminants consisting of gypsum, calcite, and beta-cloth fibers) and fine-grained angular fragments. Spherical particles are restricted primarily to the smaller size ranges; about 1 percent of the particles is spherical at 0.7-μm diameter compared with 10 percent at less than 0.2-μm diameter. Ninety percent of the total mass is within the size range of 0.3 to 3 μm. Very few lunar particles larger than 4 μm exist; some of these may be aggregates.

Sources of the dust on the mirror, and thus also of many other surfaces, include that disturbed by the abnormal Surveyor landing and by manipulations of the surface sampler scoop. It has been demonstrated that more dust exists now than at the time of the Surveyor 3 mission. Thus, the approach of the LM and/or natural lunar transport processes contributed additional material to the mirror's optical surface.

Spectral reflectance, gonioreflectance, spectral transmission, and ellipsometry measurements conducted on various components of the camera indicate that the following changes occurred on the lunar surface:

(1) A nonparticulate coating of unknown composition and origin was deposited on the mirror and possibly other surfaces. The coating is insoluble in acetone and benzene. Tests are continuing in an attempt to identify the coating.

(2) The thickness of the nonparticulate coating is not uniform, and is estimated as approximately one-half wavelength ($\lambda = 550$ nm).

(3) One or more particulate layers were deposited by at least two of the following events:
  (a) Abnormal landing of Surveyor 3.
  (b) Manipulation of the surface sampler scoop.
  (c) Normal transport processes.
  (d) Approach and landing of the LM.
  (e) Redistribution and/or contamination during camera retrieval and return.

(4) Distribution of dust on all surfaces is not of uniform thickness.

(5) Increase in spectral transmission of the blue and green filters may be due to partial dissipation of the Inconel coating.

(6) Dust on the filters caused a 25-percent decrease in transmission.

(7) Radiation darkening caused a decrease in transmission of the clear filter.

(8) Mirror acquired a pit density of approximately 1 pit per 2 mm^2 on cleaned areas.

The exterior camera surfaces showed discoloration patterns produced by lunar surface particles that were eroded and entrained on Surveyor by the LM exhaust during landing. The particles were ejected almost horizontally at 40 m sec^{-1}, struck the camera, and partially whitened its already dusty and radiation darkened surface.

Exterior surfaces of the scoop were discolored by the presence of lunar soil, but most prominently discolored by exposure to solar radiation on the Moon. The degree of discoloration, which was made apparent by a change of the original light blue paint to a whitish blue, depended upon the duration and angle of surface exposure to the Sun. Adhesion of lunar soil varied with the type of surface. Lunar material adheres more readily, in order, to (1) painted surfaces (approximately 10^4 dyne cm^{-2}), (2) Teflon, and (3) metallic surfaces (10^3 to 10^4 dyne cm^{-2}).

The Surveyor spacecraft moved from its landed configuration sometime between May 1967 and November 1969. It is conjectured that the movement occurred as a result of a sudden failure of the leg 3 shock absorber. The movement at footpad 2 was in the amount of 5° of tilt and 7 to 8 cm of lateral translation in the form of a rotation about footpad 1, which was embedded in the lunar soil.

Organic Contamination Analysis

Determination of the presence of organic contaminants was considered important in order to understand the discoloration process and to help identify possible sources of contaminating gas. Parts of the mirror and exterior camera surfaces were washed with solvents, and the residues were analyzed. Major components of the extract residue from the mirror was dioctyl phthalate and silicone oil. LM descent engine products are evident only in trace amounts.

Extracts were taken from the middle shroud on the side facing the LM and the side away from the LM. Major constituents found are hydrocarbons, dioctyl phthalate, and silicones. Several other species, thought to be derived from the Surveyor 3 vernier engine exhaust, were observed. The LM descent engine products are twice as abundant in the leeward sample; this difference in abundance is believed due to erosion of the side facing the LM by entrained lunar dust particles.

Sources of the various organic contaminants are hydrocarbons from lubricating or vacuum pump oils and general terrestrial contamination, silicones from sources as oils, outgassing of electronics and plasticizers, copolymer of vinyl alcohol and styrene from electronics insulation, and nitrogenous compounds from LM and possibly Surveyor 3 engine exhaust. The organic contamination levels do not contribute significantly to the discoloration of the various surfaces. Analyses for organic contaminants and identification of their sources, even if low in concentration, should be recognized as an important criterion for the design of optical or other active instruments for future spacecraft.

Micrometeorite Impact Analyses

A major effort in the analysis of Surveyor 3 parts has been the search for hypervelocity impact features—an effort roughly analogous to the search for the needle in the haystack. A great number of low-velocity features exist that were caused by lunar particles striking the surfaces due to Surveyor and Apollo landing events, handling of the material, and natural phenomena. The 1- to 4.5-μm size of the surface features prohibited the effective use of optical instruments. However, all participating investigators concluded that no material or surface features were found that definitely could be stated to be meteoritic in origin. Consequently, determinations of the flux rate of hypervelocity particles at the Surveyor 3 site were based on the absence of diagnostic features; as such, the flux rates represent upper limits only. In each instance, the determinations were in general agreement with those obtained from Pioneers 8 and 9, Cosmos 163, Pegasus satellites, and others.

The optical filters were inspected for primary impacts with the same results. However, because of the spatial orientation of the filters, the well-defined field of view of space for each filter and the nature of their finish provided an excellent opportunity to determine an implied impact rate of secondary particles. Particles 1 μm and larger with velocities high enough to produce plastic flow in glass were found to be about 10^3 times the cratering rate expected for primary micrometeoroids. The rate is approximately 800 impacts cm^{-2} yr^{-1} (2π sterad)$^{-1}$ for impacts ≥ 1 μm.

Comparison of pictures of the lunar surface taken 31 months apart by Surveyor 3 and Apollo 12 show no meteorite craters ≥ 1.5 mm in diameter.

Radioactivity and Radiation Damage Analyses

The camera visor was examined for an alpha radioactive deposit formed by the decay of radon isotopes diffusing from the lunar surface. The conclusion reached is that the gross activity on the visor is due to the activity of the paint. However, the amount of ^{210}Po activity expected on 1 cm^2 of the lunar surface after an infinite time at Oceanus Procellarum was estimated to be $(0.88 \pm 4.43) \times 10^{-3}$ disintegrations sec^{-1} cm^{-2}.

The cosmogenic radionuclide ^{22}Na was measured in painted and unpainted aluminum tubes, camera support collars, brackets, scoop, soil removed from the scoop, and in the mirror. The average galactic cosmic-ray flux incident on Surveyor 3 was about 4 ± 1 protons cm^{-2} sec^{-1}. Detailed radionuclide production rate calculations based on satellite data of solar flares were used to estimate the contribution of solar flare protons to the total ^{22}Na produced in Surveyor 3. Galactic cosmic-ray production of ^{22}Na in aluminum derived from the Lost City meteorite agrees with the galactic cosmic-ray production rate in Surveyor 3, indicating almost identical cosmic-ray fluxes at 1 AU and at 2.35 AU. The ^{26}Al and ^{22}Na content of lunar soil recovered from the Surveyor 3 scoop indicates that the soil originated from an average depth of 3.5 cm in the lunar surface.

The tritium content of painted aluminum samples removed from the camera shrouds was measured to be 0.48 ± 0.005 dpm cm^{-2}. This activity is more than a factor of 3 larger than would be expected if it had received the same average cosmic-ray flux and solar flux as the top of Apollo 12 lunar rock 12002. It is thought that an excess of tritium existed which was due to artificial contamination; there was a correlation, however, of tritium content with exposure to sunlight, indicative of solar wind tritium.

There was no evidence of microstructure effects caused by particle bombardment from the solar wind, solar flares, or cosmic radiation. The size and appearance of precipitate particles of Mg$_2$Si indicate appreciable thermal aging (which possibly occurred during fabrication). Elevated lunar temperatures may have been sufficient to result in thermal diffusion of trapped solar wind He and Ne in a high density of dislocations occurring to a depth of 10 µm.

Solar Wind Rare Gas Analysis

The polished aluminum tube contained trapped solar wind He and Ne with a ^{4}He-to-^{20}Ne ratio of 295. This value is lower than the ratios measured from the Apollo 11 and 12 solar wind composition (SWC) experiments. This could be due to ^{4}He diffusion loss or to a small residual dust contamination. The ^{4}He distribution around the aluminum tube is in agreement with the theoretically expected distribution and corresponds to an average solar wind ^{4}He flux of 7×10^6 cm^{-2} sec^{-1}. If ^{4}He diffusion loss had occurred, the average ^{4}He flux could be as high as 13×10^6 cm^{-2} sec^{-1}. Neglecting the small influence of possible dust contamination or of diffusion loss, table 1 shows the average isotopic composition for the solar wind during exposure of Surveyor 3 material and the Apollo 11 and 12 SWC experiments. Compared with the Apollo 11 and 12 results, the ratio of ^{4}He to ^{3}He is unexpectedly high. The differences may reflect time variations in the composition of the solar wind.

Particle Track Analyses

The energy spectrum of iron-group solar cosmic-ray particles was determined for the first time over the energy range 1 to 100 MeV/nucleon using the optical filter glass. The difference between the observed spectrum and the limiting spectrum derived previously from tracks in lunar rocks gives an erosion rate of 0 to 3 Å/yr. High-energy fission of Pb, induced by galactic cosmic-ray protons and alpha particles, was observed.

Soil Property Analyses

The soil sample returned in the scoop provided a unique opportunity to evaluate earlier, remotely controlled, in-situ measurements of lunar surface bearing properties. Assuming the lunar regolith at Surveyor 3 has a bulk density of 1.6 g cm^{-3} at 2.5-cm depth, then the agreement is good. The bearing capacity varied from 0.02 to 0.04 N cm^{-2} at bulk densities of 1.15 g cm^{-3} to 30 to 100 N cm^{-2} at 1.9 g cm^{-3}.

TABLE 1.—*Average isotopic compositions for the solar wind during exposure of Surveyor 3 material and Apollo 11 and 12 SWC experiments*

Ratio	Surveyor 3	Apollo 11	Apollo 12
^{4}He:^{3}He	2700 ±130	1860 ±140	2450 ±100
^{20}Ne:^{22}Ne	13.3±0.4	13.5±1.0	13.1±0.6
^{22}Ne:^{21}Ne	21 ±5	....	26 ±12

Pictures taken by the Surveyor 3 television camera and photographs by the Apollo 12 astronauts of identical areas have provided the opportunity to evaluate changes in the lunar regolith during the 31 months, and have helped to dispel the impression that the lunar soil may have a thin surface "crust" that breaks into flat "tiles." The impression of "tiles" and "crusting" is an illusion. Rather, the lunar soil deforms and cracks in the same manner as homogeneous, isotropic terrestrial soils of moderate bulk density, with a small amount of cohesion. Photographs viewed stereographically clearly show the three-dimensional character of the disturbed material.

No changes in the lunar soil that can be attributed to natural processes have been identified.

A previously unreported feature of lunar fines is the existence of filamentary whisker-like objects attached to individual particles in a manner resembling sea urchins. Twenty particles were found on the red optical filter with whiskers averaging 10 µm long and 0.1 µm wide. It is hypothesized that these whiskers grew on the particles during impact events on the lunar surface. If this explanation is correct, then determination of the fraction of lunar particles that contain whiskers may allow setting limits to theories that predict migration of dust over the lunar surface by various processes. These features presumably have not been observed before because of their friability.

Microbe Survival Analyses

A bacterium, *Streptococcus mitis*, was isolated from a sample of foam taken from the interior of the camera. Available data suggest that the bacterium was deposited in the camera before launch. Lyophilizing conditions existing during pre-launch vacuum tests and later on the lunar surface may have been instrumental in the survival of the microorganism.

A piece of electrical cabling also was subjected to microbiological analysis with negative results. The absence of viable microorganisms could be due to natural dieaway and dieoff caused by vacuum and heat.

Conclusions

The analyses presented and discussed in more detail in chapters IV through XI may be credited with the following achievements:

(1) Collection of a wealth of technical information applicable to the design and fabrication of future spacecraft.

(2) General agreement in the upper limit of micrometeoroid fluxes on the Moon for primary particles less than 1 μm to several millimeters in diameter and larger with values from other sources.

(3) Establishment of the sources and types of organic contamination from Surveyor and Apollo.

(4) Establishment of an almost identical cosmic-ray flux at 1 and 2.35 AU.

(5) An indication of a varying isotopic composition for the solar wind with time.

(6) Discovery of a new active erosion process on the lunar surface.

(7) Discovery of "whiskers" on lunar dust particles.

(8) Demonstration of the ability of a bacterium species to survive the rigors of the lunar environment.

Although the return of additional general hardware from the Moon or from space under similar conditions does not appear to be warranted, specific items (i.e., solar cells) or equipment from specific environments (i.e., high-energy radiation environments, the asteroid belt, etc.) could be valuable. Possible future return of space hardware should be accomplished in a controlled manner in order to preserve the effects of exposure to be examined. The value of scientific investigations on engineering hardware is severely limited by the lack of suitable controls, standards, or documentation of initial conditions. The size, shape, surface texture, and composition of engineering hardware is selected for functional performance, and therefore does not lend itself to scientific analyses. Engineering materials are typically selected for minimum response or change due to environmental factors and are therefore usually less than optimum subjects for evaluation.

In order to accommodate scientists in the future with material suitable for analysis, it is recommended that a set of coupons consisting of different types of material of interest be placed on all spacecraft regardless of the present intent of obtaining or revisiting the spacecraft. Such devices presently exist that are light in weight (several kilograms), have replaceable coupons, can be remotely deployed, and are inexpensive.

III. Returned Surveyor 3 Hardware: Engineering Results

W. F. Carroll, P. M. Blair, Jr., E. I. Hawthorne, S. Jacobs, and L. Leger

This chapter is a summary of the engineering evaluation of returned hardware performed by the Hughes Aircraft Co. Results of the engineering investigations were essentially "nonspectacular"; the primary value lies in the fact that no failures or serious adverse environment effects on the hardware were uncovered that, to some degree, had not been anticipated. The absence of detected major effects and the resulting implications for future space vehicles are significant. However, the absence of effects should *not* be construed to indicate that the problems associated with material and component selections, test, design, assembly, and systems test can be ignored.

Electronic components, including the vidicon tube, optics, materials, mechanisms, and lubricants, were in generally good condition. No identified failures or anomalies, with the exception of those resulting from thermal cycling, were caused primarily by the lunar environment.

Although not necessarily the most technically significant, the most interesting results were the external surface effects observed. The darkened color of the originally white surfaces, as observed by the astronauts, was due to expected radiation damage and to the coating of lunar dust. Although the Lunar Module (LM) landed 155 m from the Surveyor spacecraft, debris disturbed by the LM "sandblasted" the Surveyor.

All anomalies associated with lunar operations of the Surveyor 3 television camera have been resolved; however, there remain several questions regarding retrieval operations and the condition of the returned hardware.

Detailed results of the engineering evaluation, interpreted by specialists in various technical disciplines, can have an important impact on the complexity, cost, and reliability of future space vehicles. There are many implications to material and component selection, subsystem design, and assembly and test criteria.

Hardware removed from Surveyor 3 by the Apollo 12 astronauts in November 1969 and returned to Earth was subjected to intensive engineering evaluation in order to obtain information on the hardware characteristics that could be of value to the design, test, and operation of future spacecraft. No attempt was made to verify or evaluate the Surveyor design, as such, except to the extent that such an evaluation would yield information of value to future designs.

The returned hardware contained representative samples typical of many current and future spacecraft engineering subsystems and included a wide variety of electronic components, optics, functional mechanisms, materials, lubricants, and thermal-control coatings and devices. The only major spacecraft subsystems for which no meaningful hardware was obtained were propulsion and "secondary" power (solar cells, batteries, etc.).

Although the hardware remained on the Moon for 31 months before return, the electronics and mechanics subsystems functioned only during the first 2 weeks. Radiation, thermal cycling, vacuum, etc., were continuous through the remaining 30½ months, but only on nonoperating equipment.

The sequence of disassembly, engineering analysis, and incorporation of science investigations was planned and executed to maximize total technical return. The timing and scope of some of the engineering investigations were constrained by science studies and by a requirement to preserve the integrity of parts and materials for possible second-generation tests. These constraints affected the quantitative and statistical validity of some of the data. At this time, however, there is no indication that the constraints were responsible for loss of any significant information or for failure to identify any potential problem areas.

No attempt was made to conduct an "exhaustive" investigation into any component, system, or technical discipline. The tasks were structured to identify and retrieve the significant technical information, with emphasis on changes or absence of changes induced by lunar operations and storage. The scope and approach of each task were reviewed frequently to achieve optimum technical return for resources available without sacrificing significant information.

Some effort on parts of the investigation was not justified by technical return, but instead was expended because the opportunity was considered unique and because of the irreversible nature of the procedures involved. For example, it was ultimately shown that the failures of a transistor, the camera shutter, and the vidicon photoconductor were interrelated and the result of weakness induced by pre-launch testing, with lunar exposure playing only a secondary role. Extensive investigation was necessary to reach this conclusion and to preclude primary lunar effects or effects of ground command procedures.

Electronic Components

The returned Surveyor television camera contained over 1500 resistors, capacitors, diodes, and transistors. Some of these components were tested in assembled circuits and as individual components. These tests verified their general integrity after 31 months of lunar exposure. A complete description of the electronic component test program and detail results are presented in reference 1.

Surprisingly few of the electronic components failed. It was known that many of the components which were found to have failed, such as the shorted tantalum capacitor in the video amplifier circuit (described in ref. 1), were sensitive to cryogenic cycling.

There were some components with cracked glass envelopes, which were the result of thermal stress cracks in the conformal coating. Some of these exhibited malfunction due to internal damage; others were functionally satisfactory. During development tests, this effect was identified and is a material and process problem rather than an electronic component problem.

A unique failure in the returned hardware occurred in the shutter drive circuit of the television camera. A failed transistor, which acted as the shutter drive switch, caused the failure of the shutter solenoid, and indirectly, damage to the vidicon. This transistor, which had been stressed before launch by a defective test circuit, functioned satisfactorily during subsequent tests and during Surveyor 3 lunar operations. The initial failure probably was caused by a short induced by thermal stress during the lunar night. During the second or subsequent lunar day, a voltage spike from one of several possible sources (see ref. 1) caused the shutter to open and produced an overload on the shutter solenoid coil. The solenoid insulation charred; this reduced the resistance, causing an overload on the transistor and causing it to "open." Subsequent failure of the vidicon is discussed below.

Minor shifts in characteristics were observed in some of the electronic components. For example, a platinum resistance thermometer showed a change of 0.4 percent in temperature coefficient of resistivity. However, these changes are insignificant for most applications (see ref. 1).

Vidicon

When the camera was disassembled and the vidicon examined, there was no evidence of the photoconductive coating that had been on the faceplate, and the final beam control grid (grid 5) immediately behind the faceplate was ruptured. It was established subsequently that these effects were secondary failures caused by the open shutter. Solar radiation, diffusely reflected from the mirror and focused on the faceplate through the optics, caused a temperature rise sufficient to evaporate the photoconductor. During the investigation, the failures were duplicated on a spare vidicon in the laboratory.

Part of the evaporated photoconductor condensed on the adjacent grid. Subsequent diffusion into the copper grid formed an intermetallic compound with gross changes in physical strength and thermal coefficient of expansion. The actual rupture may have been due to thermal cycling or physical shock as a result of retrieval or return to Earth. The equivalent grid in the spare vidicon used to duplicate the failure was found to be ruptured when the unit was removed from the furnace.

As part of the evaluation program, the vacuum level of the vidicon was determined and found to be equivalent to that of a vidicon maintained in storage for the same period of time. This was in spite of the fact that the camera was subjected to a physical shock sufficient to cause two large dents in the camera hood some time during the recovery or return (probably during splashdown in the Pacific Ocean).

Detailed examination and partial functional tests (see ref. 1) indicated no other failures or anomalies in the vidicon tube. The observed failures emphasize the temperature sensitivity of this type of vidicon design and also demonstrate the need for configuration or mission constraints regarding solar illumination of such vidicon tubes.

Materials

The materials used in the Surveyor 3 camera had been selected for stability in the space environment. With the exception of the minor cracking and apparent loss of strength in the FEP Teflon cable wrap exposed to solar radiation, no unexpected degradation of functional performance was observed. An examination of the aluminized FEP Teflon used to wrap the cable bundles revealed surface cracks at wrinkles in the wrapping. Physical tests showed a clear decrease in tensile strength and elongation, although the change could not be established quantitatively because of limited sample size. As the Teflon was used only for thermal control, performance was not adversely affected in this stationary cable. Because FEP Teflon is used extensively as a spacecraft material, the effects of stress, radiation, and thermal cycling should be investigated more completely.

The Teflon dust seal between the mirror assembly and the camera body was discolored and curled, probably a result of dimensional change and radiation darkening of the excess adhesive used in installation.

The conformal coating used on electronic circuit boards produced the cracked envelopes described previously. Similar failures were observed during the development phase of the Surveyor program, and the observation on the returned Surveyor 3 camera was no surprise. The effect is the result of differential thermal expansion and excess thickness of application of the coating.

Peeling of the wire insulation observed in several of the cable bundles seems to be the result of physical stress imposed by the tie cords. Peeling of the polyimide overlayer had been observed during pre-flight laboratory testing of Surveyor equipment.

As expected, there was significant radiation discoloration of epoxy adhesive, nylon ties, glass fabric, and cable insulation.

Microhardness of the returned polished aluminum tube had increased, which was due to the thermal environment experienced by the tube on the Moon.

Optics

The need to protect optical elements from dust contamination was obvious during Surveyor 3 lunar operations in 1967 and was confirmed during the analysis of returned hardware. All other optical performance information gained from post-return analysis is secondary to this conclusion.

Lunar dust accumulated on the mirror during Surveyor operations was considered the primary cause of the veiling glare. (Another theory was pitting by impacting lunar particles.) Dust as the principal contributor was verified by photographs taken during operations at the Surveyor site before and after a small area at the top of the mirror was wiped by the astronauts.

Post-return analysis has demonstrated that there are at least two distinct degrees of adhesion of dust on the mirror (and other parts of the returned hardware). The area wiped by the astronauts and areas subsequently peeled for replication show remaining material adhering to the mirror. As described in references 1 and 2, there are several potential sources of the dust that contaminated camera surfaces. The differences in adhesion may be associated with the source, the time, or the condition of deposition. The analysis of the mirror is not yet complete; the relative importance of the sources remains an unanswered question. It is reasonable to assume that the material removed by the astronauts and by replication peels represents that deposited by the LM approach and descent and that the remaining material represents that deposited by the abnormal landing of Surveyor 3.

Measurements by Rennilson (see ch. IV, pt. E) of the clear filter transmission show a radiation-induced transmission loss. This is not a surprising result because radiation stability was not a criterion in selection of the clear filter. However, such radiation damage could be important to optical elements in future space missions.

Transmission and resolution of the returned lens assembly were measured in a way similar to that used before the mission. The slight decrease in measured transmission can be explained by the small amount of dust present on an outer surface of the front element and by the condensed contaminant on the beamsplitter. The decrease would not have been significant for the Surveyor camera, but could be detrimental to other instruments. The dust effect stresses the importance of particulate cleanliness during pre-launch and mission operations. The contaminant on the beamsplitter (probably from the shutter solenoid) emphasizes the importance of controlling condensable outgassing products.

Mechanisms and Lubrication

With one possible exception, no instances of cold welding were identified in any of the returned hardware. The shell of one of the electrical connectors on the front of the camera appeared cold-welded to the camera shroud. As galling during installation is possible, this single cold weld is not considered significant.

Selected mechanical subsystems were tested functionally both in air and in vacuum; frictional values obtained were nominal (refs. 1 and 3). Removal torques were measured for all accessible threaded fasteners, again with no evidence of cold welding.

Included in the returned hardware were seven mechanical subsystems with independent drive motors and gear trains; six of these subsystems were on the camera and one on the scoop of the surface sampler. During lunar operations, these were exposed to different vacuum conditions. (This was considered in planning and executing the test program.) The scoop door and camera filter-wheel drive mechanisms were exposed to space and operated in the ultra-high-vacuum conditions of the lunar surface. The three drive mechanisms associated with the lens were located inside the camera body and, because of limited egress paths (the camera was not sealed) and outgassing of adjacent components, probably never approached lunar vacuum conditions during operations.

No lubricant failure, abnormal friction values, or cold welding were detected. There was no evidence of differences as the result of the lunar exposure vacuum levels described. However, one lubricant did appear to be marginal for the application, as pre-launch tests had indicated.

Lubrication of potentiometer windings was incorporated in the design of later Surveyor cameras, but was not included on Surveyor 3. The absence of lubricant on the Surveyor 3 filter-wheel position potentiometer contributed to the failure of a substandard part. The potentiometer failed to function as the result of a broken guide block, which had been fabricated from an incorrect or substandard piece of material with a physical strength substantially lower than normal. The remaining, unlubricated potentiometers functioned during the 14 days of camera operation on the Moon, although wear was observed during the post-return evaluation.

During Surveyor 3 operations, there had been intermittent failure of azimuth step command response. The failure occurred primarily during thermal transients and in certain azimuth positions. Differential expansion during thermal transients and the gravitational side load that resulted from the angle at which Surveyor 3 rested on the Moon were assessed correctly as contributing to the problem. Lubricant failure and mechanical obstruction by lunar dust, considered contributory factors, were not evident during post-return analysis. Instead, the large azimuth drive gear had damaged teeth in positions that corresponded to positions at which step failure occurred. The damage to the gear teeth probably occurred during pre-launch vibration testing, but may have occurred during spacecraft launch.

For mechanical requirements and duty cycles equivalent to the 2 weeks of Surveyor operations, the dry film lubricants such as those used on Surveyor 3 seem to be more than adequate. For more severe thermal, torque, or duty cycle requirements, the Surveyor results can serve as a valuable baseline for design and test criteria.

Surface Effects

Studies of surface discoloration effects conducted as part of the engineering investigation are described in detail in chapter IV, part A, of this publication (also see refs. 1, 2, and 4). The overall discoloration is due to expected solar radiation darkening and a heavier than expected deposit of lunar fines. The cause of discoloration varies from all dust to all radiation, depending on location on the camera; most of the surface area has contributions from both dust and radiation. The degree of radiation darkening is proportional to the extent of solar exposure, as expected. While the magnitude of the change is somewhat greater than predicted from laboratory simulation, the discrepancy is within the uncertainty of simulation results.

Considerably more lunar dust was found on the surfaces than expected. It was known that the abnormal Surveyor 3 landing disturbed lunar material, which affected the camera mirror and presumably other spacecraft surfaces. There is substantial evidence (see ch. IV, pt. A) that the approaching LM disturbed lunar material, depositing it on the camera surfaces. Lunar material disturbed by the LM during final stages of landing "sandblasted" the Surveyor, even though the landing site was 155 m away. Details of the sandblast effect are described in references 5 and 6.

There have been no high-velocity meteoroid impact sites positively identified on any of the returned hardware. As described in chapter VI of this report, this finding describes an upper limit for meteorite distribution.

Mission Anomalies

During Surveyor 3 operations, some anomalies were noted in spacecraft performance. Three of these were associated with camera equipment, and all three have been resolved. None of the anomalies were due directly to the lunar environment. The anomalies are summarized here for the reader's convenience:

(1) Image contrast attenuation and veiling glare caused by dust on the mirror, which was deposited during the abnormal landing.

(2) Intermittent failure of response to azimuth step command caused by damaged azimuth drive gear teeth.

(3) Failure of the filter-wheel position potentiometer caused by a broken guide block. An incorrect or substandard piece of material had been used to fabricate the block.

Unresolved Questions

Several unresolved questions remain regarding the Surveyor spacecraft, Apollo 12 astronaut operations, and the returned hardware. Although some questions may be answered directly at a future time, or inferred from current investigations, others may never be resolved. These unresolved questions are discussed in the subsequent paragraphs.

Polished Tube Cutting

The astronauts were unable to cut the section of the polished tube originally designated for retrieval. The tube from the radar altimeter and doppler velocity sensor (RADVS) support strut "appeared to be more brittle and easier to cut than the tubes used in training." (See ref. 7.) Post-return analyses showed an increase in hardness of the returned tube of a magnitude that would be expected from the thermal environment. No assumption can be made regarding a change in characteristics of the originally designated tube that would prevent it from being cut. Comments made by the astronauts during debriefing and photographs taken on the Moon verify that reflected sunlight from the astronauts' suits provided sufficient illumination to insure that they were not attempting to cut the solid end fittings. Although the tube was in the shade of the spacecraft and thus would be cold, available cryogenic data indicate no change in properties that would cause an inability to cut the tube.

Cable Wrap

The astronauts reported that the cable insulation shredded and behaved like "old asbestos." This observation presumably applied to the glass fabric-wrapped cable that runs from the front of the camera to the mirror assembly. There was no evidence during the evaluation to confirm or support this observation. Cutting the fabric would produce short fiber fragments, and the observation may have been an artifact of such fragmentation in the collimated lunar sunlight.

The possibility remains that the cable wrap was highly friable and disintegrated when cut but that, upon return, absorbed atmospheric gases restored the flexibility and durability. Nylon has been demonstrated to exhibit such an effect associated with absorbed moisture; no documented evidence of a similar effect for glass fiber has been identified.

Camera Power

The interrelationship and sequence of failures including the drive circuit transistor, the shutter, and vidicon have been identified. The question of which of the possible sources provided the voltage is still unanswered. There is no evidence in telemetry that the spacecraft responded to turn-on signals or that the necessary additional signal to turn on the camera was sent. The condition of the returned hardware clearly demonstrates that the camera was powered from some source after the first lunar day. It is reasonable to assume that the spacecraft did turn on as commanded, but that response telemetry was not received. With the spacecraft on, power to the camera could result directly from some internal malfunction or from an incorrectly translated command.

Lunar Dust Contamination

It has been possible to determine quantitatively the contribution of lunar dust contamination to the total discoloration and to identify at least two sources of dust contamination: Surveyor and LM. It has not been possible to determine, except qualitatively, the relative contribution of the dust from the Surveyor and LM landings. From the results of current and planned investigations and intercorrelations, it may be possible to improve our understanding of the dust origin.

Organic Contamination

Results of the discoloration study have indicated that, from an engineering standpoint, organic contaminants are insignificant to the total observed discoloration. The presence and relative importance of organic contaminants to optics and the possible implications to science instruments on future spacecraft remain unknown at this time.

Conclusions and Recommendations

(1) Analyses of the returned Surveyor 3 hardware have produced information on the performance of typical spacecraft materials, components, etc., that can have a significant impact on future space vehicles. The major finding is the absence of significant effects of lunar exposure.

(2) Return of additional general hardware from the Moon or from space under similar conditions does not seem warranted. Specific items (i.e., solar cells) or equipment from specific environments (i.e., high-energy radiation belts around the Earth, the asteroid belt, etc.) could be of value. Possible future return of space hardware should be accomplished in a controlled manner in order to preserve the effects of exposure to be examined.

(3) Some engineering investigations were limited by the availability of controls or of documentation regarding initial conditions; however, the spare cameras and hardware in storage proved extremely valuable. Materials were selected because of their minimum response or change as the result of exposure to environmental factors. Systems were designed to allow for some variations within reasonable tolerances. Results of the comprehensive pre-launch testing to guarantee satisfactory engineering performance permitted the identification of the presence or absence of major changes. Pre-launch testing or characterization of all components to the degree necessary to identify subtle, but potentially important, changes was technically unnecessary and economically impractical. Based on requests for control parts and pre-launch information

from investigators, science investigations were similarly limited.

(4) Several continuing or potential problems for future missions have been identified or verified by this investigation. They are:

(a) Transport of lunar dust induced by landings and surface operations and the effects of such dust on optics, mechanisms, and temperature control will provide a significant constraint on future lunar operations. Results of the analysis of the returned Surveyor hardware provide valuable information on the magnitude of this problem and should be the basis of additional research.

(b) Changes observed in the physical properties of FEP Teflon and the widespread use of this material for current spacecraft indicate the need to investigate the effects of stress, radiation, and thermal cycling on these properties.

(c) Radiation discoloration observed, although expected, emphasizes the continuing need to improve the stability of thermal-control coatings and/or constrain the thermal design to allow for degradation and its uncertainty.

(d) Results of the evaluation of mechanisms and lubricants provide a significant baseline for analysis and conduct of friction and lubrication research.

(e) Cracking of conformal coating and failure of wire insulation are recognized as preventable problems. This program has emphasized the need for correct material selection and installation or application procedures.

References

1. *Test and Evaluation of the Surveyor III Television Camera Returned from the Moon by Apollo XII*, SSD 00545R, Hughes Aircraft Co., Culver City, Calif., 1970.
2. CARROLL, W. F.; AND BLAIR, P. M., JR.: "Discoloration and Lunar Dust Contamination of Surveyor III Surfaces." *Proceedings of the Second Lunar Science Conference*, MIT Press, 1971.
3. *Surveyor III Parts and Materials: Evaluation of Lunar Effects*, P-70-54, Hughes Aircraft Co., Culver City, Calif., 1971.
4. BLAIR, P. M., JR.; CARROLL, W. F.; JACOB, S.; AND LEGER, L. J.: *Study of Thermal Control Surfaces Returned from Surveyor III*, Paper 71-479, presented at AIAA Thermophysics Conference, Tullahoma, Tenn., April 1971.
5. JAFFE, L. D.: "Blowing of Lunar Soil by Apollo 12; Surveyor 3 Evidence." *Science*, vol. 171, Feb. 1971, pp. 798-799.
6. COUR-PALAIS, B. G.; FLAHERTY, R. E.; HIGH, R. W.; KESSLER, J. D.; MCKAY, D. S.; and ZOOK, H. A.: "Results of Examination of the Returned Surveyor III Samples for Particle Impacts." *Proceedings of the Second Lunar Science Conference*, MIT Press, 1971.
7. *Apollo 12 Preliminary Science Report*, NASA SP-235, Washington, D.C., 1970, p. 36.

IV. Spacecraft Changes

PART A

LUNAR DUST AND RADIATION DARKENING OF SURVEYOR 3 SURFACES

W. F. Carroll and P. M. Blair, Jr.

One of the most conspicuous features noted by the astronauts during examination of Surveyor on the Moon and later during examination of the returned hardware in the Lunar Receiving Laboratory (LRL) was the change in color. The overall tan color was in sharp contrast to the stark white paint and shiny metallic surfaces of Surveyor before launch (and to that on the model used by the astronauts during training).

Discoloration due to radiation darkening of the paint and to accumulated lunar dust had been expected. However, the expected patterns of radiation damage and conjectured patterns of dust accumulation were not evident on the returned hardware. The investigation to establish the causes of discoloration and the apparent absence of expected patterns has yielded information, primarily on the effects of lunar fines, which will be of value to future lunar operations.

The white paint used on Surveyor was known to be subject to radiation darkening. The nature and rate of discoloration had been measured in simulation tests (refs. 1 and 2), and the effect verified from temperature measurements on Surveyor 1 (ref. 3). Patterns of discoloration related to solar illumination geometry were expected because the magnitude of discoloration increases with total solar irradiation.

The abnormal landing of Surveyor 3 resulted in veiling glare and substantial loss of contrast in the pictures taken during spacecraft operation. This effect was attributed to dust on the mirror; the upper part of the mirror was significantly more affected than the lower, recessed part. It was reasonable to expect a similar coating of lunar dust on other surfaces of the camera, and with comparable variations in quantity. The astronauts observed dust contamination on the Surveyor, but detected no directional pattern associated with the Lunar Module (LM) landing (ref. 4). No effects from the LM had been expected, as there was "... preflight consideration that the landing occur outside of a 500-foot radius of the target to minimize contamination of the Surveyor vehicle by descent engine exhaust and any attendant dust excitation" (ref. 5).

Summary

Measured spectral reflectance, evidence obtained from photographs, scanning electron microscopy, and the work of other investigators on Surveyor hardware have been used to develop an understanding of the observed discoloration and its meaning to future space and lunar operations.

Measured reflectance data have been analyzed to separate and understand the effects of lunar dust and radiation damage and to conclude that organic contamination is not a major contributor to the discoloration.

Radiation-induced discoloration on the various surfaces has been found to be proportional to the degree of solar illumination. Photobleaching of the radiation damage was observed and is responsible for a gradual change in the color of the camera's surface during the evaluation program.

Organic contamination, although undoubtedly present, does not seem to be a significant factor in the observed discoloration of the external surfaces.

Almost all exposed external surfaces on the

camera are partially covered with a fine layer of lunar dust. The distribution of lunar material indicates significant contributions from fines disturbed by the initial Surveyor landing and by the approach and landing of the LM. The approaching LM apparently disturbed lunar surface material (which reached the Surveyor) over about the last 300 m of its ground track, in addition to the observed dust cloud immediately before touchdown. Some of the disturbed surface material contributed to the contamination; some of the dust cloud impacted the Surveyor and produced observable surface changes.

Lunar material, even in very small quantities, can have a significant effect on temperature control and optical performance of hardware on the lunar surface.

Examination Evidence

When the returned camera was examined in the LRL, the exterior was a dirty gray-to-tan color, with varying shades and tones and with considerable evidence of disturbance caused by handling during retrieval and return. There was no evidence of the expected contrast in radiation discoloration between surfaces with extensive solar exposure and those with little or no exposure. All external surfaces of the camera were discolored or contaminated in varying degrees.

The only obvious discoloration pattern was a series of shadows that did not correspond to solar illumination or other identifiable spacecraft geometry. In all cases, these sharply defined darker regions were found on the side of the camera that faced northwest, toward the LM landing site. Each shadow was associated with a protruding or raised surface located on the camera and near the dark region. These patterns have been shown by Jaffe (ref. 6) and Cour-Palais (ref. 7) to be the result of "sandblasting" of the camera surface by lunar material disturbed by the descending LM.

When the support collar was removed from the camera, a quantity of dark, particulate material was found inside the collar recess. (See fig. 1.) A bright spot on the camera body appeared to be an image of the inspection hole (fig. 1), but alined with the inspection hole (fig. 2) at a peculiar angle. The displacement of the image subsequently was shown to correspond exactly to the angle of incidence of material disturbed by the landing LM. Thus, the dark, particulate material trapped in the recess "sandblasted" the surface inside the clamp and produced the bright spot. It represents a sample of the LM-disturbed lunar material that "sandblasted" the Surveyor.

The first surface mirror of the camera has a diffuse appearance and light tan color. Visual examination with correct lighting, infrared photography (see fig. 3), and subsequent reflectance measurements by Rennilson (see ch. IV, pt. E) showed retention of partial mirror quality. The diffuse appearance is the result of light scattering from a partial layer of lunar fines. The mirror's surface appeared brighter in the area wiped by the astronauts as part of their examination. A small region near the top of the mirror, apparently rubbed by the plastic bag some time before release from quarantine, appeared brighter and cleaner than the region wiped by the astronauts. After the mirror was removed from the camera housing, the gradation in coverage by lunar fines from top to bottom was clearly evident. The upper protruding end had substantially more lunar material on the surface.

During subsequent examination, acetate replication peels were taken by other investigators from selected areas of the mirror to remove the adhering lunar material for study. The peeled areas showed a distinct improvement in specularity, verifying that the major source of light scattering was a readily removable layer of lunar fines. However, the protruding part of the mirror retained a slight, but distinct, diffuse character while the lower, recessed end of the mirror appeared more nearly restored to its original condition.

A geometrically sharp, curved line was identified near the bottom of the mirror. This line was a perfect projection image of the front opening of the mirror assembly from a direction in front of and below the camera. Following replication peels, a part of a second, less distinct, but geometrically sharp, similar image line was identified. Low-power, optical microscopic examination showed the upper line to be a demarcation in population of small-scale, light-scattering sites, either small pits or adhering particulate material.

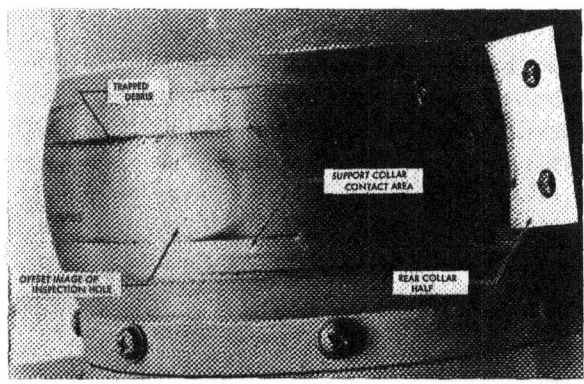

FIGURE 1.—Surveyor 3 television camera with front half of support collar removed. Back half has been displaced toward the right and upward from its original position.

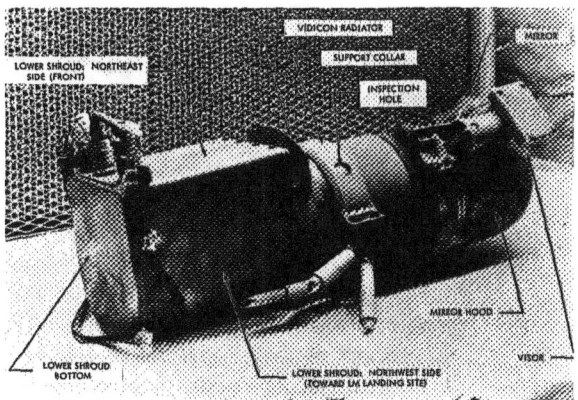

FIGURE 2.—Returned Surveyor 3 television camera.

Examination of peels using the scanning electron microscope showed a difference in small-scale (~1 μm) surface features across both of these lines. Although other explanations are possible (i.e., highly directional contamination during pre-launch vacuum testing), these lines most likely represent the effects of debris from two points on the lunar surface near the camera. The geometry associated with these images and location of the probable points on the lunar surface are described by Nickle. (See ch. IV, pt. D.)

During the evaluation program, the discolored white paint on the camera's exterior surface seemed to be fading, which was first attributed to gradual loss of lunar fines from the surface. It has been demonstrated since that the effect was due to photobleaching of radiation damage in the paint and that no loss of lunar material had occurred. The photobleaching of this paint had not been identified previously because of its slow rate; however, the effect is not surprising, as this bleaching of induced optical damage is well known (ref. 8).

Reflectance Measurements and Analysis

During the evaluation, spectral reflectance was measured in the 0.4- to 2.5-μm wavelength range on samples from representative areas of the camera surfaces. Description of the method and complete data are contained in reference 9. It has been possible to analyze these data, correlate the results with other investigations, and reach conclusions regarding the contributions of dust, or-

FIGURE 3.—Returned Surveyor 3 television camera photographed with infrared film. Note the clarity of the mirror compared with figure 2 (ch. I) and figure 26, (ch. IV, pt. E) of this document.

ganic contaminants, and radiation damage to the total discoloration.

The white surfaces showed a decrease in reflectance at all wavelengths in the range measured. Laboratory tests (refs. 1, 2, and 10) have shown that neither ultraviolet radiation nor low-energy protons cause optical damage of this paint in the near infrared (wavelength >1.0 μm). Thus, the observed reduction in reflectance at wavelengths greater than 1 μm is attributed to the presence of lunar dust; the magnitude of the reduction is proportional to the quantity of lunar dust present.

The expression developed to analyze the effects of dust and radiation is shown by

$$\rho_{m_\lambda} = \rho_{D_\lambda} K_\lambda A_D + \rho_{P_\lambda}(1 - a_\lambda A_D)^2$$

where

ρ_{m_λ} = measured sample reflectance at wavelength λ

$\rho_{D_\lambda} K_\lambda A_D$ = first surface back reflection from dust particles (negligible quantity for the white paints)

ρ_{P_λ} = reflectance of paint surface at wavelength λ ($\rho_{P_\lambda} = \rho_{0_\lambda}$ if no radiation damage has occurred; ρ_{0_λ} is original paint reflectance).

a_λ = proportionality constant related to absorptance and scattering of dust at wavelength λ

A_D = fraction of surface area covered by lunar dust

This expression shows that the reduction in reflectance is proportional to the fractional area covered by lunar fines and the spectral absorption and scattering of the lunar fines. (The incident and reflected energies pass through the dust "filter," thus the squared term.)

This expression, with the knowledge that radiation does not produce near-infrared damage and with information on the spectral properties provided by Nash [1] permits separation of the effects for all wavelengths. The radiation degradation then can be compared to laboratory simulation results, both in spectral character and total magnitude.

Similarly, the calculation permits comparison of the relative quantities of lunar material on various areas of the camera. The relative quantities so determined are shown in table 1.

Transmission measurements were made by Rennilson (see ch. IV, pt. E) before and after removing the layer of lunar fines from the clear filter of the camera. For this measurement, the detector senses only that energy in a small, solid angle in the forward direction; the energy that encounters lunar particles is either absorbed or scattered out of the forward direction of the beam. Thus, the measurement becomes a good estimate of the fractional area of the filter covered by lunar fines. The fraction 0.25, thus calculated, has been verified by Nickle (see ch. IV, pt. D) from data given by Robertson et al. (See ch. IV, pt. B.) Comparable, but somewhat different, measurements of the clean and dusty areas of the filter were made as part of this investigation. For these measurements, the filter was mounted at the entrance port of an integrating sphere so that both the forward scattered and direct transmitted energy were detected. Comparison of data from these two measurements makes it possible to estimate the value of spectral absorptance of the lunar fines on the clear filter. The accuracies of these measurements warrant only an estimate of the magnitude of the absorptance; however, such an estimate permits a reasonable assumption of the quantity of lunar material on the painted surfaces from reflectance data and the equation presented.

Other Evidence

Examination of metal surfaces (screws and washers) from the camera, using the scanning electron microscope (SEM), provided the first direct indication that lunar dust was responsible for a major part of the discoloration observed. Similar examinations permitted determinations of the quantity and particle size distribution of the lunar material on metallic camera surfaces.

It was not possible to obtain direct images of the lunar fines on the painted surfaces by using the SEM. Anderson (see ch. IV, pt. F) measured relative quantities of lunar material on the painted surfaces using a microprobe attachment for a SEM. These results show similar agreement with determinations made from reflectance data (calculated in a way similar to that described).

The relative quantities of lunar material in various surfaces were determined by Schaeffer [2] and Satkiewicz (see ch. IV, pt. H) and are given for comparison in reference 9. Schaeffer measured the quantity of trapped solar wind helium on samples from selected areas on the camera. The helium content, dominated by that trapped in the lunar fines, provides a measure of the relative quantity of lunar material. Satkiewicz, using an ion microprobe, traced the composition of sputtered materials with depth. Tracing the change in content of materials unique to the

---

[1] D. Nash, JPL, personal communication.

[2] O. A. Schaeffer, State University of New York, personal communication.

TABLE 1.—Comparison[a] of amount of lunar dust on various painted surfaces of the camera

Sample or measurement	Location	Relative quantity[a] of lunar dust
906	Top of visor	[a] (1.0)
907	Mirror hood: south[b] side (away from LM)	.5
908	Mirror hood: north side (toward LM)	1.0
898	Lower shroud: northwest side (toward LM)	.9
900	Lower shroud: southeast side (away from LM)	.4
T-3	Lower shroud: southeast side (small area adjacent to camera power cable)	<<.1
893	Lower shroud: front (facing northeast)	.7
T-7	Lower shroud: rear (facing west)	1.1
T-8	Lower shroud: rear (facing south)	.8

[a] Normalized to visor top (906).
[b] Lunar direction; for spacecraft orientation on the Moon, see ch. I.

lunar fines and to the paint permits an estimate of area coverage and effective thickness of the lunar material.

Discussion and Conclusions

Radiation Damage

Discoloration caused by radiation damage has been shown to be proportional to the solar illumination, as expected. The spectral character of the damage matches that obtained from simulation tests conducted in the laboratory. The magnitude of the damage is in reasonable agreement with laboratory simulations.

The observed photobleaching was not surprising, although it had not been observed previously on this paint. The observation emphasizes the need to return and subsequently handle hardware under controlled conditions.

The major value of the successful confirmation of expected radiation damage lies in the resulting conclusions regarding dust effects and organic contamination. The observed damage also emphasizes the need to consider degradation of thermal-control surfaces and the corresponding uncertainty in the thermal design of space and lunar vehicles.

Organic Contaminants

From analyses of reflectance data, it was concluded that organic contaminants, although most likely present, were not significant contributors to the observed discoloration. This conclusion is substantiated by the work of Simoneit. (See ch. V.) Effects of organic contaminants, although not significant to the discoloration of the thermal surfaces, may be a factor in the condition of the optics.

Lunar Dust

Adhering lunar dust radically changed the optical properties of the thermal-control surfaces and degraded the performance of the optics on the Surveyor camera. Veiling glare and contrast attenuation experienced during the Surveyor 3 lunar operations was due to lunar fines adhering to the mirror.

The distribution of lunar material on the various parts of the camera is summarized in table 1. These values are relative and normalized to the fractional area on top of the visor. The samples measured on the north and northwest side facing the LM landing site (samples 908 and 898), exposed to the "sandblast" effect, indicate a substantially higher coverage by lunar material than the opposite side. Because the sandblasting produced a lighter color by removing material, the earlier coverage was even higher. Although deposition of the heavy coating on the north and northwest surfaces may have occurred during the Surveyor landing, such an explanation is inconsistent with the amount found on the northeast (front) side.

Almost as much lunar material appeared on the front (sample 893, facing northeast) as on the side toward the LM landing site (northwest). During the Surveyor landing, deposition on the front was unlikely; deposition without some shadowing and light/dark contrast caused by protruding cable connectors would have been impossible. Deposition during final stages of the LM landing (when detected by the astronauts) also would have produced contrasts that were not evident.

The camera surface showed considerable evidence of scuffing and disturbance as the result of unavoidable handling during retrieval and return. This handling undoubtedly resulted in some redistribution of dust from one area to another. However, because the "sandblast" patterns remained so evident, redistribution was not sufficient to cancel the contrasts discussed above.

Rennilson reports evidence of more dust on the returned mirror than during Surveyor operations in 1967. (See ch. IV, pt. E.) In order to reach the mirror, dust disturbed directly by the LM exhaust must have occurred while the LM was about 300 m or more from its landing site (assuming line-of-sight trajectories for particles and assuming negligible effect from secondary material disturbed by surface impact of particles blown by the LM exhaust).

Thus, a major fraction of the lunar material on the northeast (front) and northwest sides must have arrived from a diffuse (multi-directional) source, disturbed by the approaching LM somewhat uniformly over most of the last 300 m or more of its ground track.

Some areas of the camera not in "sight" of the approaching LM also have a covering of lunar dust; this probably is due to the abnormal Surveyor 3 landing, which is known to have affected the camera mirror. The lunar material on the returned polished tube was oriented in such a way that it must have been deposited during the Surveyor landing.

Long-term deposition, such as lunar surface debris disturbed by meteorite impact, probably would produce uniformity on all sides; this was not observed. If the lines observed on the mirror are a result of secondaries produced by meteoroid impacts on the lunar surface in the vicinity of the Surveyor, such secondaries would be expected to contribute to the dust discoloration of the camera, but to an insignificant degree (<10 percent of the total lunar material).

The observed dust, therefore, originated from both the Surveyor and LM landings, with each contributing a significant amount to various surfaces. "Lunar transport" seems to be relatively insignificant, if evident at all.

From reflectance data and filter transmission measurements described, it is possible to show that the dust contaminant on the camera is in the range of 10^{-5} to 10^{-4} g of lunar fines per square centimeter of surface area. This small quantity radically alters the reflectance of the critical reflective thermal-control surfaces, increasing the absorbed solar thermal energy by a factor of 2 or 3. The quantity is small compared to the approximately 10^{-3} g/cm^2, which arrived at the Surveyor from the LM landing 155 m away. Because of the size and velocity of arriving particles, the primary effect of this final "blast" was to clean, rather than to contaminate, the surface. However, fines disturbed earlier in the LM approach contributed to the contamination of the Surveyor camera surfaces.

Clearly, lunar material disturbed by ascent or descent rockets can have a major effect on equipment on the lunar surface, even at a substantial distance from the flight path.

References

1. BLAIR, P. M.; AND BLAIR, G. R.: *Summary Report on White Paint Development for Surveyor Spacecraft*, TM-800, Hughes Aircraft Co., Culver City, Calif., 1964.
2. GILLIGAN, J. E.; AND ZERLAT, G. A.: *Study of Insitu Degradation of Thermal Control Surfaces*, IITRI-U 6061, 1969.
3. HAGEMEYER, W. A., JR.: "Surveyor White Paint Degradation." *J. Spacecraft & Rockets*, vol. 4, 1967, p. 828.
4. BEAN, A. L.; CONRAD, C., JR.; AND GORDON, R. F.: "Crew Observations." *Apollo 12 Preliminary Science Report*, NASA SP-235, Washington, D.C., 1970, pp. 29-38.
5. *Apollo 12 Mission Report*, MSC-01855, Houston, Tex., 1970.
6. JAFFE, L. D.: "Blowing of Lunar Soil by Apollo 12; Surveyor 3 Evidence." *Science*, vol. 171, 1971, pp. 798-799.
7. COUR-PALAIS, B. G.; FLAHERTY, R. E.; HIGH, R. W., KESSLER, J. D.; MCKAY, D. S.; AND ZOOK, H. A.: "Results of Examination of the Returned Surveyor III Samples for Particle Impacts." *Proceedings of the Second Lunar Science Conference*, MIT Press, 1971.

8. SCHULMAN, J. E.; AND COMPTON, W. D.: *Color Centers in Solids,* The Macmillan Co., New York, 1962.
9. *Test and Evaluation of the Surveyor III Television Camera Returned From the Moon by Apollo XII,* vols. I and II, SSD 00545R, Hughes Aircraft Co., Culver City, Calif., 1970.
10. BLAIR, P. M.; PEZDIRTZ, G. F.; AND JEWELL, R. A.: *Ultraviolet Stability of Some White Thermal Control Coatings Characterized in Vacuum,* Paper 67-345, presented at AIAA meeting, New York, April 1967.

PART B

CHARACTERIZATION OF DUST ON CLEAR FILTER FROM RETURNED SURVEYOR 3 TELEVISION CAMERA

D. M. Robertson, E. L. Gafford, H. Tenny, and R. S. Strebin, Jr.

Surveyor 3 landed on the Moon in April 1967. Part of the spacecraft was returned to Earth in November 1969 by the Apollo 12 astronauts.

A stripping film containing dust removed from the camera light filter was received for study by Battelle-Northwest (BNW) from the Jet Propulsion Laboratory (JPL). The study conducted involved the characterization of the dust; the results of the study are presented here.

Individual particles of dust from the Surveyor 3 camera light filter were examined. The dust particles (from 2 to 40 μm) were released from a (stripping) cellulose film, isolated, and analyzed by optical microscopy, electron microprobe, and X-ray diffraction. The analytical results indicate that the dust is of lunar origin. While the average composition and characteristics are in agreement with other lunar fine analyses (see ref. 1), this study clearly shows significant composition variation from particle to particle in the micrometer-size range.

Handling of Primary Samples

The samples, three cellulose films, were taken consecutively from one-half of the clear filter. The sample package was opened in the front section of a laminar air flow clean bench; the samples were immediately transferred into the bench work area. (See fig. 1.) The films were taped to clean microscope slides with the particle-containing surface facing up. (See fig. 2.) The samples remained in the bench until packaged for return to JPL.

Analytical Processing

Examination of "As Received" Cellulose Films

Figure 3 shows the particle content of the three films and a blank. This blank may not be the same lot of film used to strip the particles.

It is apparent from the photomicrographs that the first strip (ND-1) removed much more dust than succeeding strips (ND-2 and ND-3). Film ND-1 was used to obtain the particles for study. No additional work was performed on ND-2 and ND-3.

General Procedure for Individual Particles

The general procedure for analysis of an individual particle involves the steps described below. Particle 5 was photographed at various steps to help visualize the procedure. (See fig. 4.)

Step 1: Locate or select a particle in the cellulose film for analysis. (See fig. 4(a).)

Step 2: Cut a square of film (about 100 by 100 μm) containing the particle and remove the square to a clean microscope

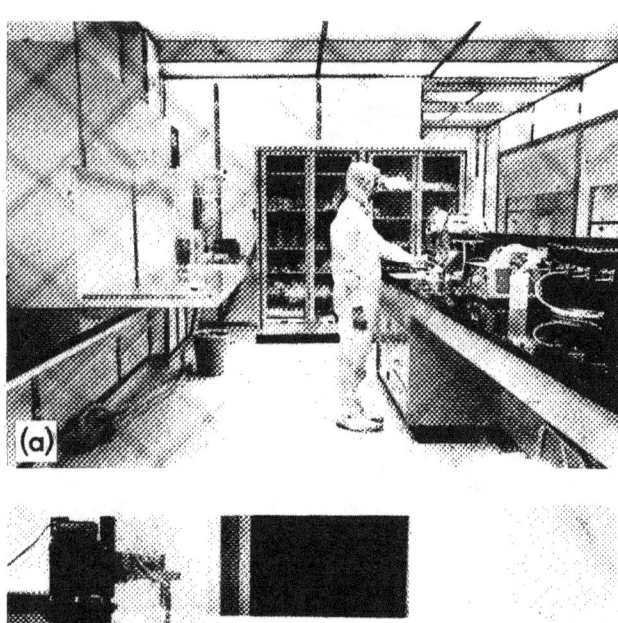
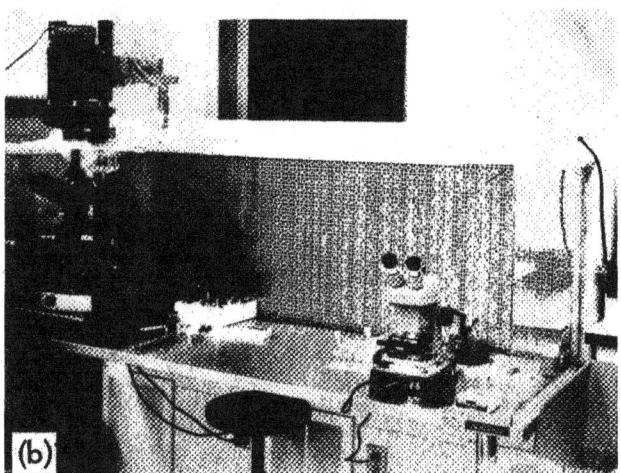
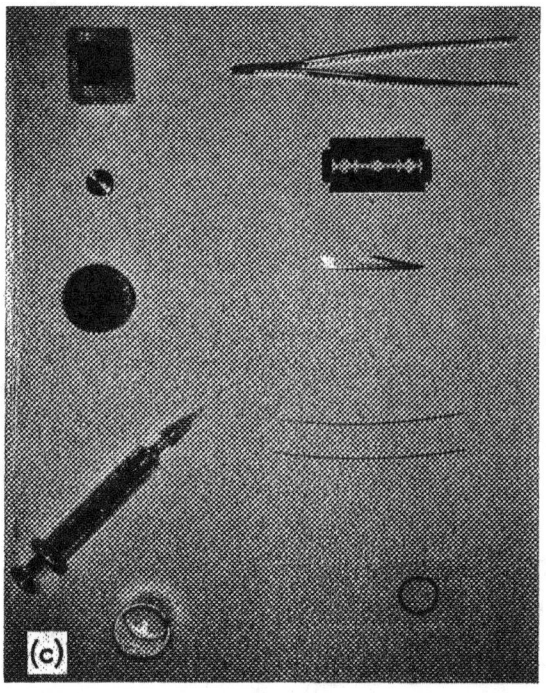

FIGURE 1.—(a) Clean laboratory and clean bench. (b) Typical optical clean bench with stereomicroscope and research microscope. (c) Particle tools. Left, bottom to top: microbreaker, micropipet, microprobe, X-ray and mass spectrometer mounts. Right, bottom to top: slide with circled work area, tungsten needles, surgical blade, razor blade, and forceps.

FIGURE 2.—As-received cellulose film placed on microscope slide.

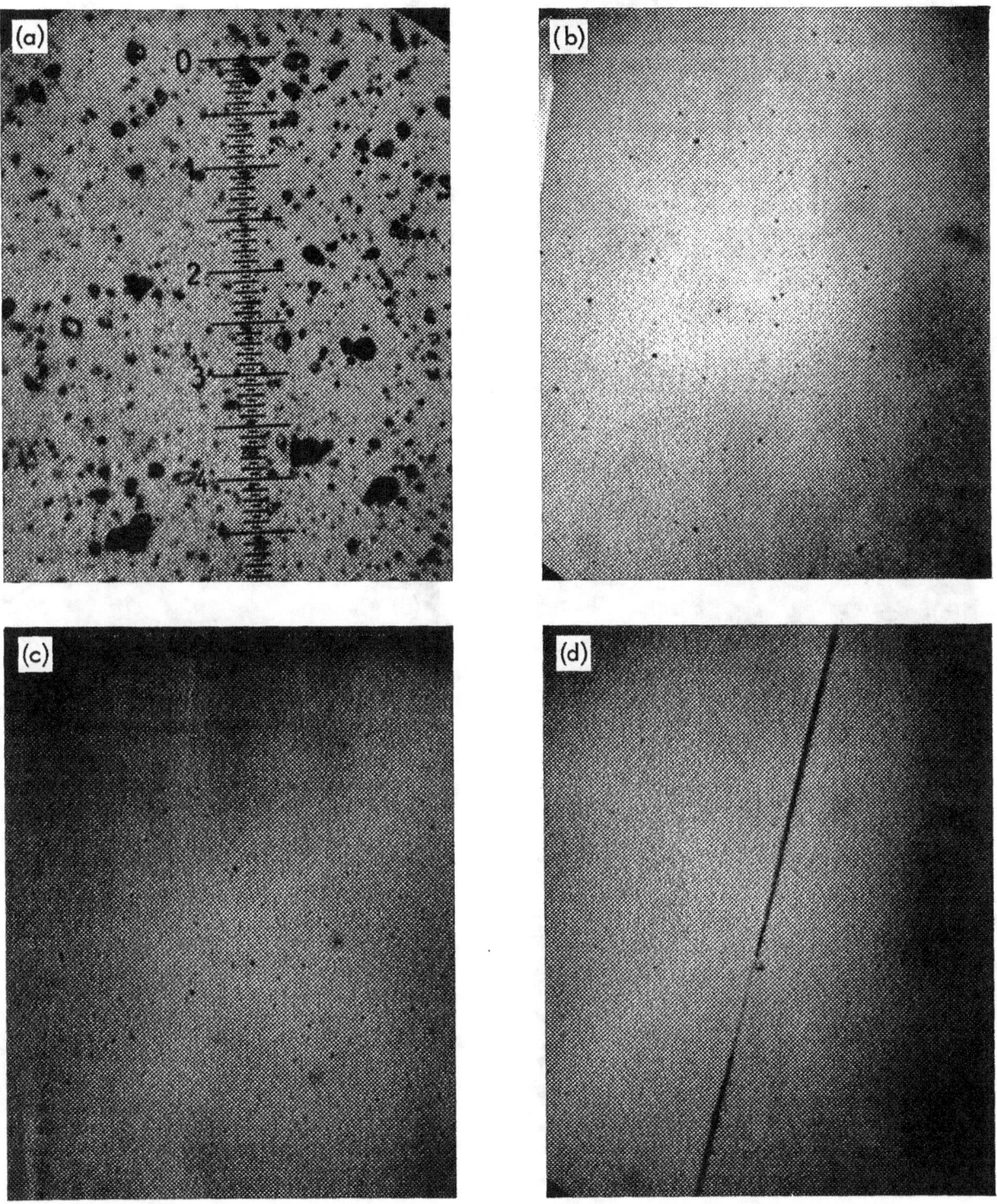

FIGURE 3.—(a) Film ND-1. Transmitted light (at 400 ×) in polacolor. (b) Film ND-2. Transmitted light (at 400 ×). (c) Film ND-3. Transmitted light (at 400 ×). (d) Blank cellulose film.

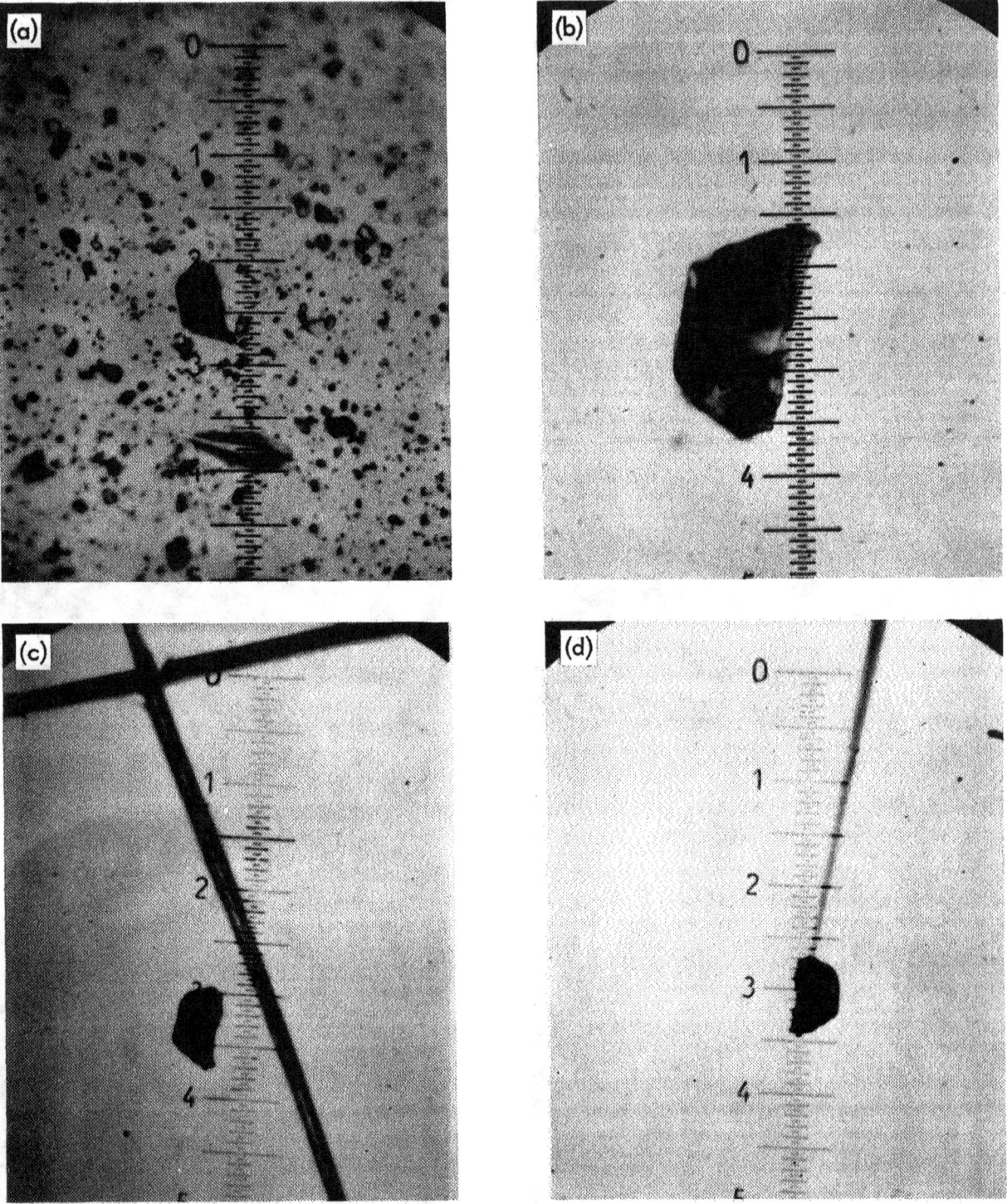

FIGURE 4.—(a) Particle located in as-received cellulose film (at 400 ×). (b) Isolated particle. Transmitted light (at 800 ×). (c) Particle mounted for microprobe analysis. Incident light (at 400 ×). (d) Particle mounted for X-ray diffraction. Transmitted light (at 400 ×).

slide. (Fig. 10 is an example of such a square.)

Step 3: Dissolve the square film (see table 1); isolate the particle from other particles in the square and wash it free of film material. (See fig. 4(b).)

Step 4: Transfer the particle with a tungsten needle to the grid of an electron microprobe mount and map the location. (See fig. 4(c).)

Step 5: After microprobe analysis, recover the particle and mount it on a glass fiber tip for X-ray diffraction. (See fig. 4(d).)

The clean laboratory, a clean bench with optical equipment, and particle tools are shown in figure 1. Figures 5 and 6 show the electron microprobe and the specially developed X-ray diffraction camera. Figure 7 shows some primary particle 5 data obtained with the microprobe. Three-stage, thermal ionization mass spectrometry is also used in particle studies when isotopic abundance data are needed.

Optical Examination

Optical examination indicated the following general morphological characteristics. (See table 2.) About 90 percent (number base) of the visible material was small (<10 μm), transparent, clear to pale yellow, slightly angular, flattened, and glassy. The remaining particles were larger, more intensely yellow, and more equant and rounded. Unique shapes included spheres and

TABLE 1.—*Analysis procedures*

Dissolution of cellulose film
In order to isolate individual particles, a solvent with rapid dissolving properties and a moderate evaporation rate was needed to dissolve acetyl cellulose stripping film. After screening 14 possible solvents, acetonitrile and N-N dimethylformamide were found to be the most promising. By combining half acetonitrile and half N-N dimethylformamide, the solution and evaporation rate allowed the solution of micro squares of acetyl cellulose on a microscope slide in small droplets of solvent.
Electron microprobe X-ray analyzer procedure and equipment
Isolated particles were analyzed on polished cobalt substrates with a Materials Analysis Co. Model 400–S electron microprobe. The emitted X-rays were resolved and measured by a cooled, lithium drifted silicon energy dispersive detector.[a] (See fig. 7.) This detector has a resolution of 300 eV for 6.4–keV X-rays and is equipped with a 1-mil Be window. Polished metal surfaces were used for standards except for sodium, potassium, chlorine, and sulfur. Carbon-coated single crystals of KNO_3, NaF, and NaCl were used as standards for potassium, sodium, and chlorine, respectively. Carbon-coated sulfur was also used as a standard. The elemental composition of the particles was determined from the X-ray spectra by a weighted least-squares fit obtained with a "GEM" computer program. (See ref. 3.) A 20-keV electron beam of 1×10^{-9} A was swept across each particle, a secondary electron image of the particle was produced on an oscilloscope. The beam then was centered on the particle and the emitted X-rays were counted for 5 min. The data were printed on punched paper tape. The computer program was used to obtain the analytical results.
X-ray diffraction procedure and equipment
The X-ray diffraction of individual lunar particles was accomplished on a Rigaku Denki rotating anode generator (RU–3V). Nickel-filtered copper Kα radiation was used with the tube operated at 45 kV and 45 mA. A 2.58-cm-diameter powder diffraction camera (fig. 6) was used for the analysis. This camera was designed and built at BNW to determine X-ray spectra on micrometer-size particles. The camera is evacuated to a pressure of 50 μm during the exposure time. Each individual particle was mounted on a glass fiber that had been drawn out to a 2- or 3-μm point. Lunar particles were held to the fiber with a small amount of rubber cement. During exposure, samples were rotated at 1 rpm. The X-ray diffraction spectra were recorded on Kodak No-Screen Industrial X-Ray film and processed in a normal manner.

[a] Sodium was determined by wavelength dispersion and a flow counter.

FIGURE 5.—Electron microprobe X-ray analyzer.

rods with somewhat bulbous ends. Opaque material was less than 5 percent of the total. Birefringence was present in less than one-half of the material and was generally weak. Scanning electron microscope photomicrographs of typical particles (67, 77, and 52) are shown in figure 8.

A number count and estimated size of particles were made at four locations in the film corresponding to filter locations marked in figure 9. Squares, nominally 100 μm on a side, were cut from the film. (See fig. 10(a).) The square was dissolved and the particles allowed to separate over a restricted area to facilitate counting. (See fig. 10(b).) The separated particles were counted and sized at about 500 $\times$ magnification in transmitted light. Size was estimated to the nearest micrometer with a calibrated reticule. (See fig. 11.) No depth estimate was made. Sizes up to about 5 μm were recorded as a single dimension; i.e., diameter of an "equivalent" area circular particle. The average estimated lengths and widths were recorded for larger particles. The data are shown in figure 11 and are presented in table 3.

Electron Microprobe Elemental Composition Analysis

Seventy-five individual particles were analyzed using the microprobe. Only particle 60 (a stainless steel) appears to be man-made and may be a piece of the Surveyor 3 spacecraft.

Table 4 lists the "average" composition of the particles analyzed. This composition is compared with the wet chemical analysis of bulk fines (ref. 1).

The microprobe data for the 75 individual particles are listed in tables 5 and 6 according to increasing percentages of silicon, the most prevalent element. Table 5 lists the weight percent for each element. Table 6 lists calculated and normalized data, with the assumption that certain elements are present as oxides. Oxygen could not be measured with the present detector system.

There is good agreement between bulk and our averaged individual particle values for sev-

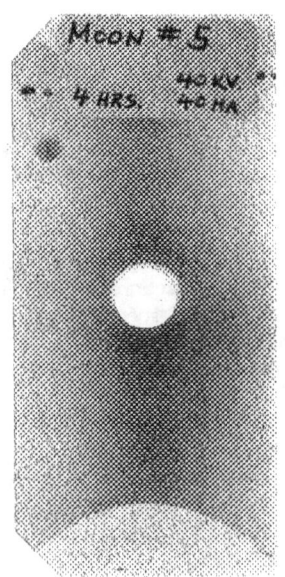

FIGURE 6.—(a) X-ray diffraction unit showing camera (center) in place. (b) X-ray diffraction spectrum.

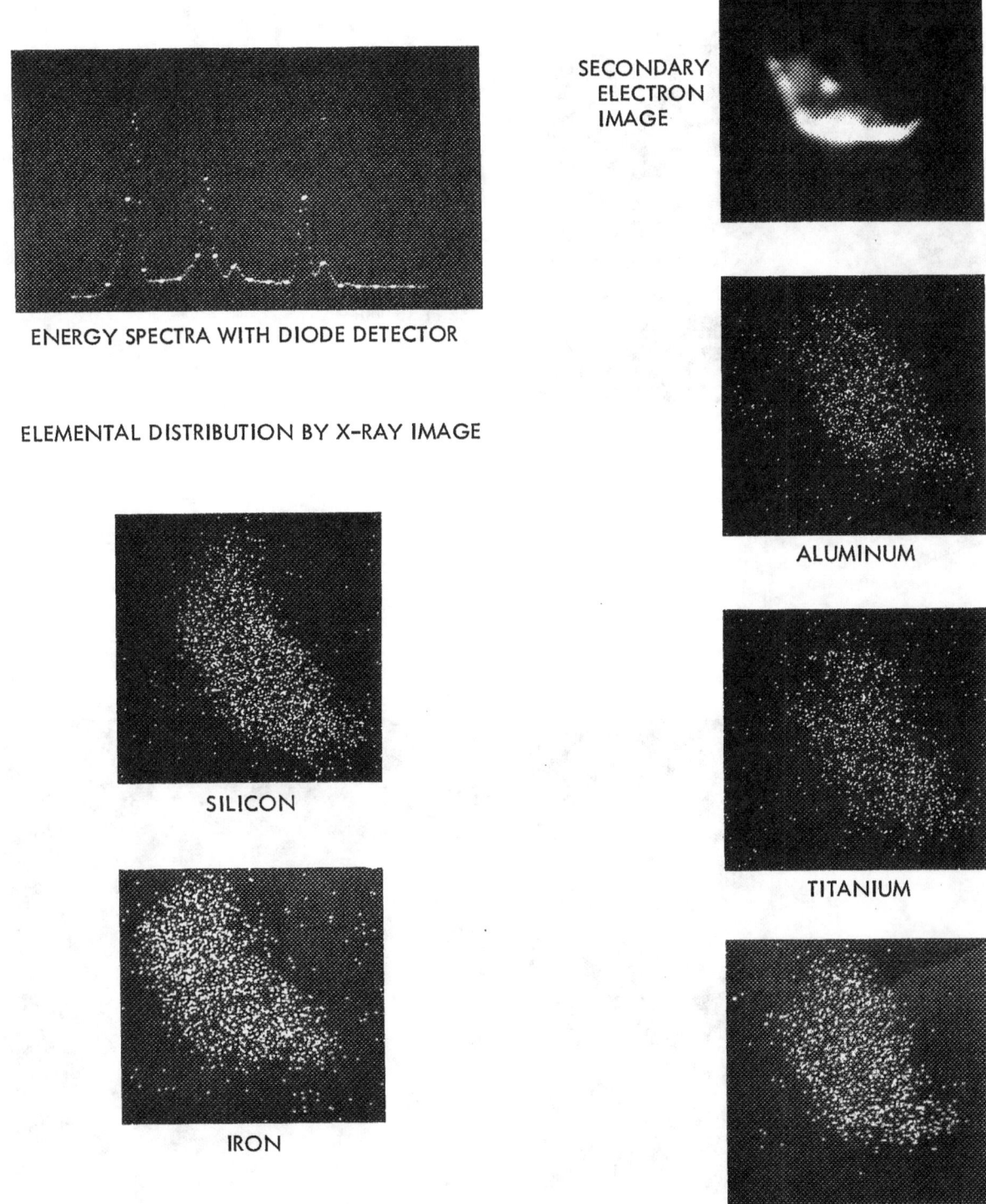

Figure 7.—Electron microprobe data for particle 5.

FIGURE 8.—Scanning electron microscope photomicrographs of typical particles. (a) Particle 67 at 4000 ×. (b) Particle 67 at 10 000 ×. (c) Particle 67 at 25 000 ×. (d) Particle 77 at 2000 ×. (e) Particle 77 at 15 000 ×. (f) Particle 52 at 4000 ×. (g) Particle 52 at 15 000 ×.

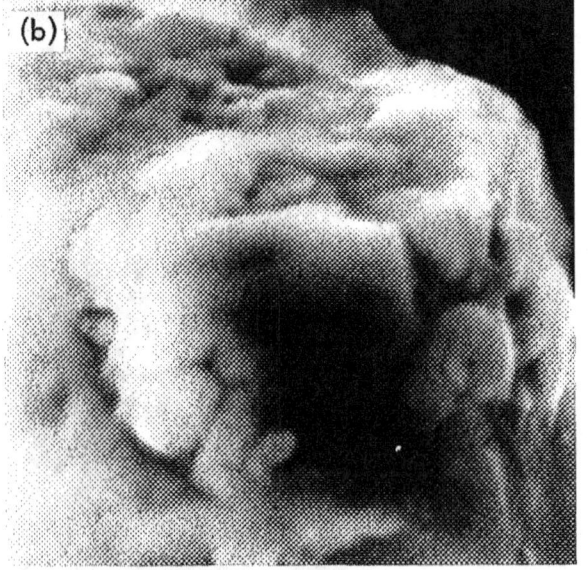

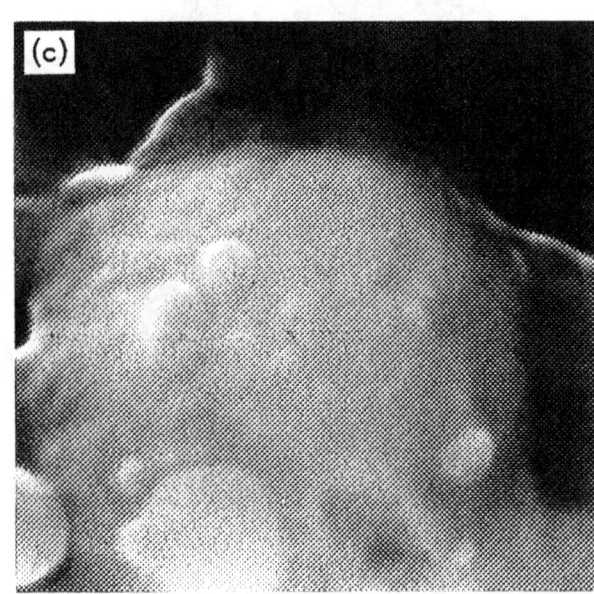

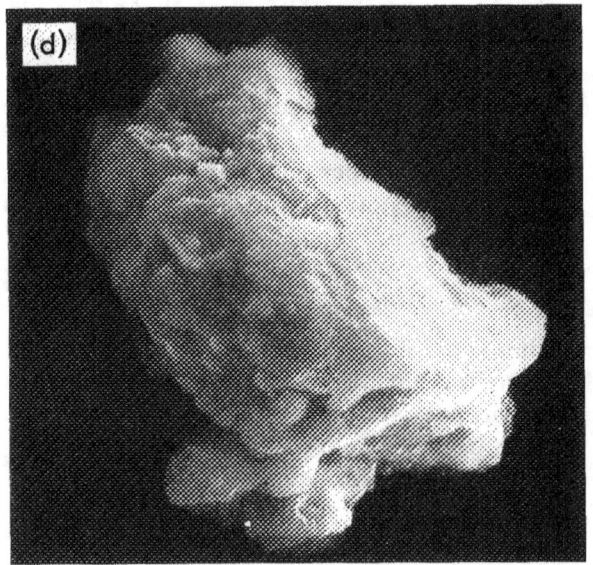
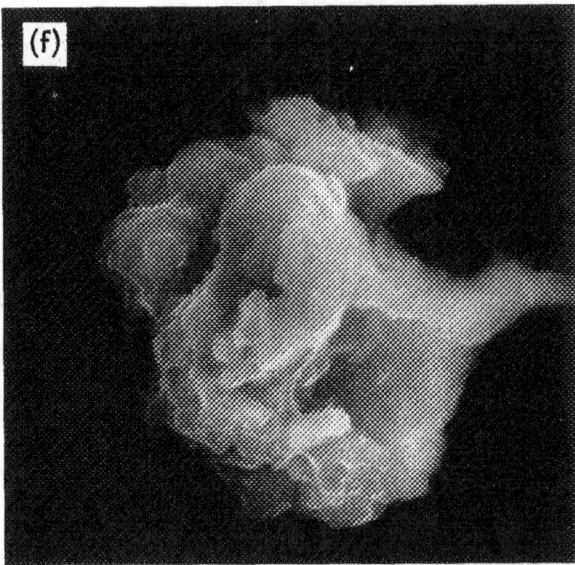

FIGURE 8.—Concluded.

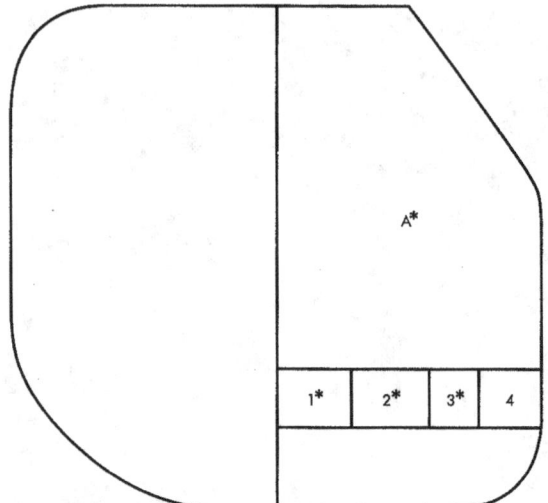

*LOCATIONS ON CLEAR FILTER CORRESPONDING TO POSITIONS ON THE STRIPPING FILM (SEE FIGURE 2), WHERE SQUARES WERE EXTRACTED FOR PARTICLE SIZE AND COUNT DETERMINATIONS.

FIGURE 9.—Diagram of clear filter. Strip films were taken from right half.

eral oxides; e.g., SiO_2, FeO, and Na_2O. The differences that appear are perhaps to be expected because our composite was small (75 individual particles). This is supported by the fact that differences between our average value and our single particle compositions showed even greater variations. Thus, analysis of individual particles can be important when dealing with fines and dust.

X-Ray Diffraction Data

The X-ray diffraction results of 30 dust particles are listed in table 7. About 57 percent of the dust particles are amorphous or glassy material. This appears to be consistent with previously examined lunar fines and soils. (See ref. 2.)

Of the crystalline material examined, there are two major mineral phases present: plagioclase and clinopyroxene. Bytownite, anorthite, and labradorite members of the plagioclase group were found. Augite and pigeonite clinopyroxenes were the other major minerals identified. Tridymite also was found.

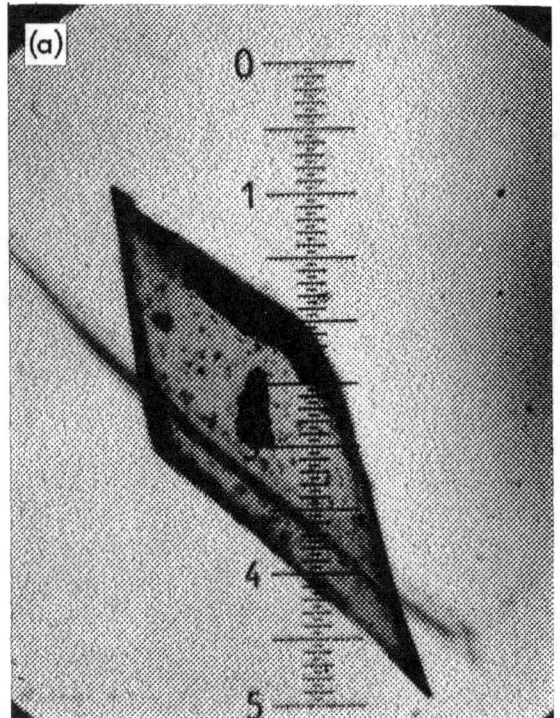
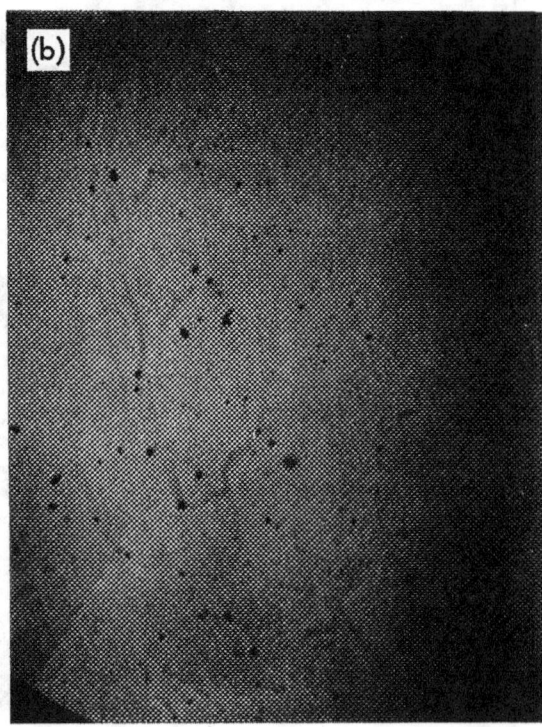

FIGURE 10.—(a) Square of film containing particles to be counted (at 400 ×). (b) Separated particles from square of film (at 50 ×).

TABLE 2.—*Morphological data on individual particles*

Sample	Size, μm L×W×H	Color Transmitted	Color Incident
76	10×8×9	Yellow	Yellow
56	15×7×6	Colorless	Yellow
78	11×6×8	Black	Yellow
66	12×6×6	Yellow	Yellow
67	10×8×10	Colorless	Yellow
61	25×8×6	Colorless	Colorless
75	19×8×7	Yellow	Yellow
59	12×12×6	Yellow	Yellow
83	5×3×3	Yellow	Yellow
52	12×10×9	Brown	Yellow
72	20×19×13	Opaque	Yellow
69	15×7×6	Colorless	Yellow
73	28×16×8	Yellow	Yellow
68	10×8×9	Yellow	Yellow
57	20×17×15	Yellow	Yellow
4	Diameter=5	Colorless	Colorless
62	15×12×6	Yellow	Yellow
74	29×19×8	Brown	Yellow
54	18×10×10	Dark yellow	Yellow
50	22×18×15	Dark yellow	Yellow
58	12×10×8	Yellow	Yellow
79	57×31×24	Opaque	Dark yellow
51	17×17×10	Yellow	Yellow
70	11×6×8	Yellow	Yellow
80	21×13×9	Yellow	Yellow
55	20×15×15	Brown	Colorless
81	16×9×7	Yellow	Yellow
64	10×8×10	Colorless	Colorless
53	18×10×8	Yellow-brown	Yellow
60	30×6×6	Opaque	Metallic silver

During the diffraction studies, a significant feature was observed that may bear upon the origin of the dust. Even though the general particle size was in the micrometer-size range, these individual dust particles were not small, homogeneous pieces of larger single-phase material. Most of these particles were mixtures of more than one mineral. The mineral name applied to each particle in table 7 was the major or dominate spectrum that could be identified.

That these dust particles were mixtures of more than one type of material is indicated by the microprobe data. Individual chemical analyses deviated markedly from theoretical values of identified crystalline phases.

Two spheres (82 and 4) were X-rayed and found to be amorphous.

Test for Fissionable Material

Two pieces of ND–1 film were subjected to thermal neutron irradiation. (See fig. 12.) The film was placed on a solid-state fission track detector plastic and irradiated to 10^{15} neutrons/cm². Examination of the plastic after etching revealed no fission fragment damage tracks. Figure 12 also shows the final condition of film ND–1 after our analytical sampling.

Conclusions

It seems reasonable to conclude that the dust examined is of extraterrestrial origin. The following points support this statement:

(1) Mineralogy indicates a similarity with

TABLE 3.—*Particle counts on light filter strip ND-1*

Location (see fig. 9)	1	2	3	4
Size,[a] μm:				
<1	1405	1443	983	911
<1 to <2	476	335	395	523
<2 to <3	210	190	183	388
<3 to <4	65	49	52	121
<4 to <5	21	20	16	29
5×4	4	2	5	
5×5	2	5	5	
5×6		7	8	4
5×7		1		
6×2	1			
6×4	4	2		
6×6				4
6×7			1	
6×8	2	4	2	
7×7	3	2		
8×8				3
8×10	2	2		2
8×14	1	1		
10×10			1	2
10×20		1		
12×14				1
Total	2196	2064	1651	1988
Square size, μm	14 000	12 350	8500	11 550
Particles/10 000 μm (100×100 μm square)	1569	1670	1942	1721

[a] Sizes with single dimension were estimated average diameter (of equivalent circular area). On larger particles, both average length and width were estimated.

TABLE 4.—*Comparison of Surveyor 3 dust and lunar fines*

Component	SiO_2	FeO	CaO	Al_2O_3	TiO_2	MgO	K_2O	Cr_2O_3	S	ZnO_2	Na_2O
(A)[a]	39.0	15.8	17.3	19.8	2.6	3.2	0.7	0.06	0.3	0.07	0.4
(B)[b]	41.8	15.98	11.68	13.68	7.42	8.38	.13	.36	.10	.05	.41

[a] (A) "Average" composition of analyzed dust (this article).
[b] (B) Average fines (p. 450 of ref. 1).

TABLE 5.—*Electron microprobe elemental composition of lunar samples (in weight percent)*

Particle	Si	Fe	Ca	Al	Ti	Mg	Cl	K	Cr	S	Pb	Sn	Zn	Na	V	Mn	Ni
17		47		1			1										
60		75														1	9
13		1	35	6													
36			18	10		4											
35	5		21	20	25												
32	4		12	19													
16	5	35	2	4	15				16								
8	4		35	6	.5				.5								
29	5	23		1		4			.6				0.5	6	0.4	.5	
11	3	1	8	14			3										
34	8		7	14													
40	14	27	.9	15	4	9											
14	13	4	15	12	.5												
21	8	2	5	8													
20	10	12	6	4	.8	3		2			1						
12	10	14	6	17	1												
15	12		11														
33	7	7	9	9	1												
24	9	6	6		1												
76	5	7	5	7		10	.5			0.5							
42	13	24	.6	7	2	3											
31	13	9	8	13	.5												
22	14	5	12	13													
37	12		11	5		5											
56	7	2	8	2		11											
78	11	17	2														
18	8	8	10	11	.9		1		.9					1			
39	13	2	8	11	.6					2							
66	18	31	15	3													
67	18	.9	.8	16	2					1							
61	3		10	8		5	4	.7	.6								
48	17	11	11	3	2			.8									
75	14	2	11	11	2												
19	14	13	5	3	2					1							
43	14	12	8	5	2												
59	13	9	8	4	1				.5								
5	11	8	7	3		2			.2								
83	13	3	9	9							.5						
52	18	11	8	7	1	3	.6			2							

TABLE 5.—Electron microprobe elemental composition of lunar samples (in weight percent)—Concluded

Particle	Si	Fe	Ca	Al	Ti	Mg	Cl	K	Cr	S	Pb	Sn	Zn	Na	V	Mn	Ni
72	14	4	7	7	0.3	2	4										
69	13	.9	11	9										0.7			
46	18		14	14			1										
23	12	10	3	3	2	4											
41	16	11	8	5	.6					0.7							
73	13	3	8	8			1										
30	14	10	3	3	2	7											
68	17	10	8	6						.5							
57	17	5	9	9													
4	16	3	10	9	.6												
62	16	11	5	3	.5	3	.7		.4								
74	14	7	7	5	2		.7										
54	17	6	8	7	.9	3											
77	17	17	8	2	1		.7		.5	.6							
50	17	13	4	2	1	5	.6	5									
45	21		7	13	.9												
47	21	13	16	3	2	7			.4	.6							
58	19	11	6	5	1					.6							
79	14	8	6	4	.7												
1	15	8	7	6	1	4			.4	.6							
27	15	8	6	5	.8			.7									
51	19	9	7	5	1	6	1	.7									
70	15	7	7	5	1												3
63			16	7	7												
26	14	6	5	6	2	3	.6		.3	.6							
82	17	7	7	6	.8	3											
80	17	15	6		1	9											
9	16	10		3		3											
55	19	7	7	5	1	3	.6			.8							
81	11	8	3		.8			6									
10	18		1	10	1			10									
64	23			9									2				
28	25			8				.6									
53	19		.7		.6		.8							.8			
25	25			8													
38	25			8													
44	37																

TABLE 6.—*Composition of lunar samples*[a]

Particle	SiO$_2$	FeO	CaO	Al$_2$O$_3$	TiO$_2$	MgO	Cl	K$_2$O	Cr$_2$O$_3$	S	Ni	PbO$_2$	Sn	ZnO	Na$_2$O	VO	MnO
17				3			2										
13		95		18													
36		2	79	41					1							1	1
35			54	53					2				2				
32			40	68													
16	9	45	32	12	42												
8	11		3														
29	12	40	77	15	33												
11	15	4	44	8	3												
34	28		20	52													
40	30	41	15	36		7											
14	30	6	27	49	8												
21	30	.6	15	26	1	13											
20	31	26	14	14	2		9	6									
12	31	31	14	47	3	7											
15	31		22	45													
33	32	24	16	34	4												
24	32	16	26		7	19											
76	33	33	1	19		6	7										
42	34	46	16	34	4												
31	34	19	23	41	1												
22	34	8	25	26		11											
37	34		32	7		37											
56	35	6	5														
78	36	40															
18	37	26	22	33	3												
39	37	5	13		1												
66	38	48	25	36													
67	38	1	7														
61	38		18	19	4		.9	6			14						
48	39	18	23	33				1									
75	39	5	11	10	5	10					2						
19	40	27	17	14	5												
43	40	23	19	14	4												
59	40	21	19	11											24		
5	40	19	19	30		6						1.1					
83	40	7	23	17	3		31		.6		3						
52	41	18	14	17	.8	5	<1										
72	41	8	17	21		5	7										

TABLE 6.—Composition of lunar samples[a]—Concluded

Particle	SiO$_2$	FeO	CaO	Al$_2$O$_3$	TiO$_2$	MgO	Cl	K$_2$O	Cr$_2$O$_3$	S	Ni	PbO$_2$	Sn	ZnO	Na$_2$O	VO	MnO
69	41	2	27	30													
46	41		25	33			1										
23	42	26	8	12		10											
41	43	21	17	14	5												
73	43	7	20	28	2												
30	44	23	7	11		16											
68	44	19	16	16	4	.7											
57	44	9	19	25	1	.5											
4	44	6	22	27	1												
62	45	23	11	10	5	6	<1										
74	45	16	18	16	4		1				1						
54	45	12	17	20	3	6	<1										
77	46	28	14	5	3		1		<1	<1							
50	46	26	9	6	3	9			<1								
45	46	13	31		2												
47	47	22	28	8	4	12	<1		<1	<1	3						
58	47	19	11	17	3					1							
79	47	19	15	14	2												
1	47	18	18	19	4												
27	47	18	14	12	2	7			<1	<1							
51	47	17	14	16	3												
70	47	16	17	18	3			1									
63	47	15	17	25	3												
25	47	15	13	15		7											
82	48	12	13		4	6											
80	48	31	13		2	19	.8		.5	.8							
9	48	22		10		6											
55	50	13	15	13	3	10	<1			<1							
81	51	26	10		2												
10	55			33				12									
64	56		1	23				16									
28	71			24											5		
53	74		22		2												
25	75			25													
38	75			25	2												
44	100																

[a] Normalized microprobe data, in percent.

TABLE 7.—*X-ray diffraction data on individual lunar dust particles*

Particle	Compound
53	Plagioclase-bytownite
55	Plagioclase-anorthite
56	Plagioclase-anorthite
73	Plagioclase-anorthite
76	Plagioclase-labradorite
80	Plagioclase-anorthite
5	Clinopyroxene-augite
50	Clinopyroxene-augite
74	Clinopyroxene-pigeonite
77	Clinopyroxene-pigeonite
81	Clinopyroxene-augite
64	Tridymite
78	No identification (crystalline)
2	Glass
3	Glass
4	Glass
11	Glass
12	Glass
51	Glass
52	Glass
54	Glass
57	Glass
58	Glass
63	Glass
64	Glass
72	Glass
75	Glass
79	Glass
82	Glass

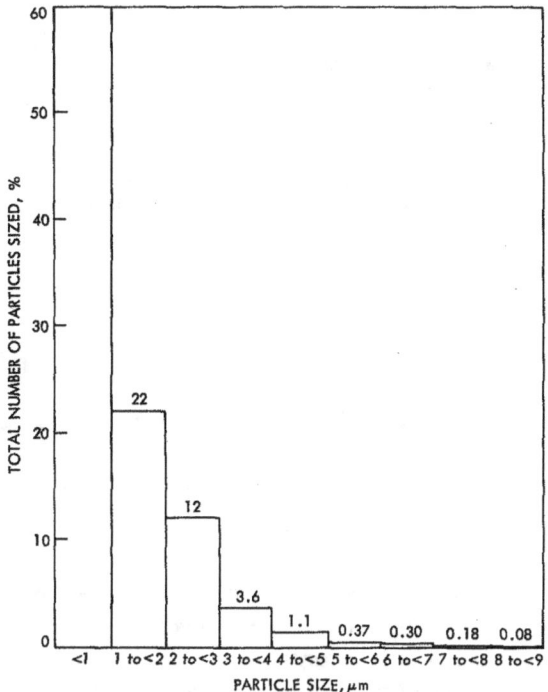

FIGURE 11.—Particle number distribution as a function of size.

FIGURE 12.—Final condition of ND-1 film.

bulk mineral phases found in lunar rocks and soils.

(2) Reasonably high percentage of glassy or amorphous material is typical of lunar solids examined to date.

(3) Presence of glass spheres is a feature that is typical of lunar rocks and soil.

(4) "Average" chemical composition of the particles approaches that reported for other lunar material. However, there are significant differences among the compositions of individual particles. These differences can be seen only by analyses of the type conducted in this study.

The origin of this dust appears to be from a fine-grained rock or soil. The X-ray examination shows that the majority of the particulates are complex mixtures of more than one crystalline phase and not merely micrometer-size pieces of single-phase minerals. Therefore, the most logical parent material of this dust is a fine-grained breccia or a soil from such a rock type.

References

1. FRONDEL, C.; KLEIN, C., JR.; ITO, J.; AND DRAKE, J. C.: "Mineralogical and Chemical Studies of Apollo 11 Lunar Fines and Selected Rocks." *Proceedings of the Apollo 11 Lunar Science Confer-*

ence; vol. 1, Mineralogy and Petrology, 1970, pp. 445–474. Also Geochim. et Cosmochim. Acta, suppl. 1, 1970.
2. DUKE, M. B.; WOO, C. C.; SELLERS, G. A.; BIRD, M. L.; AND FINKELMAN, R. B.: "Genesis of Lunar Soil at Tranquillity Base." Proceedings of the Apollo 11 Lunar Science Conference; vol. 1, Mineralogy and Petrology, 1970, pp. 347–361. Also Geochim. et Cosmochim. Acta, suppl. 1, 1970.
3. NICHOLSON, W. L.; SCHLOSSER, J. E.; AND BRAUER, F. P.: "The Quantitative Analysis of Sets of Multicomponent Time Dependent Spectra from Decay of Radionuclides." Nucl. Instr. Methods, vol. 25, 1963, pp. 45–66.

PART C

DEBRIS ON THE SURVEYOR 3 MIRROR

M. H. Carr and S. J. Proudfoot

This article describes work performed on debris that adhered to the Surveyor 3 camera mirror after it was returned from the Moon during the Apollo 12 mission. The chemical and morphological natures of the debris are described and some fine-scale features of the mirror surface are discussed. Almost all of the debris is from the Moon. Astronaut Conrad wiped part of the mirror before removing it from the lunar surface; the wiped area was clearly visible when our samples were taken from the mirror. It is suspected that much of the material had been on the mirror since the Surveyor 3 landing and that it was the main cause of the veiling glare encountered during the Surveyor 3 mission. No new conclusions regarding the nature of lunar fines are presented here, nor were any anticipated when the work began. The main intent was to provide supplementary information on the adhering debris so that the causes of the optical degradation of the mirror could be determined more accurately. The data are, therefore, presented with a minimum of discussion.

Sampling

A standard peel technique was used to remove the debris from the mirror. A preliminary examination indicated that most of the debris was below the limit of resolution for optical microscopy. It was clear that the material had to be removed from the mirror in such a way as to allow for subsequent examination in the electron microscope. Removal in a plastic replicating tape softened with acetone was decided upon because it is efficient and because normal electron microscope procedures for sample preparation could be followed. Before the peels were made, three large particles visible to the naked eye were removed with a needle. These particles later were found to be contaminants.

Several areas of the mirror were sampled (fig. 1). Most of the mirror appeared to be covered with dust, but some slight shading was apparent at one end. This may have resulted from shielding by the mirror housing. Also some interference bands were visible when the mirror was viewed under oblique light. Samples were taken along a strip that crossed both the shading bands and the interference bands. At each location, approximately 1-cm² pieces of acetyl cellulose tape (0.0034 cm thick), moistened with acetone, were placed on the mirror; they were removed after the acetone had dried. The debris was molded

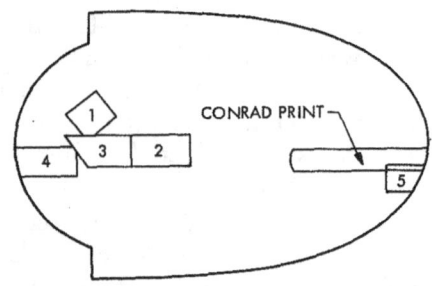

FIGURE 1.—Location of the sample areas of the mirror. Several peels were taken at each sample location.

into the soft plastic and removed from the mirror when the peel was lifted. Duplicate peels were taken at each sample location. Subsequent examination showed that each peel removed 95 to 98 percent of the material on the surface. (This was contrary to the experience with the Surveyor 3 aluminum struts on which there was still adhering material after several peels.) As only a small portion of each peel was used, the peels are available to other experimenters.

After several peels had been taken in areas 2, 3, and 4, the interference bands were still visible. One possibility was that the bands were caused by material adhering to the surface, so more severe steps were taken. The strip that had been sampled was rubbed vigorously with a Q-tip to remove any remaining material, then additional peels were taken. Examination of the peels showed that almost all of the material had been removed from the surface of the mirror by the initial peels, so that the interference bands could not be attributed to dust on the surface.

Electron Microscopy

Samples were prepared for electron microscopy in two ways. The first and more simple technique placed the sample directly in the microscope for a check on the second and more complex replication technique and possibly for electron diffraction work. The mounts were prepared by shadowing the tape containing the sample with carbon and platinum, then dissolving the tape in acetone. This left the particulate debris directly on a carbon-platinum film that could be viewed in the electron microscope (fig. 2). This type of mount, while necessary for diffraction work, is unsuitable for observing particle morphology as only shadows of the particles can be seen (fig. 3). To obtain a better view of the particles, a replication technique was used. The cellulose tape containing the sample was painted with polyvinyl alcohol (PVA); the cellulose tape was dissolved in acetone to leave only the sample and the PVA (fig. 2). After shadowing with platinum, and then with carbon, the PVA was dissolved in water. The sample itself was dissolved in hydrochloric acid to leave a platinum-carbon replica of the sample which,

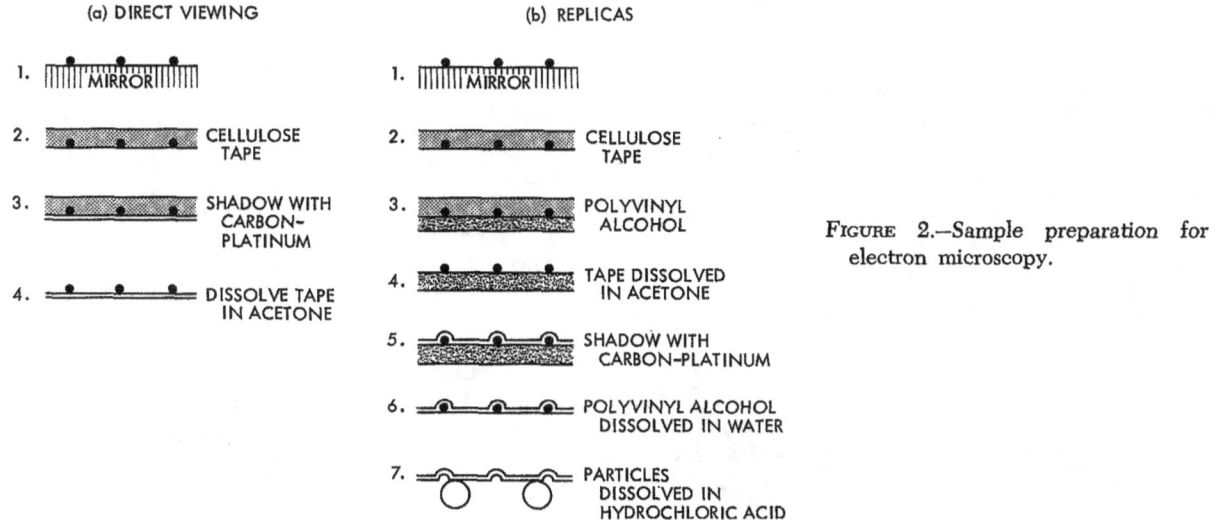

FIGURE 2.—Sample preparation for electron microscopy.

when mounted on a grid, could be viewed directly in the electron microscope. A typical replica is shown in figure 4.

The material on the mirror consists of fine-grained, angular fragments. Spherical particles are restricted primarily to the smaller size ranges; approximately 1 particle in 100 is spherical at 0.7-μm diameter compared with 1 in 10 at 0.2-μm diameter. The particles fall within a very narrow size range. The size frequency curves for different areas (fig. 5) show a steep falloff above 3 μm and few particles smaller than 0.3 μm; 90 percent of the total mass of the sample is within

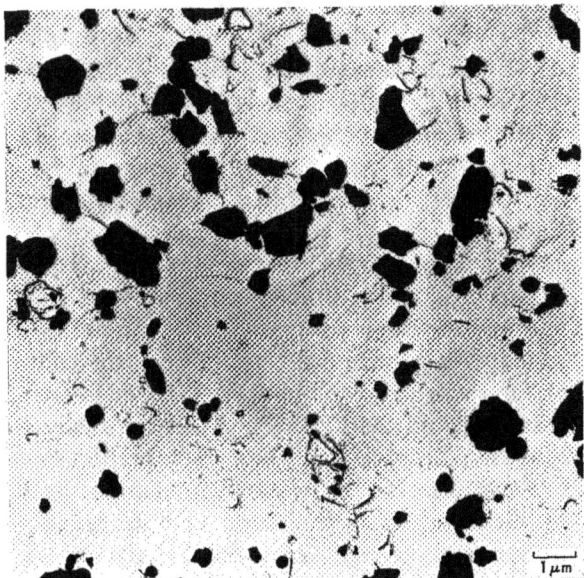

FIGURE 3.—Electron micrograph of debris from the Surveyor mirror.

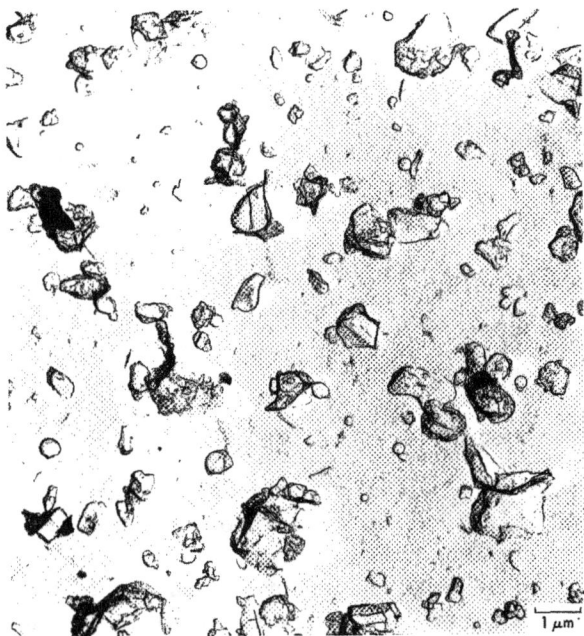

FIGURE 4.—Electron micrograph of a replica of the debris from the Surveyor mirror.

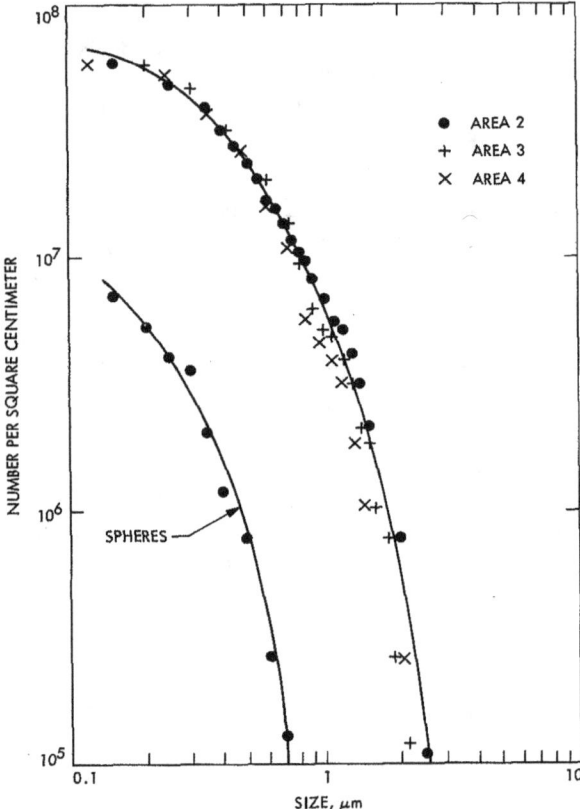

FIGURE 5.—Cumulative size-frequency distribution of debris on different parts of the mirror. Curve for spheres is average for all areas.

the size range 0.3 to 3 μm. Very few particles larger than 4 μm were observed; some of these may have been aggregates. Area 4 had a slightly higher particle frequency than areas 2 and 3 for particles larger than 0.7 μm, but the difference is less than a factor of 2. No diffraction work was attempted because of the nature of the sample and because of our instrumental limitations, which do not allow orientation of the sample or operating voltages in excess of 100 kV.

While observing the sample, a recurring defect was noted in the surface of the mirror. It was especially evident in the second and third peels taken at a particular location, as these contained virtually no masking debris. The defects are flat-bottomed, shallow depressions; they are irregular in outline, and generally less than 2 μm across. They all have a characteristically pitted floor (fig. 6). They probably indicate places in which the protective silica coating is absent. It is not known whether these defects were on the mirror before the Surveyor mission, nor whether they are a result of the mirror's manufacture or its subsequent history.

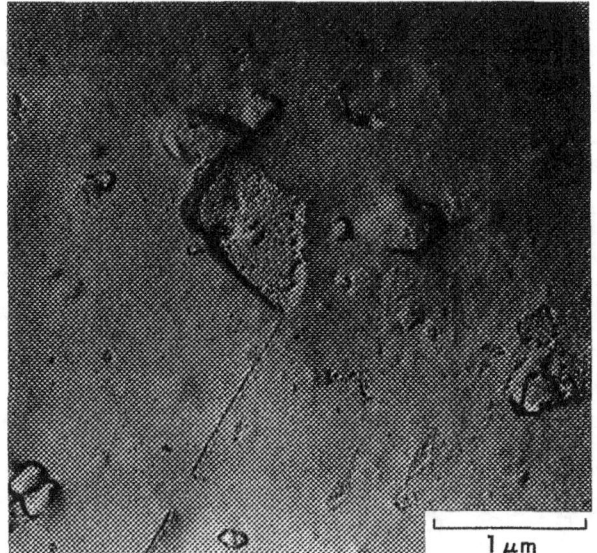 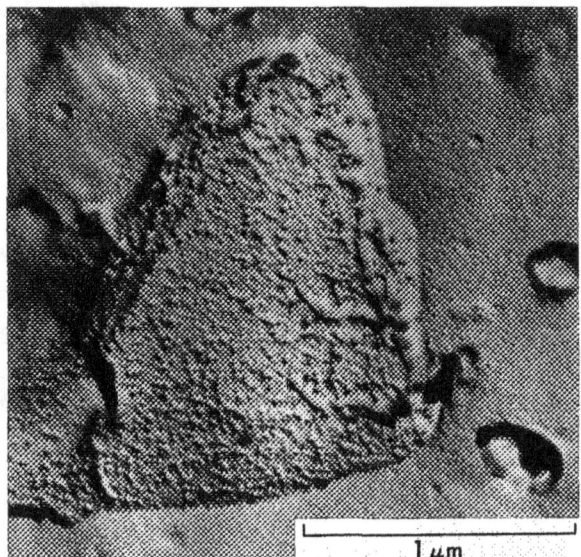

FIGURE 6.—Defects in the surface of the mirror.

Electron Microprobe Analysis

The small size of the individual particles prevented the sample from being prepared for analysis in the usual way. The cellulose tape peels containing the sample were shadowed with carbon; the tape then was dissolved in acetone. The carbon film containing the sample was floated onto the surface of water and picked up on a beryllium probe mount. After drying, the sample was ready for analysis. No attempt was made to mount particles individually for analysis, nor was any attempt made to polish particles. Generally, larger particles were selected in the probe for analysis.

Table 1 lists analyses, normalized to 100 percent, for 20 individual particles. Errors of 10 to 20 percent are probable, as the particles were small (<4 μm) and not polished. All analyses (except No. 20) are consistent with a lunar origin and very similar to analyses on Apollo 11 debris (ref. 1).

The three large particles mentioned were analyzed independently of the rest of the sample. The particles were white to light brown, irregular in shape, and extremely friable. A small fiber was attached to one. Only Ca and S were detected from microprobe analysis, but at such low levels as to indicate that the main constituents of the particles were not apparent. This was suggestive of an organic composition. X-ray analysis showed weak calcite and gypsum lines, which was consistent with the microprobe data. Darkening of the film and the weak lines suggested again that the particles were primarily organic. They are interpreted as contaminants, probably from acoustic tile or some similar material.

Reference

1. DUKE, M. B.; WOO, C. C.; BIRD, M. L.; SELLERS, G. A.; AND FINKELMAN, R. B.: "Lunar Soil—Size Distribution and Mineralogical Constituents." *Science*, vol. 167, 1970, pp. 648–650.

TABLE 1.—*Electron microprobe analyses of individual particles*

	1	2	3	4	5	6	7	8	9	10	11	12	13	14	15	16	17	18	19	20
Na_2O	...	0.5	...	...	0.3	0.6	0.6	0.7	0.5	0.8	0.7	...	0.9	0.4	...	0.9	...	0.8	1.1	...
MgO	6.9	5.5	0.6	8.8	6.4	19.5	9.6	4.5	6.7	10.0	9.6	16.3	15.3	5.6	16.7	1.1	11.5	2.9	5.9	1.3
Al_2O_3	10.3	11.2	4.6	12.5	6.5	18.2	11.0	22.3	9.4	10.0	15.6	3.0	1.9	17.1	3.2	18.7	5.8	23.2	11.4	6.6
SiO_2	39.1	50.2	89.1	44.0	45.9	39.6	43.8	48.0	8.3	52.1	45.7	51.1	52.5	47.8	52.2	66.0	47.3	53.3	60.1	87.6
K_2O	.3	.2	2.5	.1	...	...	...	.2	...	.8	.8	...	...	.4	...	.5	...	.4	1.5	1.5
CaO	12.8	7.2	2.5	12.7	10.3	11.9	14.5	16.2	6.4	7.6	12.3	11.0	7.1	13.1	9.9	10.6	8.1	14.7	7.0	1.8
TiO_2	.5	4.2	...	2.5	1.7	.4	2.7	.9	30.6	1.6	2.3	.8	1.1	1.3	.8	...	1.2	.6	...	...
Cr_2O_3	...	3.9	...	.5	...	.4	.5	...	...	.6	...	.8	.4	...	...	...	...	...	...	...
Fe_2O_3	30.0	17.1	.7	18.9	28.8	10.1	17.3	7.2	38.0	16.6	13.0	17.0	20.9	14.2	17.2	2.3	26.2	4.2	12.9	1.3

SPACECRAFT CHANGES

PART D

DYNAMIC CONSIDERATIONS OF DUST ON THE TELEVISION CAMERA MIRROR

N. L. Nickle

The mirror on the Surveyor 3 television camera is an optical device used to vary the viewing direction of the statically mounted camera. The mirror is fabricated from cast beryllium, is nearly elliptical in shape, measuring 15.5 × 10.8 cm, and is front surface plated. The various coatings that comprise the flat mirrored surface consist of:

(1) Precipitated nickel deposited on a ground beryllium blank and polished to a thickness of 50 to 80 µm.

(2) Aluminum vapor deposited to a thickness of 0.1 to 0.3 µm.

(3) Silicon monoxide vapor deposited to a thickness of 0.1 µm. The silicon monoxide coating, which contains unknown amounts of SiO_2, is optically clear and provides a protective film over the reflecting aluminum.

After return to Earth, the mirror had a coating of fine-grained particulate material adhering to its surface, which was typical of nearly all exposed surfaces. (See fig. 1.) The discovery of this material was no surprise, as the television pictures transmitted to Earth during the mission were degraded by a veiling glare caused by the presence of what was reported to have been dust deposited there during the abnormal landing sequence (ref. 1).

Figure 1 shows numerous features emphasized by the low angle of illumination. Individual particles visible in the figure are considered to be terrestrial contamination or contamination from the astronauts' tote bag. The six largest particles and, undoubtedly, many smaller ones consist of agglomerates of calcite and gypsum. (See ch. IV, pt. C, of this document). These minerals are unknown in lunar soil. Other large particles include glass fibers from the tote bag and lint.

The 7- to 8-mm-wide swath down the center of the mirror was made by astronaut Conrad before the camera was cut from the spacecraft (compare with fig. 7, which was taken before the finger swipe). His gloved finger was dirty; consequently, the swath contributes to the overall contamination of the mirror. This swipe did not compromise the integrity of the mirror for the type of tests performed.

It is believed that the smudged area at the top of the mirror occurred during the time the camera was in the tote bag. Peripheral markings above the trunnions (horizontal pivot axis) are primarily pre-flight features; marks up to 5 mm extending in from the edge were caused by the Teflon-felt seat used to seal the camera's upper shroud and all optical elements (a protective feature that was not employed during the mission); the raised portion at the edge that resembles accumulations of particulate material is residual adhesive contamination.

Two features not visible in figure 1, but which are readily visible under different lighting conditions, can be seen in figure 2. A spectral band running between the trunnions and a subtle, but distinct, shadow line running diagonally below the band are two of three features that have created the most interest in the mirror. Rennilson (see ch. IV, pt. E, of this document) has discussed the optical properties of the mirror and the probable thickness of the non-particulate coating that gives rise to this spectral band. The third feature is the dust itself.

Tests Conducted on the Mirror

The mirror has been subjected to many tests that have modified its surface (see fig. 3); the results of these tests by other investigators are presented in this document. Lunar dust has been removed from specific areas by rubbing, by acetate and metallic film stripping techniques, by rinsing (fig. 4), by scraping (fig. 5), and by inadvertently touching the surface. The acetate film stripping technique revealed a second

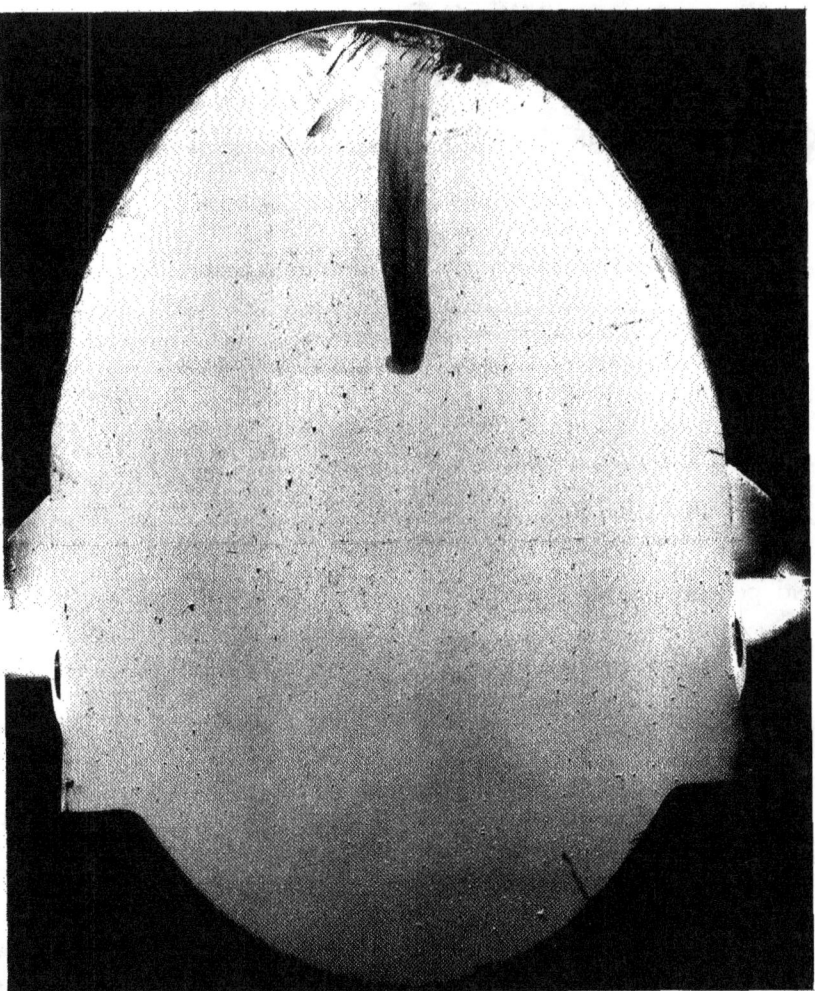

FIGURE 1.—Surveyor 3 television mirror showing a coating of fine-grained lunar dust covering its entire surface and some coarser contaminating particles of calcite, gypsum, glass fibers, and lint. The vertical swath down the center of the mirror was made by Conrad before removing the camera from the spacecraft.

shadow line, more subtle but just as distinctive as the first. It can be seen with difficulty only where the overlying lunar dust has been removed, an indication that its deposition may have preceded the more obvious one or that the lines are defined by residual dust tenaciously adhering to the surface. A closeup of the lower part of the mirror shows the persistence of the shadow lines after most of the dust has been removed (fig. 6).

Second- and third-generation peels were taken across the upper shadow line. Carroll and Devaney[1] have concluded:

(1) That the demarcation line is more obvious to the unaided eye than at higher magnifications.

(2) That no actual line exists, but that it is a sharp transition in the density of light-scattering centers.

[1] W. Carroll and J. Devaney, Jet Propulsion Laboratory, personal communication, 1970.

Please See Figure 2 (which has been moved to the end of the book)

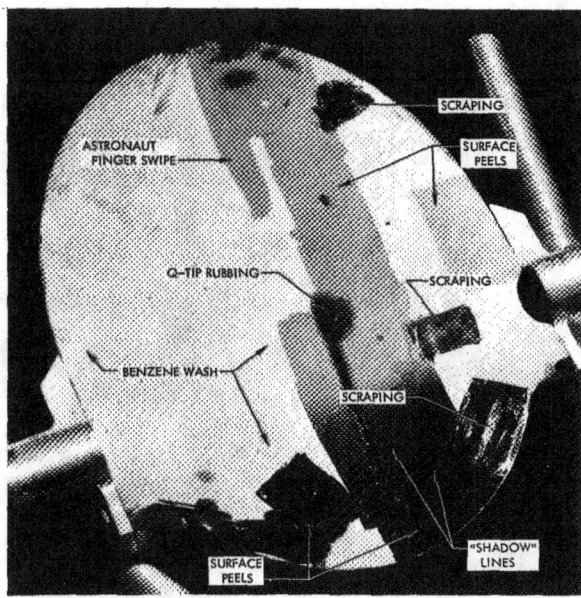

FIGURE 3.—Surface features on the mirror were created by other investigators to define the nature of the adhering material and the shadow lines, to identify the film that caused the spectral band, to search for micrometeoroid impact features, and to identify organic contaminants (compare with fig. 6).

(3) That the light-scattering centers consist of positive and negative features (adhering material and pits).

(4) That the light-scattering centers are due primarily to adhering material.

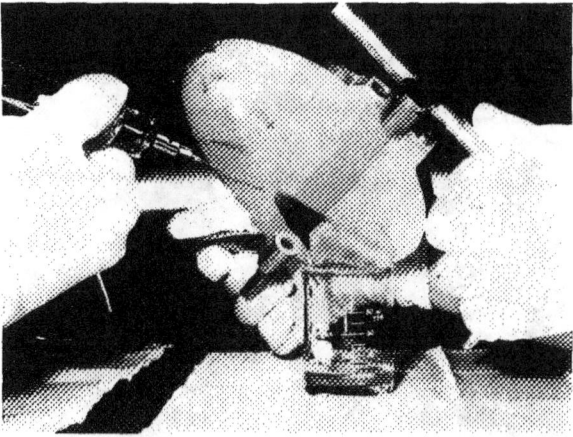

FIGURE 4.—The rinsing technique used by chemists to define the type of organic contaminants on the surface of the mirror. The nonparticulate film causing the spectral band shown in figure 2 proved to be insoluble in acetone and benzene.

Tests using the scanning electron microscope to define the shadow lines quantitatively were not complete when this work was prepared.

The primary objective of this article is to define the source(s) or event(s) responsible for creating two shadow lines that occur on the lower part of the mirror. Figure 7 shows the relative orientation of the mirror with respect to the front opening of the camera and the lunar surface as it existed when the camera was removed

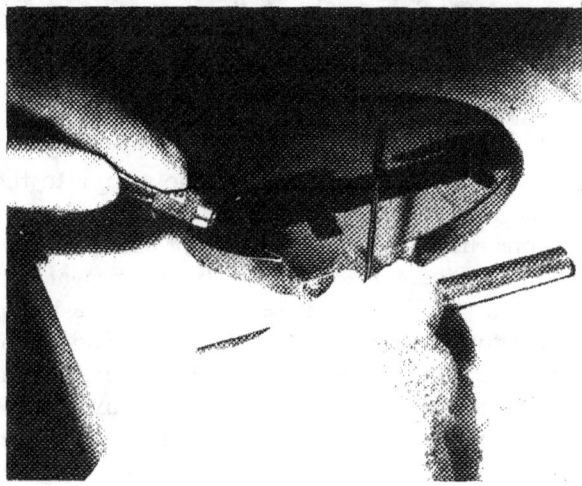

FIGURE 5.—The scraping technique used by chemists to define the composition of the nonparticulate film. Results of this test were not available at the time of this writing.

FIGURE 6.—A closeup photograph of the lower part of the mirror showing some of the features observable in figure 3. The two shadow lines can be seen easily; the lower line can be seen only where overlying dust has been removed, and then only in bright light.

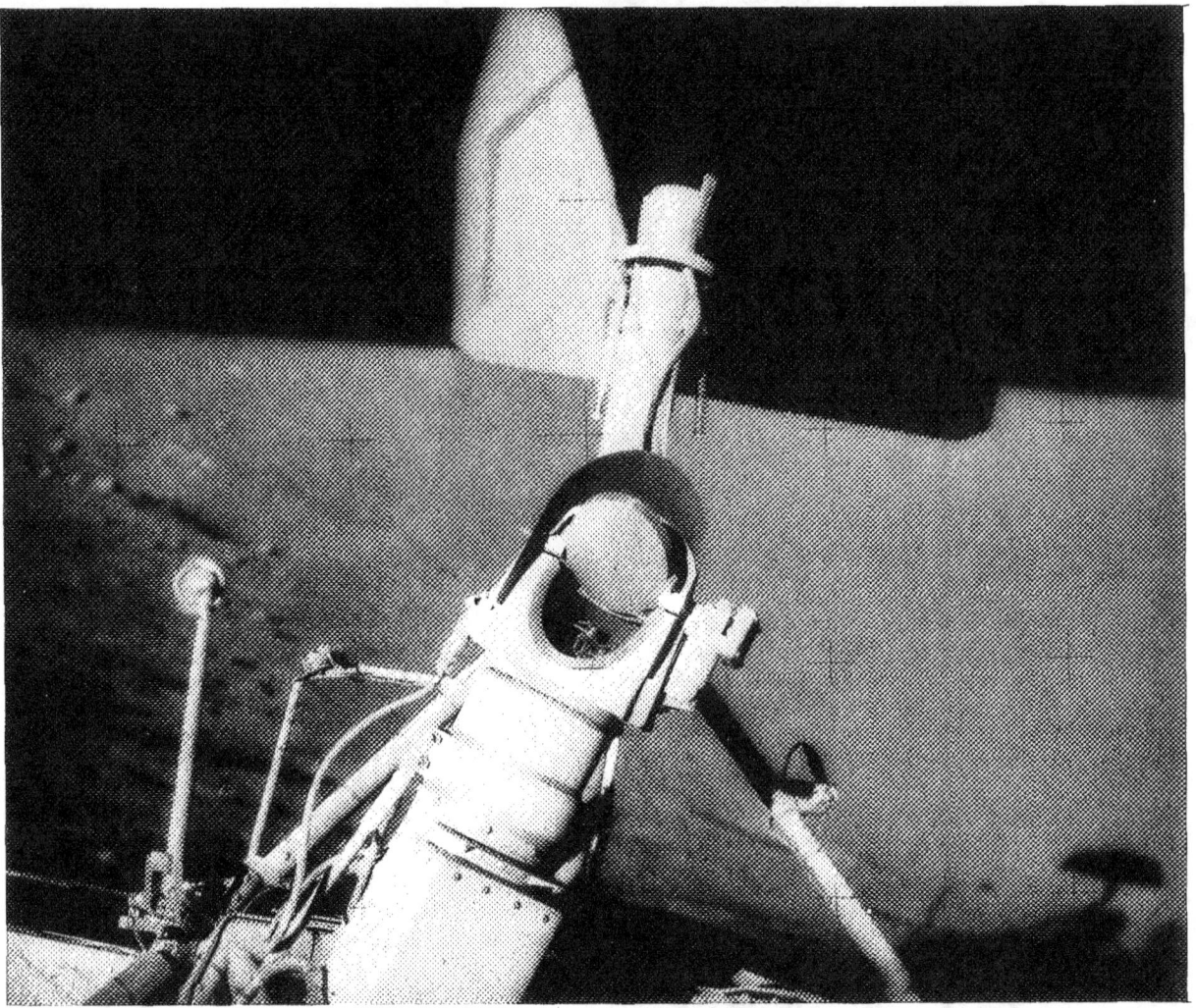

Figure 7.—The Surveyor 3 television camera as it existed before removal by astronauts Conrad and Bean. The front opening of the camera provided the silhouette that defines the shadow lines.

from the spacecraft. (See ch. IV, pt. L, and app. A of this document.) The front U-shaped opening is beveled near the trunnions and, as such, has provided shadow lines on the mirror unique to a given source vector for a given mirror orientation. In order to find a probable source or event on the lunar surface responsible for producing the shadow lines, a mirror orientation, from which geometrical measurements could be made, had to be selected. Two orientations were chosen to make the measurements: (1) that which existed at the time of Surveyor 3 touchdown, and (2) that which existed at the termination of the Surveyor 3 mission.

The spatial relationships of the mirror to the camera and the camera to level ground were reconstructed to simulate the orientations discussed. Figure 8 shows the type-approval test camera (TAT-1), which is identical in design to the returned camera, mounted on a tripod in the configuration of the camera as it was at the end of the Surveyor 3 mission. The pivot axis of the mirror was situated 1.5 m above the floor. A reference point was located directly below the center of the mirror, and a reference line was located on the floor coincident with the bearing of the flat face of the lower shroud. The bearing of this line on the Moon, N 47° W, was determined to have pointed within 1° of the Apollo 12 Lunar Module (LM).[2] These references were used to define the locations of points on the floor that produce shadows cast by the front opening of the camera; these shadows, in turn, produce the best fit to existing shadow lines on the mirror.

The choice of orientation used in this study was based on the highest probability of one or more events occurring at a given orientation to produce the observed features. During the landing maneuver, the mirror was pointed toward leg 3. The abnormal landing sequence was considered a good candidate for producing the shadow features. That is, the vernier engines could propel a small object, impart the lunar surface, and cause secondary events with sufficient force to create the shadow lines. The orientation at the termination of the Surveyor mission also was a possibility because of the long exposure time before retrieval, and hence a greater opportunity to record secondary impacts created by primary events on the lunar surface in view of the mirror.

The orientations described were reproduced in the laboratory and the unique points determined. A paper pattern of the upper shadow line was prepared and an image of the lower line was drawn on it. The pattern was taped to the TAT-1 mirror, and a point source of light was moved about the floor until the closest match was achieved. This method was used to define the upper and lower lines for each camera-mirror orientation.

Figure 8 shows the camera-mirror orientation that represents the end-of-mission configuration. Figures 9 and 10 are closeups of the camera's head as seen in figure 8. Point a in figure 8 produced the shadow visible in figure 9; point b produced the shadow in figure 10. Similarly, figure 11 shows the camera-mirror orientation that represents the landed configuration. Point c in the figure produced the shadow visible in figure 12 and point d produced the shadow in figure 13. Comparison of figures 9 and 13 shows a slightly better fit in figure 9 of the upper shadow with the patterns in the vicinity of the trunnion and the beveled edge. The lower shadow fits equally well in both orientations.

[2] W. Carroll, personal communication, 1970.

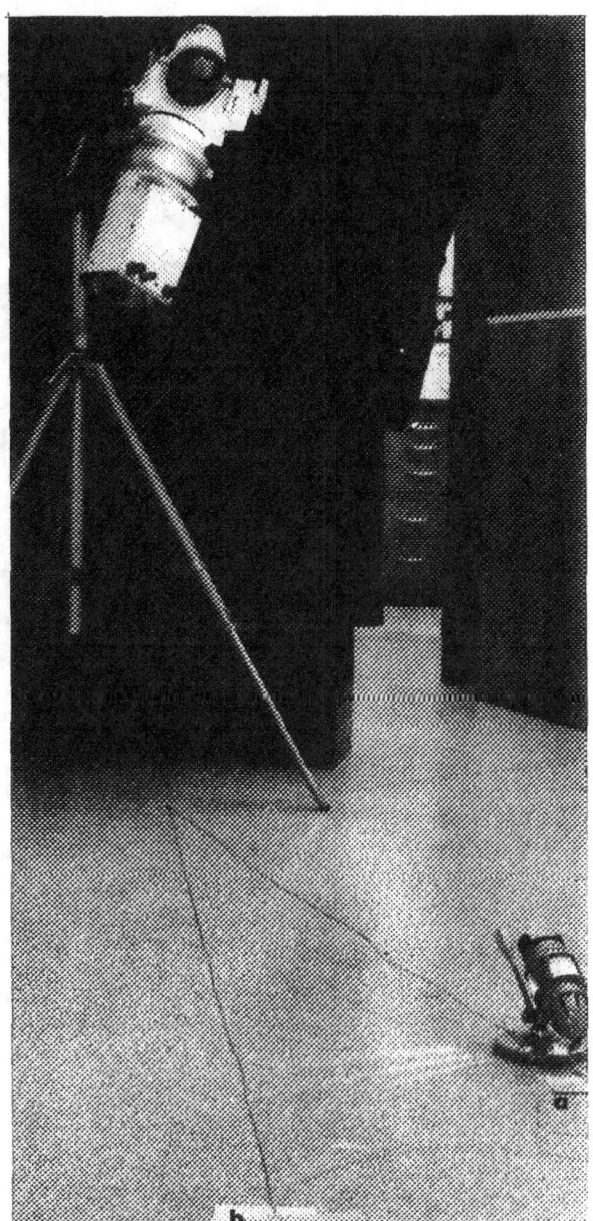

FIGURE 8.—The end-of-mission configuration used to define the vectors that account for the two shadow lines. Point a creates a silhouette approximating the upper line, and point b the lower line. The camera is TAT-1, a replica of the Surveyor 3 camera.

The geometrical relationships of the point sources (a, b, c, and d) with respect to the mirror were corrected for differences in the level floor and the lunar topography and plotted on a drawing of the spacecraft in plan view (fig. 14).

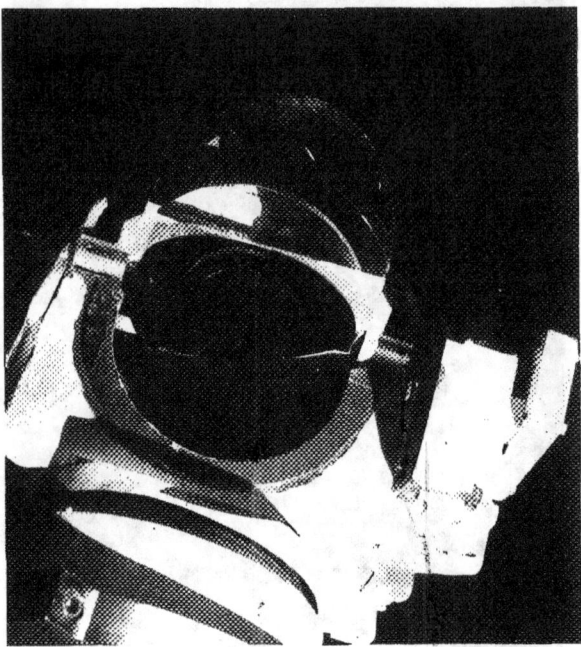

FIGURE 9.—A paper pattern of the upper shadow was prepared and an arc representing the lower line was scribed in its correct orientation. The pattern was taped to the mirror and a silhouette from point a of figure 8 projected onto the pattern. Note the unique profile cast by the trunnion and beveled edge adjacent to the mirror, and the relatively good fit.

FIGURE 10.—The silhouette projected from point b in figure 8.

The spacecraft is shown in its actual lunar orientation along with surface features created by the footpads and scoop. The rectangular areas represent four trenches dug by the surface sampler scoop; small squares, circles, and triangles represent bearing tests, contact points, and impact tests, respectively. (See ref. 2, p. 75.) The larger squares with rays joined to the camera represent the relative locations of the unique points.

Points a and b are found to coincide well with the scoop's trenching operations and impact tests. The mirror's position during many of the trenching and impact operations was within several degrees of its position (azimuth and elevation) at the end of the mission. Points c and d are situated under the spacecraft with an unobstructed view of the mirror. Either pair of points could account for the observed shadow features with only slight changes in the mirror's orientation to produce a more exact shadow-to-pattern fit.

Discussion

The objective of this study was to find one or more fresh impact craters, within view of the mirror, that could be analyzed for changes in surface properties by comparing Surveyor 3 pictures with Apollo 12 photographs. The published and unpublished works of L. D. Jaffe (see ch. VI, pt. G) have demonstrated how inactive the Moon is on this time scale, and how deceptive small-scale surface features can be in photographs taken under different lighting conditions. (See ch. X, pt. B.) It was calculated that an impact crater that could be responsible for all the dust on the mirror would have to be so small that it would be less than or equal to the resolution limit of the television pictures.[3] Consequently, the opportunity to make detailed studies of small areas was welcomed. The conclusion reached in this study is that the manipulations of the surface sampler scoop caused the impingement of lunar dust responsible for the shadow lines. This conclusion is sheltered by the absence of a way of discounting production of the fea-

[3] W. Carroll, personal communication, 1970.

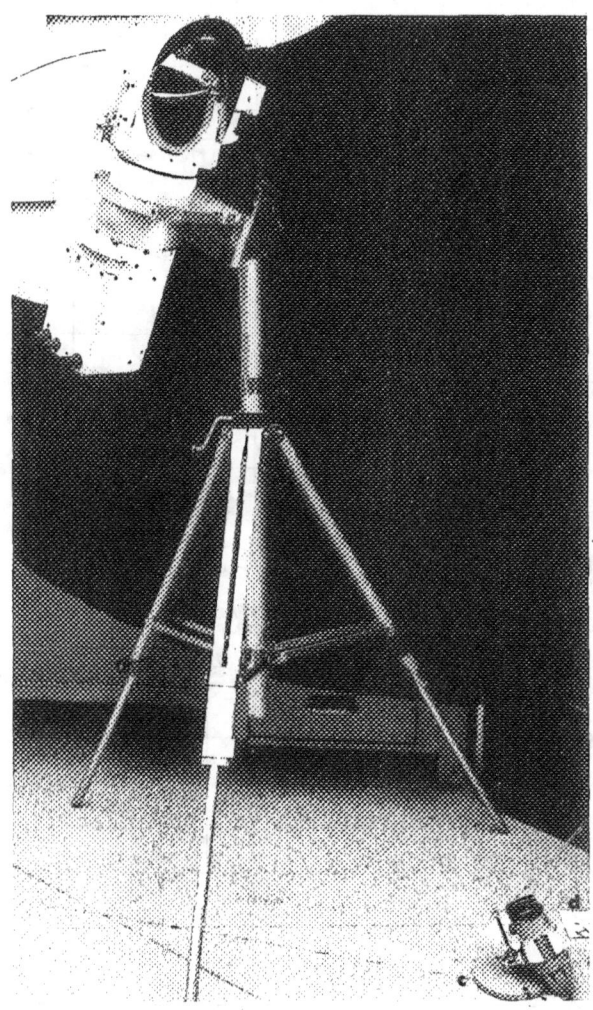

FIGURE 11.—The second camera-mirror configuration used to determine the two points (c and d) that may eventually point to features on the Moon responsible for the shadow lines. This configuration represents the one that existed at the time the spacecraft landed in April 1967.

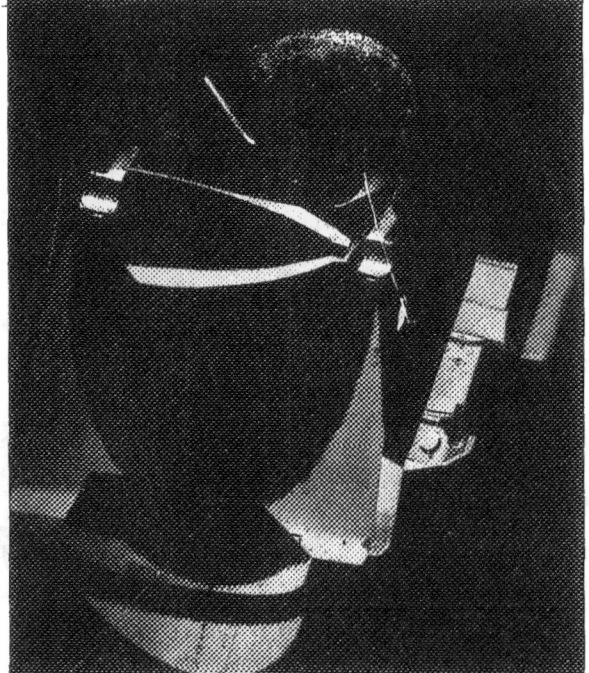

FIGURE 12.—The silhouette produced by the light projected from point c in figure 11.

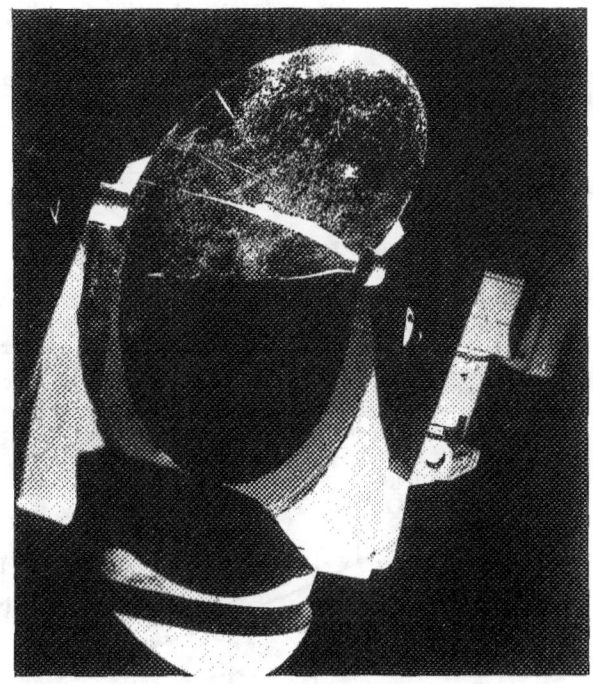

FIGURE 13.—The silhouette produced by the light source at point d in figure 11. The "fit" has been judged to be less exact than that shown in figure 9.

tures by Surveyor's abnormal landing, by micrometeoroid impact or other transporting processes, or by the approach of the LM.

It has been demonstrated that the LM is capable of entraining and eroding mechanical surfaces located 155 m away. (See ch. IV, pt. I, of this document.) It is reasonable to assume, therefore, that the same process would occur during the LM approach because of the closer pass to Surveyor than its relative position at the landing site. (See fig. 15.) At the closest point, the LM was about 67 m above a point on the ground

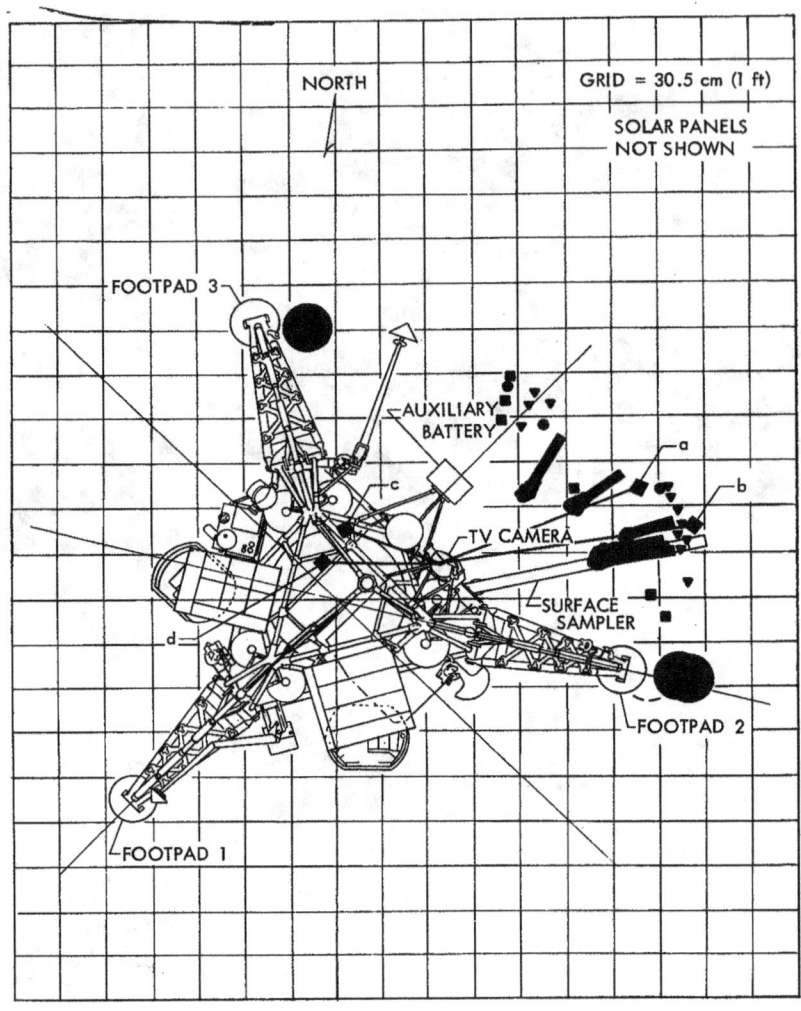

FIGURE 14.—A plot of the two pairs of determined points on a drawing showing the surface features created by Surveyor's footpad and scoop. Rectangular areas represent four trenches dug by the scoop; small squares, circles, and triangles represent bearing tests, contact points, and impact tests, respectively. Note the positions of points *a* and *b* in relation to the trenches and impact points made by the scoop. This figure was modified from a drawing prepared by L. D. Jaffe and F. I. Roberson.

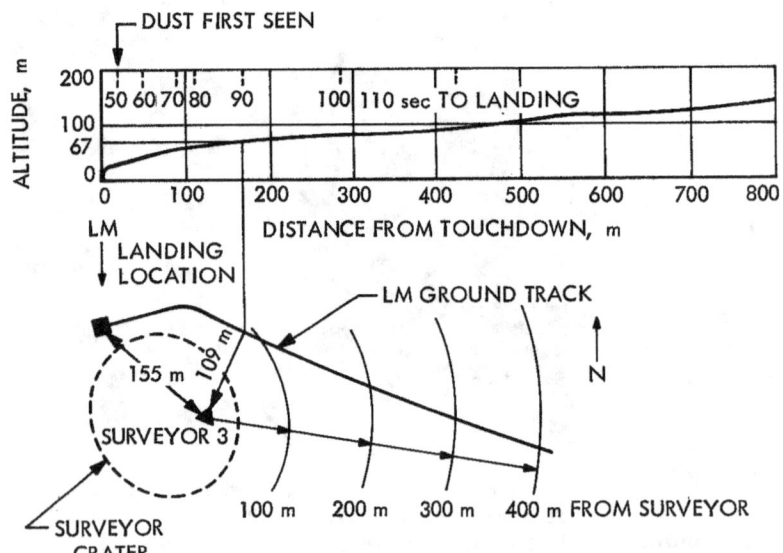

FIGURE 15.—A profile and plan view of the Apollo 12 approach trajectory and landing site in relation to the Surveyor 3 spacecraft. The closest point along the ground track was about 109 m, while the LM was at an elevation of 67 m.

FIGURE 16.—Apollo 12 photograph of the Surveyor 3 scoop showing most of the surface area available to the scoop. Points *a* and *b* coincide with those found in figure 14. No changes in these features were noted by comparing Surveyor 3 pictures with the Apollo 12 photographs taken 31 months later by the astronauts.

located about 109 m from Surveyor. It has been speculated that the LM rocket exhaust could disrupt particles and entrain them to points *a* and *b* (shown in fig. 14) with sufficient force to produce the shadow lines. The writer agrees that this is possible; however, the fact that these points coincide so well with impact points and trenches created by the scoop seems more than mere coincidence.

Figure 16 is an Apollo 12 photograph taken from the south side of footpad 2 (see fig. 14) in a northerly direction. Points *a* and *b* have been located on figure 15 for comparison. No changes were noted between this picture and a similar one taken 31 months earlier by the Surveyor 3 television camera.

The value of continuing this study with the intent of obtaining more conclusive evidence to the origin of the shadow lines seems neither justified nor rewarding. The scoop is considered to be solely responsible for the lines, and this simply serves to iterate the need to protect optical devices from activities that tend to redistribute the rather tenuous lunar soil.

References

1. CHRISTENSEN, E. M.; BATTERSON, S. A.; BENSON, H. E.; CHOATE, R.; JAFFE, L. D.; JONES, R. H.; KO, H. Y.; SPENCER, R. L.; AND SUTTON, G. H.: "Lunar Surface Mechanical Properties." *Surveyor III Mission Report. Part II: Scientific Results*, TR 32–1177, Jet Propulsion Laboratory, Pasadena, Calif., 1967, pp. 114–152.
2. SCOTT, R. F.; AND ROBERSON, F. I.: "Soil Mechanics Surface Sampler: Lunar Surface Tests, Results, and Analyses." *Surveyor III Mission Report. Part II: Scientific Results*, TR 32–1177, Jet Propulsion Laboratory, Pasadena, Calif., 1967, pp. 69–110.

ACKNOWLEDGMENTS

I thank W. F. Carroll, L. D. Jaffe, R. Davis, and F. I. Roberson for their helpful comments and suggestions and W. F. Carroll and L. D. Jaffe for their review of the manuscript.

PART E

CHANGES IN OPTICAL PROPERTIES OF THE SURVEYOR 3 CAMERA

J. Rennilson, H. Holt, and K. Moll

For 942 Earth days, the Surveyor 3 television camera was exposed to the harsh lunar environment. On November 20, 1969, 928 days after its last picture had been transmitted to Earth, the camera was retrieved for return to Earth by the Apollo 12 astronauts in order to measure and analyze the changes in the camera's optical performance.

The measurements made involved the following areas:
(1) Spectral reflectance (specular).
(2) Gonioreflectance.
(3) Ellipsometry.
(4) Spectral transmission.
(5) Contrast and modulation transfer.
(6) Photography (macro and micro).

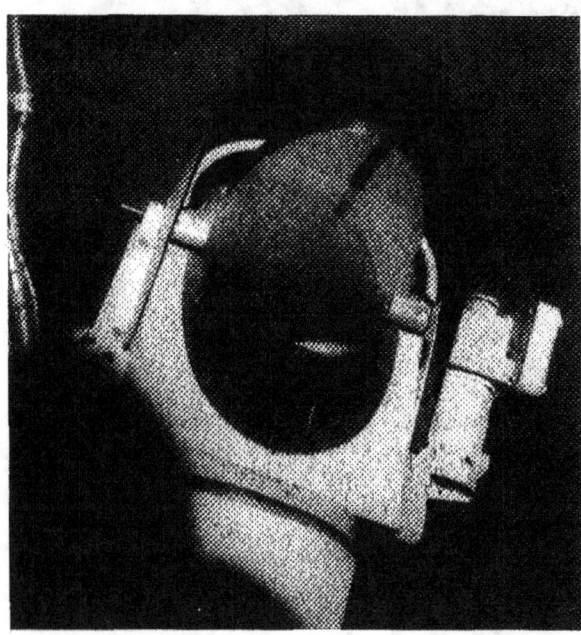

FIGURE 1.—Enlargement of the Surveyor 3 television camera. Photograph was taken by P. Conrad on the lunar surface during the Apollo 12 mission. Outline and parts of the filter-wheel mechanism are visible in the lower part of the mirror. The image of the mirror housing is visible in the upper part of the mirror (AS12–48–7132).

The optical parts of the television camera can be divided into three groups:
(1) Scanning mirror.
(2) Filter glasses (six pieces).
(3) Variable focal length lens (25 to 100 mm).

The mirror is formed of beryllium metal, polished and electrodeposited with a thin nickel coating (Kanigen), which was polished to an optical quality surface. An aluminum coat was vacuum deposited with an overcoating of silicon oxide (SiO) as a protective layer. The SiO film was deposited with a thickness of about ½ wavelength at $\lambda = 550$ nanometers (nm). The durability and optical properties of the film depend greatly on the oxygen pressure and deposition rate (ref. 1) under which they are applied.

When astronauts Conrad and Bean first examined the Surveyor 3 television camera on the lunar surface, they said: "It's no longer a mirror —it's just got a fine dust on it." However, photographs obtained by the astronauts showed that some parts of the filter-wheel assembly, especially the bearings, were visible by reflection (fig. 1). Most parts of the filter-wheel assembly were reflected by the lower part of the mirror. This agreed with the effect observed during the Surveyor 3 mission (fig. 2).

Measurements

After the camera was returned to the Hughes Aircraft Co. (HAC), it was mounted on a special bracket, and photographs of the mirror in collimated light were compared with the Apollo 12 photographs. The image of the filter-wheel assembly agreed in contrast and detail with that from Apollo 12 (fig. 3; compare with fig. 1), indicating that a relatively small amount of lunar material had been lost during the return to Earth.

The observations made after receipt of the camera at HAC indicated that a pronounced band of color occurs across the mirror at the

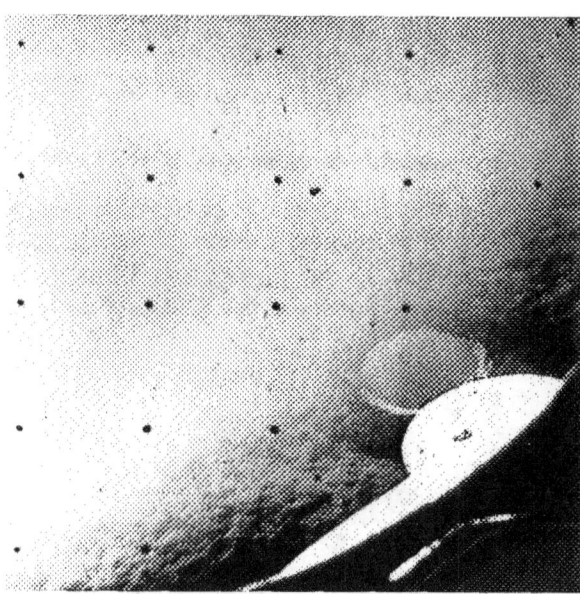

FIGURE 2.—Surveyor 3 picture taken on April 21, 1967, at 02:41:19 GMT. The upper half of the frame is almost deplete of contrast; in the center, craterlet detail can be seen. The azimuth angle was −54°; elevation angle was −48°.

approximate position of the elevation axis. The band bows inward toward the small end of the mirror. Specular reflected light in this band exhibited a violet appearance; nonspecular reflected light was yellow-green in hue. The mirror still reflected a satisfactory image with no direct illumination falling on the surface. Five specks of material, whose origin is presently unknown, were visible on the mirror's surface.

A question often asked is whether the contrast attenuation of the camera is the same now as it was during the Surveyor 3 mission. In order to answer this question, we proposed a plan of using a light box and a spare operating Surveyor camera. The spare was a type-approval test camera (TAT-2) used extensively during mission testing; thus, its characteristics were well known. The mirror assembly of the TAT camera was removed, and the Surveyor 3 assembly was placed on the TAT camera, allowing video pictures to be recorded with the same configuration as the original Surveyor camera. If the contrast attenuation had been greater than that measured during the mission, an accrual of lunar material had taken place.

To resolve this question, a light box illuminated with 1000-W tungsten lamps, powered by a variable transformer, was used. The light box was positioned in front of the TAT camera at a distance enabling one-third of the frame to be illuminated. A target, consisting of five equally spaced opaque and clear bars, was placed in front of the box. Thus, the camera was recording a scene of square-wave modulation approximating zero frequency. To simulate the Sun, a collimated beam from a xenon arc lamp was used and oriented at about the same solar elevation and azimuth that corresponded to the early Surveyor pictures (fig. 2). The source illuminated the Surveyor mirror completely.

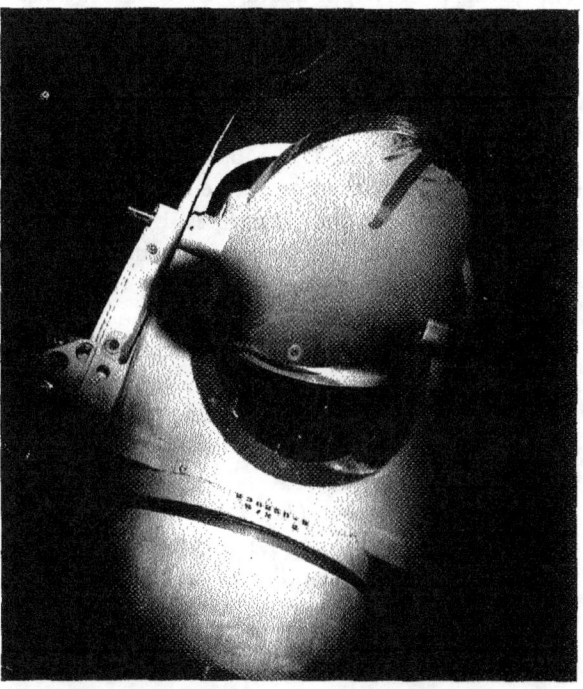

FIGURE 3.—Photograph of the camera, after its return to Hughes Aircraft Co., taken in collimated light at approximately the same geometry as figure 1. The improved resolution identifies the areas in figure 1.

The video signal was recorded by an oscilloscope camera, while the luminance level of the clear areas of the target was decreased. When no signal differences between the clear and opaque areas of the target could be seen on the oscilloscope photographs, contrast threshold had been reached. Then the luminance level of the clear areas and the illuminance of the xenon collimator were recorded.

Analysis of the video signal was made using the following relationships. Inherent contrast of the target is given by

$$C_0 = \frac{L_0 - L_c}{L_c}$$

where L_0 is the luminance of the opaque areas (about 0) and L_c is the luminance of the clear areas; thus, $C_0 = -1$. The apparent contrast "observed" by the vidicon through the mirror is given by

$$C_r = \frac{L_0' - L_c'}{L_c' + L_s}$$

where the primed parameters designate attenuated values, and L_s is the additional luminance contributed by the sunlight scattered from the dust-covered mirror. Thus, the contrast threshold is reached when the contrast transmittance approaches zero; i.e., $C_r \to 0$. A target scene of $C_0 = -1$ implies a large value of L_s, and the scattered light by the mirror is then the primary signal source. To determine whether more or less particulate matter existed on the mirror after 31 months than it did during the Surveyor mission, measurements were needed of the luminances at threshold. If the luminances were higher than those observed during the Surveyor 3 mission, more material would exist on the mirror now.

A Surveyor picture (fig. 2) was chosen that closely corresponded to the geometry of illuminating and viewing conditions of the Apollo 12 photographs. This Surveyor picture has resolvable craterlets with a background luminance of about 640 cd/m². The shadows of the crater walls are assumed to be dark (3.4 cd/m²); thus, the inherent contrast of the scene is about -1. Because the craters are still detectable, the video signals were above threshold. Laboratory conditions were then established to duplicate this Surveyor 3 scene. For threshold with an unchanged geometry or mirror condition, one has the inherent relationship

$$\frac{L_B}{E_s} = \frac{L_C}{E_{xe}K}$$

where

L_B = luminance of the Surveyor background

E_s = normal illuminance of sunlight at the lunar surface

E_{xe} = normal illuminance of the xenon collimator

L_C = luminance of clear area of the target

The factor K is put after a compensating parameter required because of a different vidicon, and different spectral power distributions of the tungsten, xenon, and solar sources. It is expressed by

$$K = \frac{\int_\lambda H_s'(\lambda)\rho_s(\lambda)S_3(\lambda)\,d\lambda}{\int_\lambda H_L'(\lambda)S(\lambda)\,d\lambda}$$

where

$H_s'(\lambda)$ and $H_L'(\lambda)$ = normalized spectral power distributions of sunlight and tungsten light, respectively

$S_3(\lambda)$ and $S(\lambda)$ = normalized spectral sensitivities of the Surveyor 3 and TAT-2 camera vidicons

$\rho_s(\lambda)$ = spectral reflectance of the lunar surface

The incident sunlight and xenon collimated light on the mirror have about the same spectral power distributions over the spectral sensitivity of the vidicons, and thus are eliminated in this equation.

The value for threshold L_C, determined by

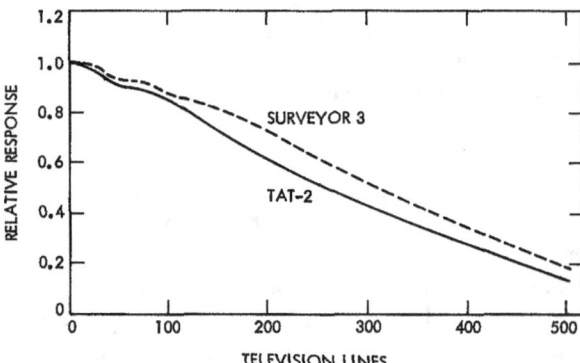

FIGURE 4.—The modulation transfer functions of the Surveyor 3 mirror and the TAT-2 mirror, used with the TAT-2 camera. The conditions were identical to preflight tests. The slight improvement may be due to positioning and flatness differences.

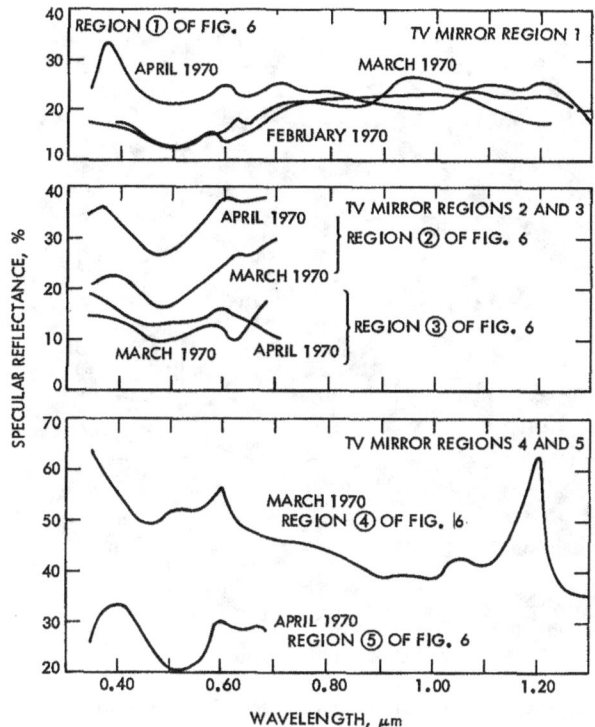

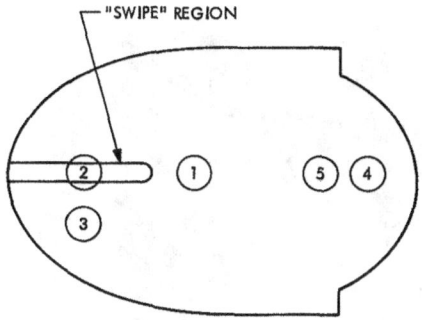

FIGURE 5.—Measurements of the spectral reflectance on the mirror made by Hughes Aircraft Co. over the period of February through April 1970. Incidence angle equals reflectance angle. (Data are through the courtesy of Hughes Aircraft Co.)

FIGURE 6.—Locations of the areas measured by Hughes Aircraft Co. on the Surveyor 3 mirror. (Data are through the courtesy of Hughes Aircraft Co.)

the laboratory simulation, was 127 cd/m²; the normal illuminance of the xenon collimator was 1290 lm/m². The parameter K was valued at 3.37, resulting in a Surveyor background luminance level of 4080 cd/m² for threshold conditions. The fact that the measured L_B was only 640 cd/m² indicates that threshold conditions of the mirror were lower at the time of the Surveyor 3 mission. This would occur if more lunar fine material were present on the mirror now than during the mission. Other evidence supports this view and attributes the additional material to the landing approach of the Lunar Module (LM). It should not be discounted, however, that some material accumulated in the 31 months the camera resided on the lunar surface. From the optical viewpoint, it is difficult to distinguish between the materials according to their age.

The modulation transfer characteristics test used during pre-launch calibrations was repeated in order to check the effect of lunar fines on the mirror. The original calibration test involved a series of discrete sinusoidal frequency photographic targets. Each target image was evaluated by recording an oscilloscope trace through the center of each target. The maximum peak-to-peak signal of the sine wave was measured relative to a gray-to-white ratio of almost zero frequency. The gray and white portions were at almost the same density level as the sinusoidal peaks in the original target. The ratios (relative responses) plotted against the frequencies result in a modulation transfer function (MTF) for the camera subsystem. Figure 4 shows the MTF measurements made on the TAT-2 camera with its own mirror assembly and that from the Surveyor 3 camera. Very little change in the MTF can be detected over the frequency range of the camera. The optical modulation may be affected at higher frequencies, but no provision was made for its measurement.

The spectral reflectance of the mirror was one of the prime types of measurements during the mirror investigation. Initial studies were made using a tungsten halogen collimator at HAC in Culver City, Calif. The detection apparatus was a EG&G spectroradiometer. Five positions were measured on the mirror at specular reflectance geometry. The positions and representative curves are given in reference 2 and are shown in this chapter as figures 5 and 6.

Some infrared photography was performed at the Lunar Receiving Laboratory (LRL) after return of the camera (fig. 7). This photograph

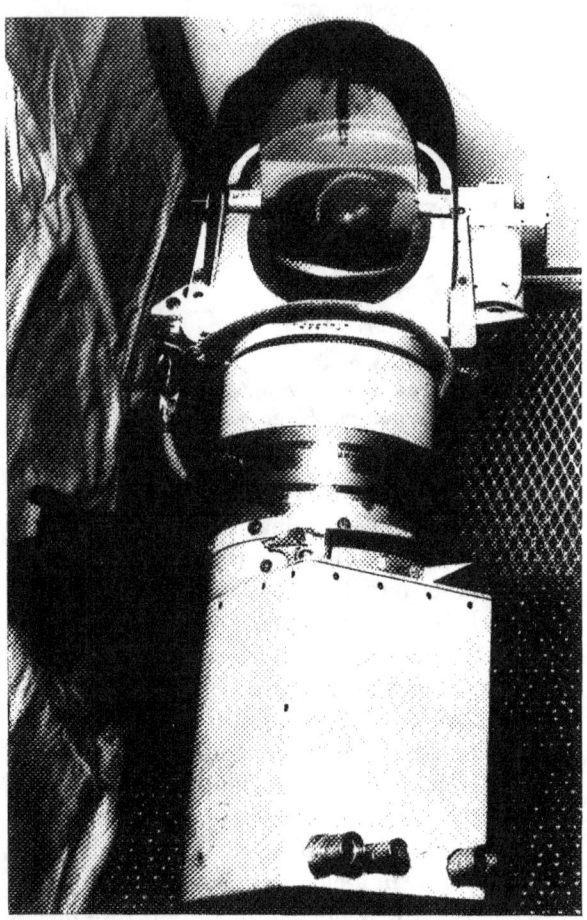

FIGURE 7.—Infrared photograph of the camera taken at the Lunar Receiving Laboratory. More of the camera and mirror housing is reflected by the mirror compared with figure 3.

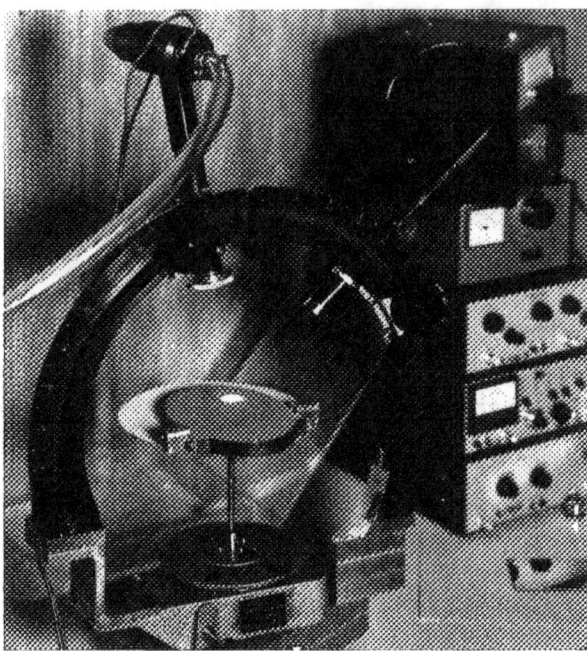

FIGURE 8.—Goniophotometer used in the photometric tests on the mirror. The detector fiber optics probe (right) is coupled to the entrance slit of a monochromator.

shows much more detail in its mirror image than visible light. Measurements were made at HAC to confirm this observation. A goniophotometer (see fig. 8) was used to make angular measurements of the reflected light flux at three narrow wavelengths. Two positions on the mirror were chosen. As figure 9 shows, much more light was scattered at the wavelength centered at 413 mm than at 625 mm. Incidence angles of 30° and 60° indicated that at least 50 percent more light at 413 nm was scattered at the larger angle. The amount of scattered light at 60° incidence appears to respond like the exponential function e^{-x}, where x is the sum of the absorption and total scattering coefficients. No estimate of the particle size distribution was made from these measurements using the methods of Grum, Paine, and Simonds (ref. 3) because of the angular nature of the particles. (See ch. IV, pt. C, of this report.) The scattering decreased for angles of incidence close to the normal.

Specular reflectance values measured with the goniophotometer indicated close agreement with the EG&G spectroradiometer data at the wavelengths indicated. These goniophotometer measurements were taken during February 1970 and repeated just before April. The reflectance had increased about 60 percent at 613 nm, indicating that the loose lunar fines were being removed by exposure to laminar air currents. This completed the measurements made while the camera mirror and filters were at HAC.

The mirror and filters were removed from their respective assemblies and transferred to the Jet Propulsion Laboratory (JPL), Pasadena, Calif. After this, the mirror was taken to the Center of Astrogeology, U.S. Geological Survey, in Flagstaff, Ariz., where the following types of measurements were performed:

(1) Spectral reflectance (specular).

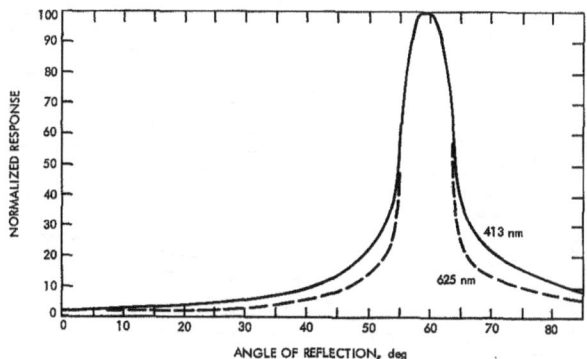

FIGURE 9.—Goniophotometric data on the reflection of the collimated light from the mirror at 2 wavelengths. Angle of incidence is 60°. The peak is normalized at 100 and is broad because of the detector acceptance angle. About 50 percent more light is scattered at 413 nm than at 625 nm. The area is beside the swipe made by Conrad.

(2) Goniophotometry.
(3) Ellipsometry.
(4) Photography.

The measurements and their analyses are discussed in the subsequent paragraphs.

Spectral Reflectance (Specular)

Because of the apparent symmetry of lunar material distributed around the long axis of the mirror, five positions were chosen for measurement. These are shown in figure 10. Area A is geometrically the lowest of these and represents a part of the mirror that has the highest reflectance with lunar material still in contact. Area B is important because it occupies the region of prominent color banding. Areas C and D show less coloration, but contain greater quantities of lunar material. Area E is important as a repeat of previous data taken earlier at HAC. The swipe made by astronaut Conrad was investigated by HAC, and no additional measurements were made on this area.

The equipment utilized consisted of a goniophotometer with a fiber optics coupled monochromator and detector. A manual scanning of the wavelength range 380 to 700 nm was obtained at incidence angles of 10°, 30°, 40°, 50°, 60°, and 75°. The results are plotted in figures 11 through 13. Upon first observation, the abso-

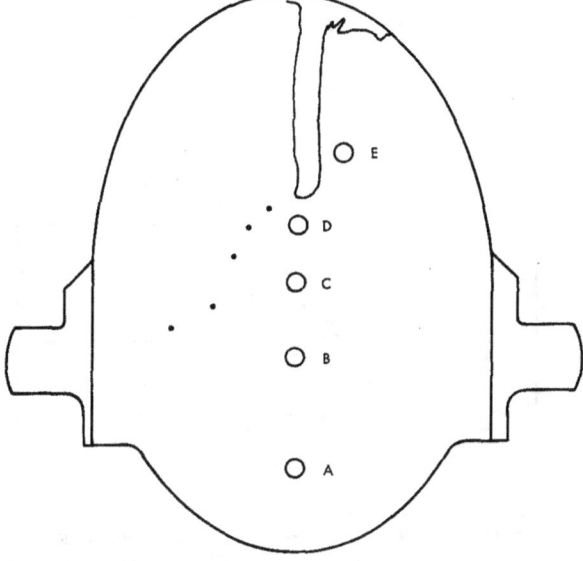

FIGURE 10.—Positions of the areas on the mirror used for spectral reflectance. The five white specks are indicated and discussed in the text.

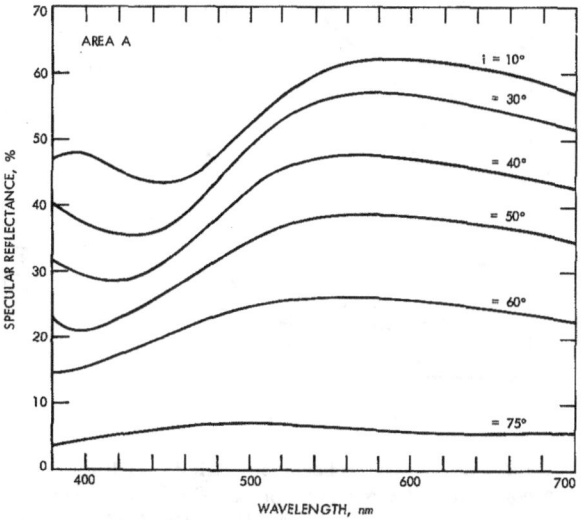

FIGURE 11.—Plot of the absolute spectral reflectance of area A at varying angles of incidence. The detector was set equal to the reflection angle.

lute reflectance curves show a large decrease with angle of incidence. For mirror surfaces in general, the reflectance is almost constant. Thus, the effect of particulate matter on the mirror is the dominant factor in decreasing the light flux reaching the detector. This result, however,

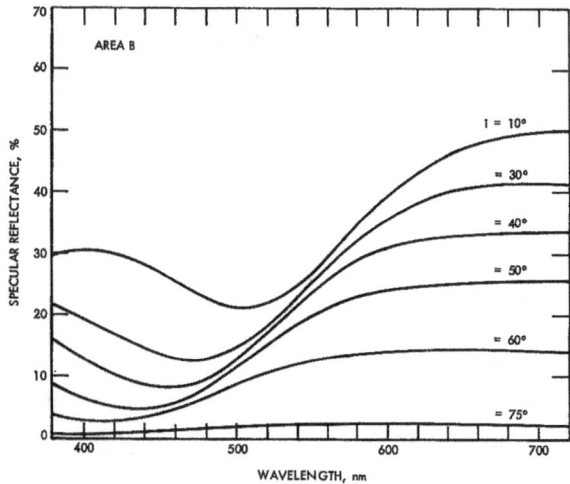

FIGURE 12.—Plot of the absolute spectral reflectance of area B at varying angles of incidence. The detector was set equal to the reflection angle.

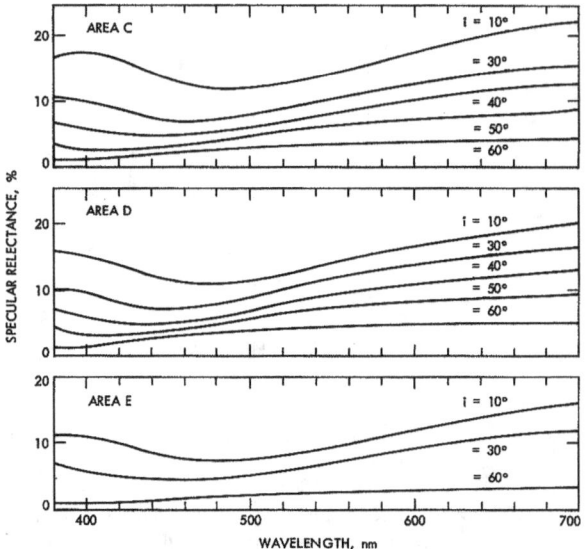

FIGURE 13.—Plot of the absolute spectral reflectance of areas C, D, and E at varying angles of incidence. The detector was set equal to the reflection angle.

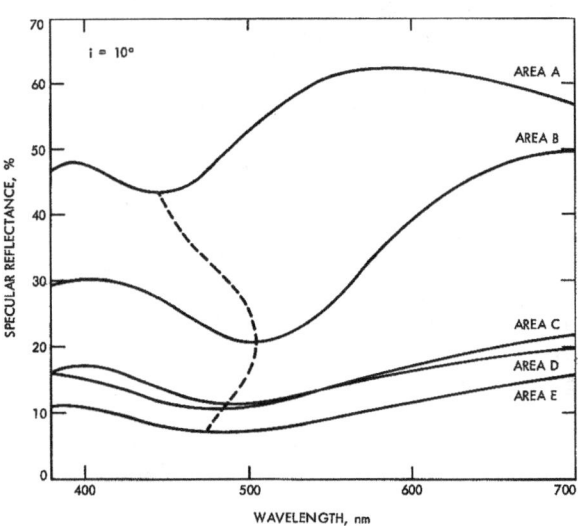

FIGURE 14.—Spectral reflectance variation with area at a constant incidence angle of 10°. The dotted curve is a trace of the minima and shows the increased film thickness at the elevation axis.

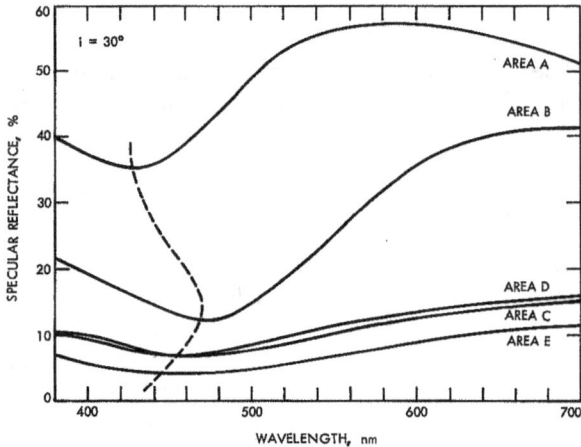

FIGURE 15.—Spectral reflectance variation with area at a constant incidence angle of 30°. The dotted curve is a trace of the minima and shows the increased film thickness at the elevation angle.

could have been predicted from the goniophotometric data taken at HAC. What is interesting is the minimum in each curve indicating an additive residual reflected color of violet, the almost normal visual impression of the color band on the mirror. The shift toward the blue or shorter wavelength as the angle of incidence increased, however, is a phenomenon closely associated with interference effects from thin films. The appearance of only one absorption band in the visible would indicate the presence of a film less than one wavelength thick. The shift in the minima with increasing angle is similar to the direction and magnitude of interference filters.

The comparison of the reflectance curves also reveals the fact that the positions of the minima change with the area on the mirror, indicating nonuniform coating of some kind of film (figs. 14 and 15). This appearance of the spectral properties, as well as the Carr and Proudfoot results (see ch. IV, pt. C), lead the authors to the conclusion that scattering or diffraction effects from the lunar material are not the cause of the colors.

Goniophotometry

The measurements of nonspecular reflected flux at different wavelengths were closely associated. Comparisons could be made concerning the changes, if any, which occurred to the mirror as the analysis program proceeded. Figure 16 is a plot of the reflection at $i = 30°$. Less material is present on the mirror than at the start of the observations.

The absolute reflectance (integrated over the vidicon sensitivity) varies over the mirror's surface. At an incidence reflectance angle of 10°, the lower area of the mirror is 49 percent, the central band is 28 percent, the middle of the mirror is 18.3 percent, the bottom of the swipe 17.1 percent, and the areas left and right of the swipe 16.1 and 12.9 percent, respectively. Thus, the right area is more heavily coated with lunar material than the left.

Ellipsometry

The general appearance of the color band and other visual observations suggested the possibility of a layer of film deposited after the standard SiO overcoating. To test this hypothesis, film measurements were made using the technique of polarized light.

Measurements on the primary mirror of the Surveyor 3 television camera were made on a Gaertner Scientific Co. Model L–119 ellipsometer with Glan-Thompson prisms and 0.01° divided circles. The light used was a mercury arc filtered at 5461 Å. The geometry of this ellipsometer precluded measurements at an incidence angle greater than 65° because of interference with the support brackets on the mirror. The measurement technique consisted of passing the light (5461 Å) through a collimator, polarizer, and quarter-wave plate and allowing it to strike the mirror at a known incidence angle, Φ. The reflected beam, also at angle Φ, passes through an analyzer and telescope to a photomultiplier photometer. (See fig. 17.)

The phase change factor, Δ, was observed to vary linearly with incidence angle, Φ, over

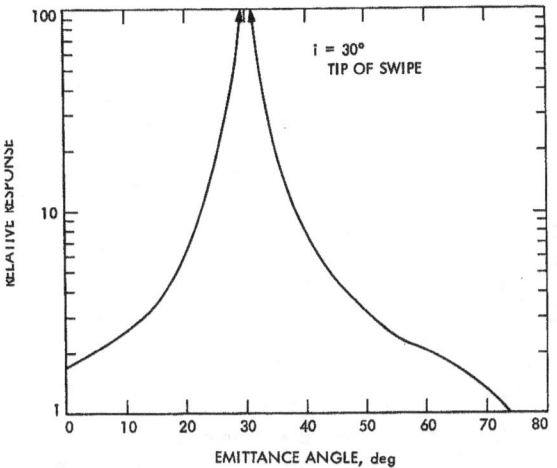

FIGURE 16.—Plot of the mirror reflectance as a function of emittance angle at $i = 30°$. It is similar to figure 9, but over the entire spectral range of the detector.

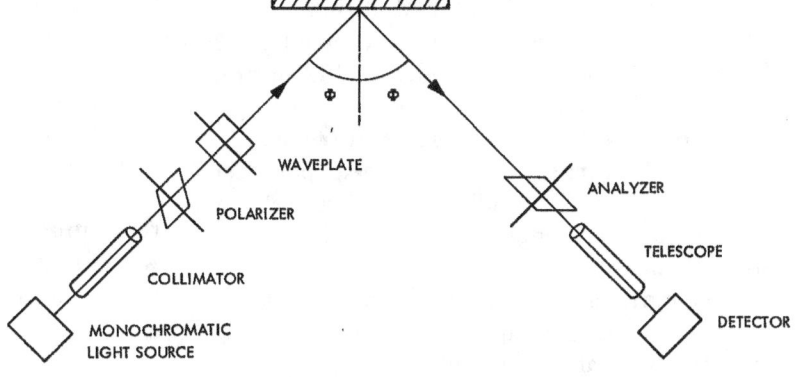

FIGURE 17.—Setup used to measure the effect of reflection on the state of polarization of incident light. The incidence angle is Φ.

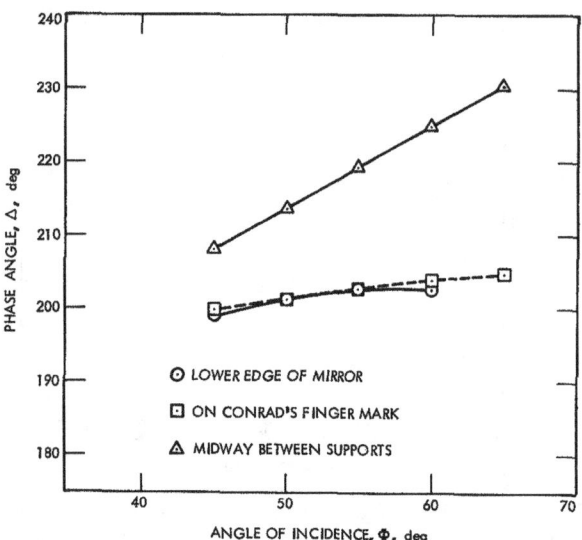

FIGURE 18.—Plot of the phase change Δ as a function of the angle of incidence. The slope of the Δ vs. Φ curve in the region adjacent to the elevation axis indicates increased film thickness. The top and bottom of the mirror are of approximately equal film thickness.

the range $45° \leq \Phi \leq 65°$. (See fig. 18.) The slope of the Δ and Φ curves increases in the region adjacent to the mounting brackets, indicating an increase in film thickness in this region. The Δ vs. Φ data also indicate that the film thickness is about the same at the top and bottom of the mirror. No calculation of film thickness and index of refraction has been made because of the lack of ellipsometric data on films deposited on silicon monoxide overcoated aluminum substrates. Work is currently in progress on developing a multilayer film theory with SiO overcoated aluminum substrates; when available, measurements on the Surveyor mirror will be interpreted and published at that time.

The physical difficulty in measuring all parts of the mirror resulted in the idea of full mirror, white light ellipsometry. Basically, this technique is similar to the standard Gaertner instrument in that white light passes through a linear polarizer by a large lens. The full mirror then reflects the light at an incidence angle of about 70°, through a large lens to a focus where an analyzer and photographic camera are located. Thus, at certain orientations of the waveplate, the polarizer, and the analyzer, all light of a given wavelength is extinguished from a uniform mirror. Because light at different wavelengths is not completely eliminated, an additive color mixture of these wavelengths is formed. For example, the light of 550 nm (green) is extinguished, allowing the blue and red to combine into violet. A phase change thus would be manifested by the appearance of different colors across the mirror. Phase change, as stated previously, is associated generally with film thickness, although differing indices would also be included in this change. Thus, green light extinguished at one location on the mirror would be prominent at another if the film thickness changed. Color photographs show the variation of film thickness as color differences. The same color would indicate the same film thickness. (See fig. 19.)

Initial observations of the mirror under polarized light conditions showed a wide variety of colors ranging from violet to green and blue. Especially important was the lack of color change between adjacent areas where replication had been performed. Long strips of the mirror had been cleaned by an application of acetyl cellulose tape moistened with acetone so most of the lunar fines were removed. Thus, the color band must be caused by a film of some kind or a removal of the SiO overcoating. To distinguish the validity of these conditions, a small Q-tip covered with lens tissue was used to rub small areas of the mirror. After each rubbing, the mirror was mounted in the polarization setup described, and new color photographs were taken.

Although no data on the Surveyor 3 mirror were available to the authors on this subject, it was felt that the edges of the mirror would be rubbed sufficiently to remove the SiO film and expose only the aluminum. The appearance of the mirror under polarized light indicated that this had been accomplished. Three areas at the top, bottom, and side showed the same yellow color, signifying the aluminum layer had been reached. The lower part of the mirror after replication gave a violet color in polarized light. A rubbing was then attempted at various areas: top of mirror, center on a previously peeled area, at the color band, and at the bottom of the mirror. Each area resulted in the same

Please See Figure 19 (which has been moved to the end of the book)

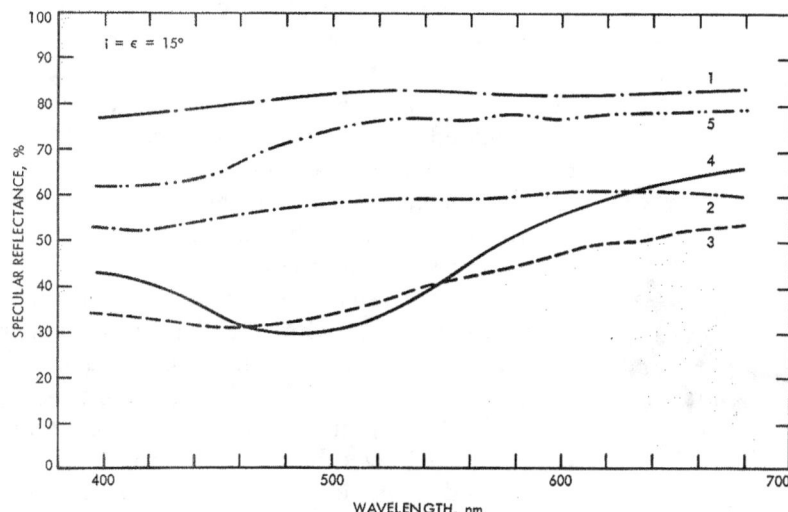

FIGURE 20. — Spectral reflectance curves of areas marked in figure 19, after replication and rubbing. The incidence and reflectance angles=15°.

violet color, indicating that the SiO film layer had been reached. Thus, the greenish color appearing around the elevation axis was replaced by violet in the rubbed locations. By this technique, the proof was established of a film layer or layers of post-launch origin. These layers were of different or graduated thicknesses. From the spectral and ellipsometry data, the maximum thickness of this film or films was determined to be less than one wavelength ($\lambda = 550$ nm). Vigorous rubbing in the central area (violet in color) also indicated that the SiO overcoat had become hard and removal was difficult, although a color change toward yellow was detectable.

The many cyclic thermal changes in the hard lunar vacuum probably had increased the durability of the overcoat beyond its original properties. Included in figure 19 is a probable cross section of the deposits on the mirror. High-resolution mass spectrometric analyses of the mirror were made by B. Simoneit and A. Burlingame. (See ch. V of this document.) Their findings indicate some contamination from hydrocarbons, silicones, and dioctyl phthalate, together with traces of exhaust products from the LM and Surveyor 3 engines. However, additional analysis is needed before the composition and possible origin of the film, or films, on the mirror can be resolved.

A repeat of the spectral reflectance data described previously was performed on five replicated and rubbed areas (fig. 20). This was measured on the goniophotometer using narrow-band interference filters. The areas are marked on figure 19. Curve 4 corresponding to area 4 closely repeats the wave at $i = 10°$ on area B, that of maximum color. The five curves were measured at $i = \epsilon = 15°$. Curves 3 and 5 are also similar to those of areas D and A, respectively. However, the rubbed areas of 1 and 2 show an almost flat response over the visible wavelength range. Curve 1 is close to that expected from a clean aluminum surface, while that of 2 indicates that some roughness or pitting is still present. Again, the spectral measurements reflect the observations made during the rubbing investigations.

Another observation of the mirror surface requires some comments. Two very faint boundaries can be seen under correct illumination conditions in the lower portion of the mirror. These are marked in figure 19. From the work by N. Nickle (see ch. IV, pt. D), these boundaries are apparent shadows of the mirror assembly hood. Microphotographs (fig. 21) of these boundaries indicate that the upper part is particulate material, and possibly includes pitting. The authors agree with Nickle's conclusions that trace these boundaries to locations of surface sampler hardness tests during the Surveyor 3 mission. The lunar fines easily removed from the mirror probably were caused by the LM descent; the remaining material had its origin

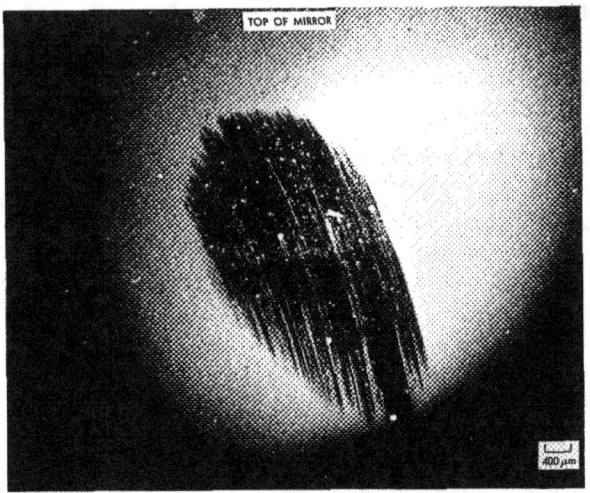

FIGURE 21.—Microphotograph of one of the faint boundaries of the mirror surface.

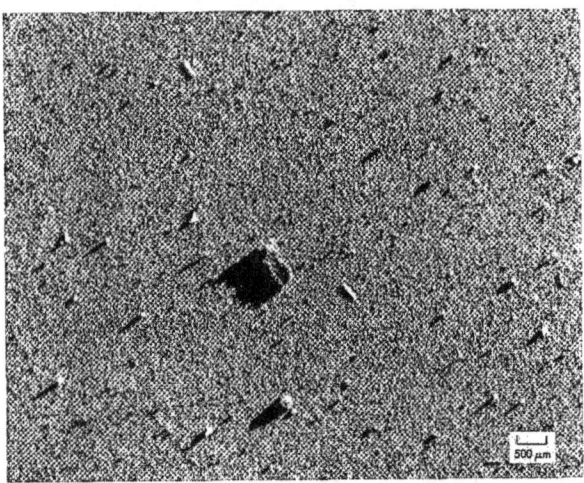

FIGURE 22.—Microphotograph of the fine material on the mirror surface. The rock is about 800 by 330 μm; the two prominent spheres are 230 and 100 μm in diameter.

in the Surveyor 3 landing and mission operations. This hypothesis also would explain the differences in the optical performance of the mirror on the lunar surface and at HAC upon return. Even after replication of the mirror, the reflectance in the central part was less than the top or bottom of the mirror. Thus, either some degree of pitting has occurred in the underlying film or some sort of molecular binding must have taken place, making complete removal of the fines difficult. How much material has been added to the original Surveyor imposed surface can be only surmised, but values of two or three times do not seem unreasonable. The authors believe that little, if any, additional lunar fines were added to the mirror surface in the 31 months between spacecraft landings.

A brief comment is warranted here concerning the condition of the mirror as determined from pictures during the Surveyor 3 mission.

As stated in reference 4, during the mission: "The broad diffuse white band over the upper half of the mirror appears to be consistent with the glare observed in the photographic images, ..." The report discounted the scattering effects caused by exhaust gases from the midcourse and main retro maneuvers. Most of the scattering was assumed to be caused by lunar particulate matter sprayed up from the interaction with the lunar surface during touchdown. The estimated amount of particulate matter in cross section was given as 30 percent or more for the upper half of the mirror.

The "present" condition may be contrasted against the 1967 conclusions by reviewing the previous results. Particulate matter is now present all over the surface and would thus change the appearance of the television image as shown in figure 4 (ch. 8) of reference 4. This particulate matter, allowing for an estimated mean particle size and an incidence angle of 10°, is less than 30 percent in cross-sectional area in the lower part of the mirror. It rapidly increases to more than 50 percent at about the center and to between 70 and 80 percent at the top of the mirror. This is much more than that concluded from the Surveyor 3 mission and is probably a result of the LM landing.

A difficulty exists in using the reflectance data to estimate the particle density because the film over the surface changes the absolute values. However, it is a correct conclusion to have stated that most of the particulate material caused the scattering. The film effect was almost impossible to detect during the mission.

Photography

Several microphotographs of the mirror were taken to examine the nature of the material covering the mirror. A wide variety of lunar

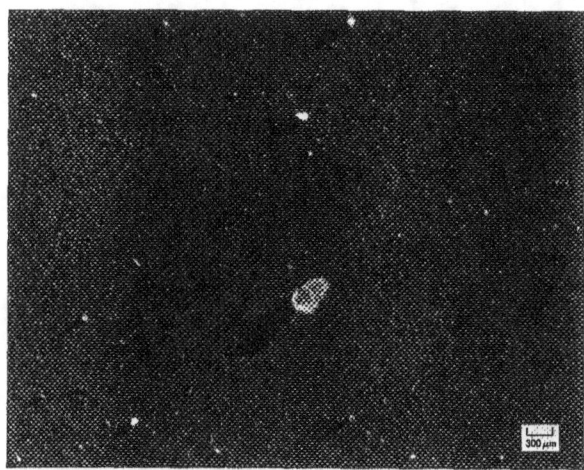

FIGURE 23.—Microphotograph of a clear feldspar grain. The fragment is about 440 by 280 μm; the small sphere, about 2 cm "up Sun" from the large fragment, is about 30 μm in diameter.

FIGURE 24.—Microphotograph of pitted sphere on the mirror's surface. The sphere diameter is about 186 μm. The pit measures about 56 by 37 μm.

material (similar to other lunar fines) was discovered by this photography. Figure 22 shows a lunar rock fragment and numerous glass spheres. The fragments in the figure are probably pyroxene grains, while those in figure 23 resemble a clear feldspar. Much of the material has pitted surfaces even to those of the spheres (fig. 24). Also of interest is the accumulation of lunar fine material around the base of the spheres. (See fig. 25.) The five prominent specks of material located on the mirror from the top of Conrad's swipe to the left elevation axis trunnion (see fig. 10) are especially interesting. These specks were on the mirror at the first unpacking at the LRL. (See fig. 26.)

A microphotograph of one of the specks (fig. 27) shows an apparent growth pattern in the material. The long needle-like object is a beta-cloth fiber from the bag that contained the camera on the Apollo return flight. Isodensitometry of the original film is in progress to determine whether these specks were present on the mirror when the astronauts arrived at the Surveyor 3 site. The origin of these specks is still in doubt, even after the analysis of Carr and Proudfoot (see ch. IV, pt. C) because all areas in the LRL exposed to the camera were devoid of ceiling tile. The scratch on the mirror is a real one and probably was there before flight.

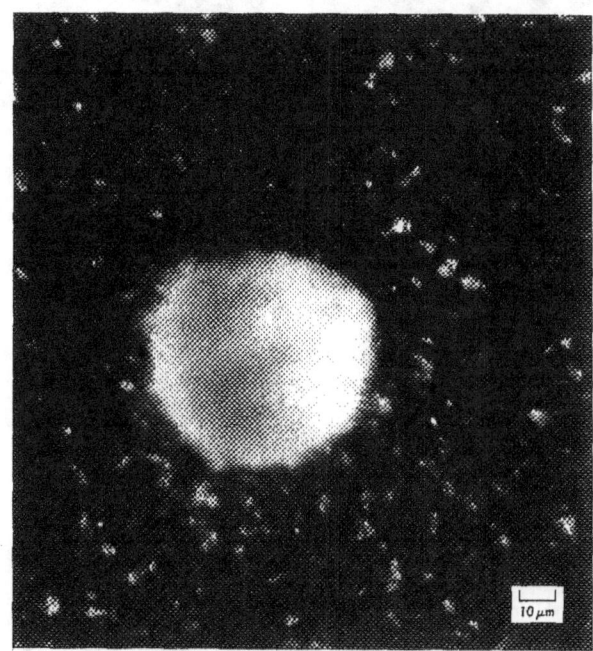

FIGURE 25.—Microphotograph in crossed polarizer illumination of a sphere with adjacent lunar material banked against the sides. The sphere is about 36 μm in diameter. The black areas are the reflecting parts of the mirror surface.

Figure 28 is an oblique illumination view of the tip of Conrad's swipe, showing the amount of lunar material that did not easily rub off.

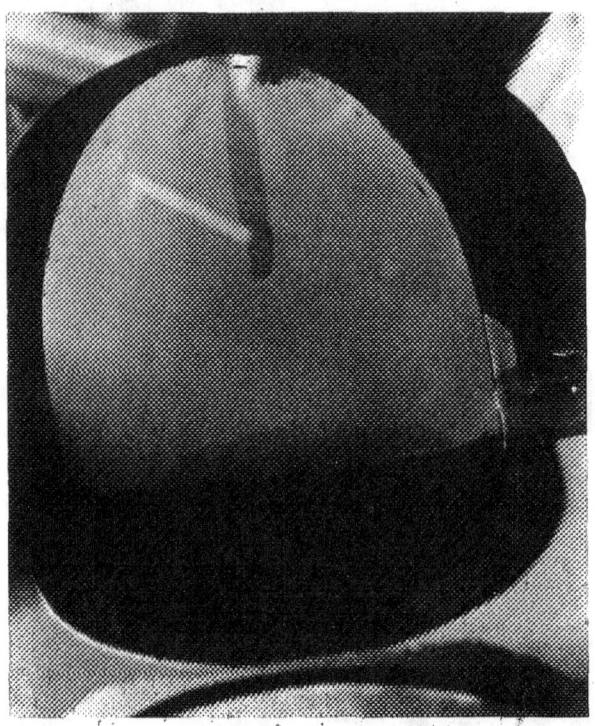

FIGURE 26.—Photograph of the mirror assembly immediately after official unpacking at the Lunar Receiving Laboratory. The five specks (dark in the photograph) are visible here.

FIGURE 27.—A microphotograph of one of the five prominent white specks on the mirror. The long needle fiber is beta cloth.

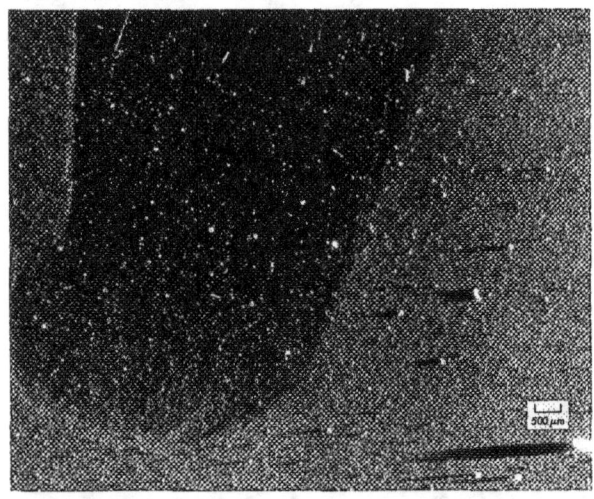

FIGURE 28.—A photograph of the tip of the swipe made by Conrad. Much lunar material still remains.

Filters

The Surveyor mirror assembly contained a filter wheel with four apertures. Two of the filter positions had double-element absorbing glass; the third and fourth positions were occupied by a single absorbing filter and a clear glass, respectively. Figure 29 shows the positions of the filters in the wheel when the last Surveyor transmission ended. The lunar fines coating the filters were not evenly distributed; the blue filter was most protected. Spectrophotometric measurements, before and after cleaning all filter assemblies, were made on the filter glass:

(1) Before cleaning all filter assemblies:
 (a) Five discrete positions through the visible.
 (b) Central position from 0.3 to 2.5 μm.
 (c) Each component of the double filters (four) at the above positions and wavelength range.

(2) After cleaning: Each component from 0.3 to 2.5 μm.

Goniophotometric measurements were made of the surface of the clear filter.

The glass filters were chosen to fit the color-matching functions $\bar{x}$, $\bar{y}$, $\bar{z}$ of the international system of color measurement (CIE) when used

TABLE 1.—*Thicknesses of spare and flight filters used on Surveyor 3*

Type	Thickness, mm					
	Clear flint	3-76	GG15	BG1	OG4	OGR3
Flight	2.98	2.97	1.96	1.19	1.28	1.74
Spare	2.91	3.50	1.79	1.17	1.16	1.78

in combination with the Surveyor 3 vidicon and optics (ref. 5). The original glass was obtained from four manufacturers: Schott, Chance, Corning, and Bell & Howell. The Schott glasses were GG15 and BG1 for the blue filter and OG4 for one element of the green filter. The Chance glass was OGR3 for the other component of the green filter. Corning glass 3-76 was used for the red or amber filter, and Bell & Howell flint glass $n_D = 1.612 v = 37$ for the clear filter. The GG15 and OG4 glasses were on the top in the blue and green filters and collected all the lunar fines. The inside surface of these glasses was coated with a thin layer of Inconel metal. This uniformly attenuated the incident light and equalized the exposure through each colored filter. The thickness of each glass filter varied in order to obtain the best possible "fit" to the functions. Each camera had two identical filter sets manufactured, and the spares were preserved in total darkness, forming an excellent comparison set. The glasses were from the same glass melt and differed only slightly in thickness. (See table 1). The transmission of the Inconel coatings on the spare set was within a few percent of the flight set.

Measurements

Two instruments were used in the spectral transmission measurements: A Bausch & Lomb Spectronic 505, and a Cary 14 spectrophotometer. Both were checked against NBS filter standards and agreed well in their overlapping wavelength range. The Bausch & Lomb 505 was used for five positional measurements over the range of 0.3 to 2.5 µm. Figure 30 is a plot of the spectral transmission of all flight and spare sets. The data are for the clean portions of the flight filters. An evident yellowing of the flint

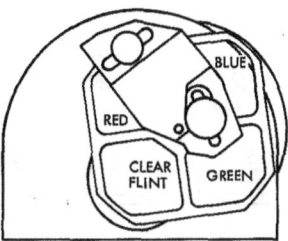

FIGURE 29.—Surveyor camera mirror assembly from the top showing the positions of the glass filters at the time of the last transmission.

glass can be seen and probably is caused by the long ultraviolet and radiation exposure. No apparent change in the anti-reflection coating (MgF) was detected. Large changes in the filter combinations over pre-flight measurements are indicated in figures 31 through 33. In all cases, the transmission of the combination increased. The red filter (fig. 33), made of a single piece of 3-76 glass, apparently underwent a bleaching effect, resulting in increased green transmission. Much more difficult to explain is the increase in the blue and green filters. If the curves for the blue filter (BG1) and green filter (OGR3) elements of these glasses are compared with the spare set, the large transmission increase cannot be explained. If, on the other hand, the uppermost elements GG15 and OG4 are responsible, then the cause must be the Inconel coatings because both glasses, uncoated, have transmissions over 92 percent. The uppermost glasses were exposed to thermal and radiation cyclic changes that eliminated the ultraviolet transmission of the one and shifted the other cut-on position toward the red. Some removal of the Inconel coating would account for the transmission increase, but such dissipation is unknown at the temperatures involved.

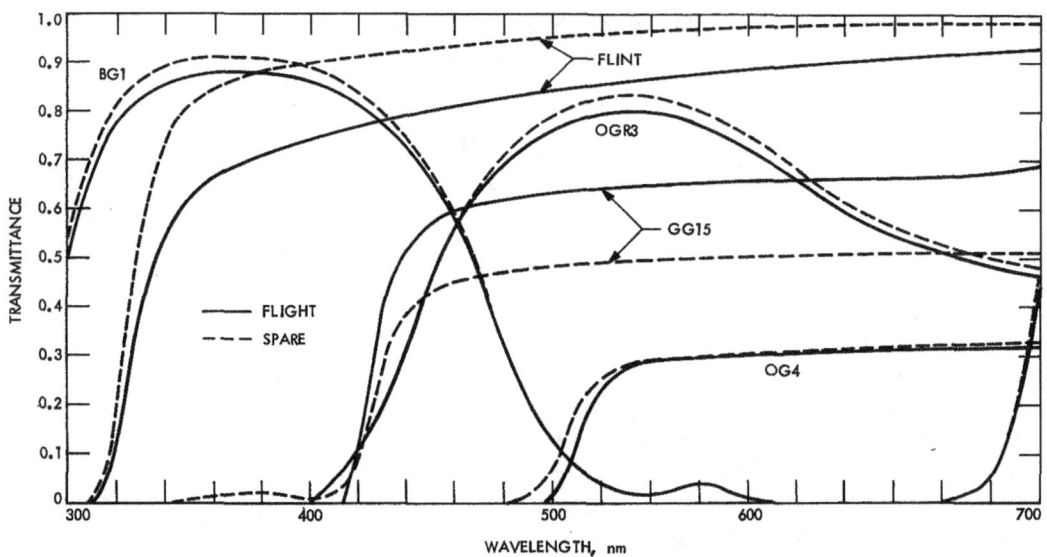

FIGURE 30.—Spectral transmission of the flight and spare set of filter glasses, with the exception of 3–76.

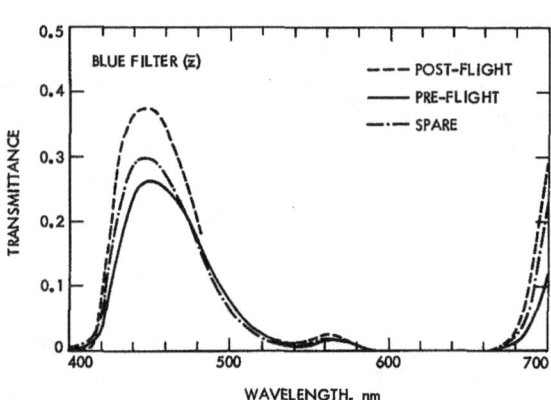

FIGURE 31.—Transmission of the blue filter for pre-flight, post-flight, and spare sets. The increase in overall transmission of the post-flight measurements is apparent.

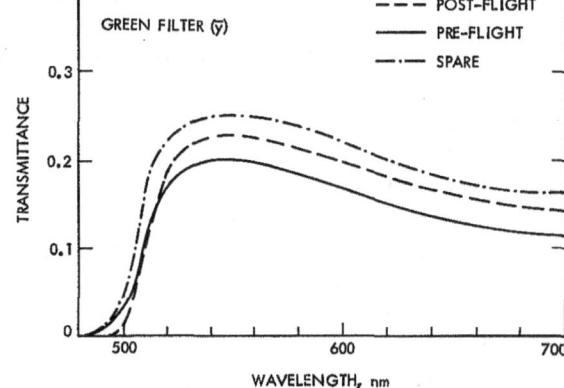

FIGURE 32.—Transmission of the green filter for pre-flight, post-flight, and spare sets. The increase in overall transmission of the post-flight measurements is apparent.

Particle Density

The filters, flint, GG15, OG4, and 3–76 possessed a coating of lunar fines to some degree. By using the transmission measurements of the dirty filters with respect to the cleaned areas of the same filters, an estimate of the particle density can be made. Five measurement positions of the filters were used to calculate the percentage loss (dirty:clean ratio) caused by the fines. If the assumption is made that the glass filter remained homogeneous over its area, then the area covered is directly proportional to the transmission loss. This has been substantiated by analysis of the peels taken from the filter surfaces (ref. 6). The areas covered at the five positions on each filter are given in figure 34. The dotted area shows the observed limits of the lunar fines when the filters were removed from the camera.

Figure 35 is a plot of the transmission ratios

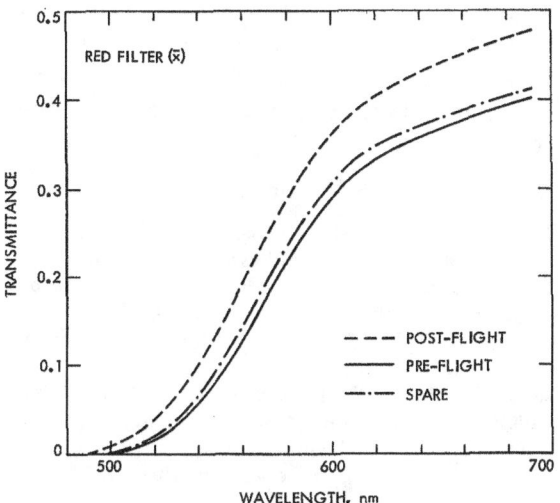

FIGURE 33.—Spectral transmission of the red filter. The increase in post-flight transmission is less than that noticed in the other two filter sets.

of the clear filter at the five positions as a function of wavelength. Data beyond 700 nm were obtained for the center only. The transmission curves are very flat with the exception of position 5, which tends toward a red increase as seen in the reflectance of the lunar fines. Position 5 was the closest to the front edge of the camera, and thus received the greatest amount of lunar fines. The large proportion of glasses in these fines undoubtedly accounts for the spectral variation. The last curve in figure 35 clearly shows the effect of exposure on the clear filter in the visible, but the almost complete transparency in the infrared. This agrees well with the data obtained from fines on the mirror. However, the gonioreflectance of the upper surface of the clear filter shows that much less material is present on the filter than on the mirror surface.

Lens

The last major optical component of the camera is the variable-focal-length lens assembly built by Bell & Howell. Shortly after disassembly of the camera, the lens was returned to Bell & Howell for an exact repeat of the pre-flight tests. Although a detailed review will not be attempted here, the appearance of a thin film on the first lens surface is worthy of mentioning. The film is present on the side closest to the front

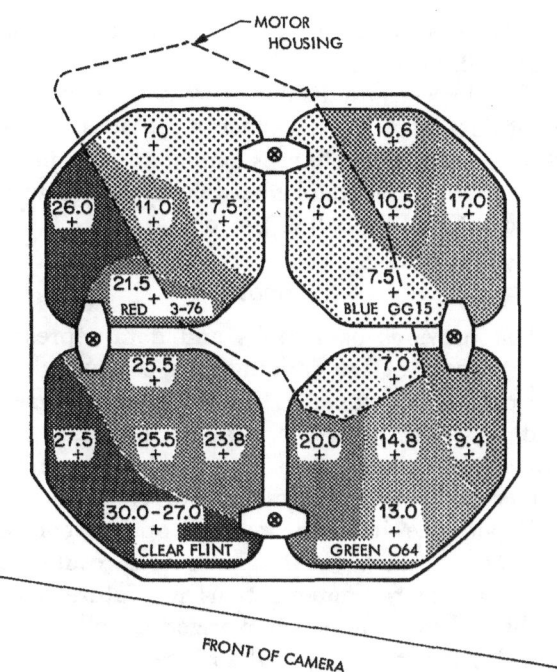

FIGURE 34.—Drawing showing the four filter positions and the extent of the coating of lunar fine material. The values by the filter positions (marked by crosses) are the percentages of area occupied by the fines.

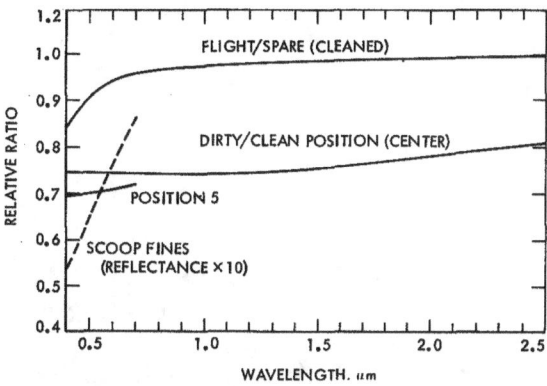

FIGURE 35.—Plot of the transmission ratios for the clear flint glass. The reflectance of lunar fine material found in the Surveyor scoop is shown as a dotted line. Its reflectance in absolute terms is one-tenth of that plotted.

of the camera. Removal of this film also was apparently easy (small area). No spectral tests were made, but the similarity of this type of film to that on the mirror surface is an important fact. The origin of these films may be diffusion pump oil, as suggested in reference 2.

The lens transmission decreased over pre-flight data, showing the effect of radiation on the glasses used. This effect is similar to that measured in the flint glass filter. Both glasses were manufactured by Bell & Howell.

Summary

The previous paragraphs and data represent only a part of the investigation conducted on the optical components. The lunar environment had its effect on the optical performance, but not seriously enough to prevent the use of the camera if the electronics had survived.

Experience in analyzing the optical components of the Surveyor 3 camera have resulted in some strong recommendations for future space flights. These may be summarized as follows:

(1) All external optical surfaces, except during the interval of measurement, should be protected from the local environment. Any contamination seriously affects the radiometric use of optical instruments.

(2) Glasses or films should be chosen which will not significantly change their characteristics during the lifetime of the mission.

(3) In the event that neither of the above recommendations can be implemented, an onboard calibration system should exist that will accurately monitor the degradation of the optical components during the lifetime of the mission.

References

1. DRUMMETER, L. F.; AND HASS, G.: *Physics of Thin Films*, vol. 2, Academic Press, 1964, pp. 305–361.
2. *Test and Evaluation of the Surveyor III Television Camera Returned From the Moon by Apollo XII*, vol. I, SSD 00545R, Hughes Aircraft Co., Culver City, Calif., 1970.
3. GRUM, F.; PAINE, D. J.; AND SIMONDS, J. L.: "Prediction of Particle-Size Distribution from Spectrogoniophotometric Measurements." *Journal of the Optical Society of America*, vol. 61, 1971, pp. 70–75.
4. GAULT, D.; COLLINS, R.; GOLD, T.; GREEN, J.; KUIPER, G. P.; MASURSKY, H.; O'KEEFE, J.; PHINNEY, R.; AND SHOEMAKER, E. M.: "Lunar Theory and Processes." *Surveyor III Mission Report. Part II: Scientific Results*, TR 32–1177, Jet Propulsion Laboratory, Pasadena, Calif., 1964, pp. 195–213.
5. RENNILSON, J. J.: "A Television Colorimeter for Lunar Exploration." *Proceedings of the 8th International Colour Meeting, Lucerne, Switzerland*, Centre d'Information de la Couleur, 23 Rue Notre-Dame-des-Victoire, Paris, France, 1965, pp. 498–506.
6. NICKLE, N.: "Surveyor III Material Analysis Program." *Proceedings of the Apollo 12 Lunar Science Conference.* To be published.

PART F

PARTICLE IMPACT AND OPTICAL PROPERTY ANALYSIS OF THE SURFACES OF SURVEYOR 3 MATERIALS

D. L. Anderson, B. E. Cunningham, R. G. Dahms, and R. G. Morgan

As part of the overall scientific investigation of Surveyor 3 materials, a study was conducted at the Ames Research Center to determine the effect of the lunar environment or some of the painted and unpainted exterior surfaces. Several surfaces were examined:

(1) White thermal-control paint on parts of the television camera (elevation-drive housing—a small 5.1- by 7.6- by 1.3-cm box—and the lower shroud).

(2) Polished surface of the unpainted aluminum radar altimeter and doppler velocity sensor (RADVS) support tube (two 2.5-cm sections, B and E, as cut from 19.7-cm length of the 1.3-cm-diameter tube).

(3) Unpainted surfaces of two stainless-steel

screws and two aluminum washers from the lower shroud.

Results of this study are presented here as they pertain to surface cratering, to changes in surface spectral reflectance, and to a better definition of the lunar micrometeoroid and secondary particle environment (lunar ejecta).

These parts were returned by the Apollo 12 astronauts in November 1969. During recovery, the parts were subjected to the high "g" loads associated with re-entry and to the jolt at splashdown in the Pacific Ocean. It is believed that the television camera became dislodged from its stowed position at splashdown, causing two dents in the primary mirror hood (ref. 1). At the NASA Lunar Receiving Laboratory in Houston, the parts were removed from the astronauts' recovery bag, given a brief examination, and sealed in polyethylene bags. No specific effort was made to maintain a vacuum or light protection around any of the parts examined. The parts have, therefore, been exposed to a variety of laboratory environments since their release from quarantine in January 1970. The examination reported in this article was conducted between July and December 1970. For a more complete description of these parts and their handling before investigation at Ames, see reference 2.

Examination and Analysis

Examination of the camera parts and tube sections was conducted using three techniques:

(1) Optical and scanning electron microscopy (SEM).
(2) Energy dispersive X-ray probe analysis.
(3) Spectral reflectance measurements.

The microscopy techniques were used to examine the surface features of each part at magnifications up to .1700 and 30 000 times, respectively. The normal practice of vapor depositing a gold film over an insulating paint surface for SEM was not permitted on the Surveyor parts because of constraints imposed by subsequent experiments to be conducted by other investigators. Therefore, some difficulty with charge buildup was encountered; this limited the useful magnification for examination of the thermal-control paints to 10 000 times. The X-ray probe,

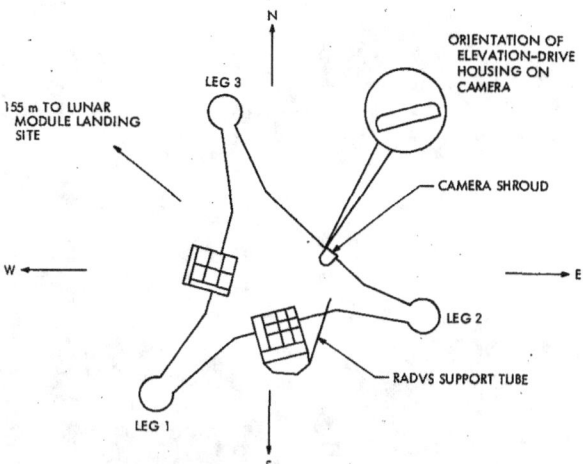

FIGURE 1.—Orientation of Surveyor 3 on the lunar surface, shown with respect to Apollo 12 LM. Locations of television camera and RADVS support tube are shown.

an accessory to the SEM, was used to obtain the elemental composition of a surface. X-ray maps of a specimen, which showed the presence and spatial distribution of each element analyzed, were also obtained. The spectral reflectance of each part was measured in an integrating-sphere reflectometer, with the sample located in the center of the sphere whenever possible. Spectral reflectance measurements were made at several locations on each part; special emphasis was given to the cleanest, dirtiest, or most contaminated areas.

Figure 1 shows the positions of the parts on the Surveyor 3 spacecraft and the relative location of the Apollo LM landing site (refs. 3 and 4). Figure 2 shows the locations of the television camera parts; the locations of sections B and E on the 19.7-cm length of the RADVS support tube are shown in figure 3. Discussion of results of the examination of these parts is divided into three categories:

(1) Physical features of painted and unpainted surfaces.
(2) Chemical composition of each type of surface before and after exposure to the lunar environment.
(3) Spectral reflectance of each type of surface.

In support of the first category, an ancillary laboratory program was conducted in the Ames

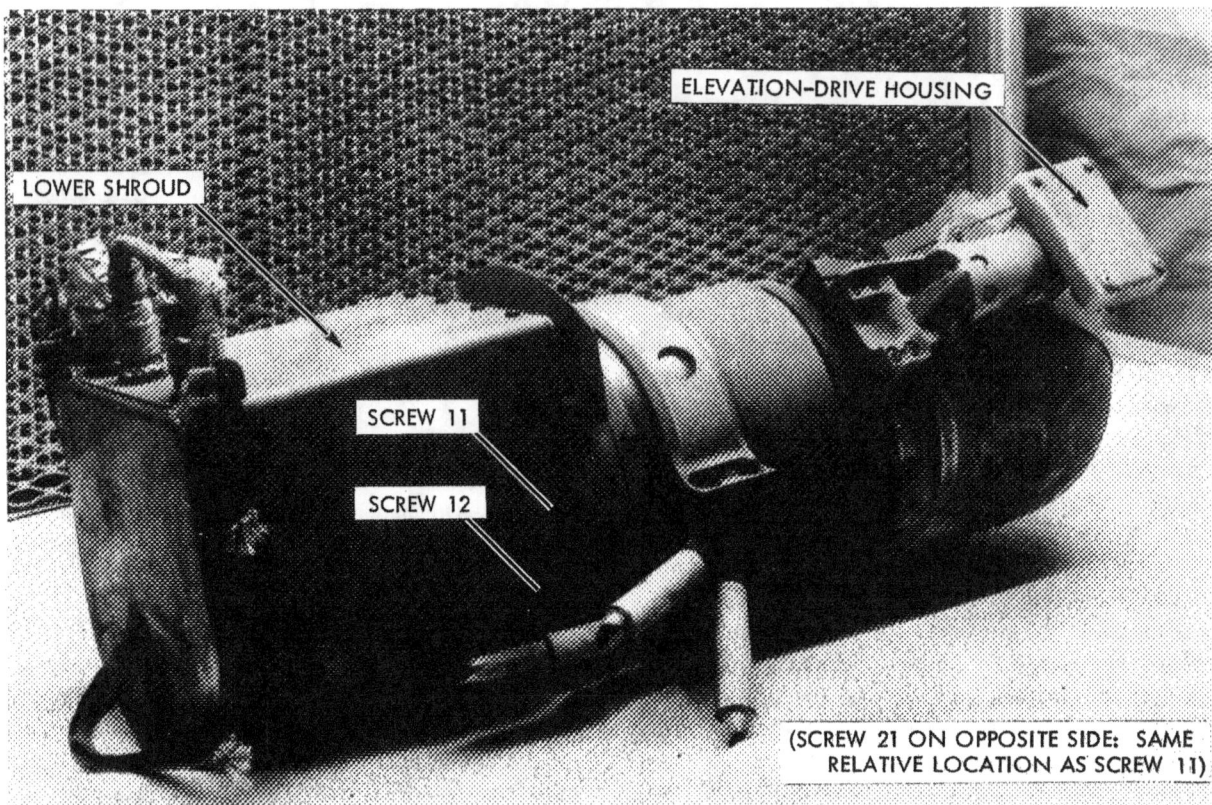

Figure 2.—Surveyor 3 television camera, showing location of specific parts of the camera examined in the study by Ames Research Center.

Space Environment Simulator (ref. 5) to obtain a better interpretation of possible impact features observed on the Surveyor 3 materials. Control specimens of these materials (polished aluminum and painted surfaces) were bombarded with micrometer-sized carbonyl-iron particles, electrostatically accelerated to velocities of up to 20 km/sec with a Van de Graaff accelerator, to produce "standards" for characterizing impact craters. This approach was used even though it was recognized that the lunar micrometeoroid environment may be comprised of particles with densities as low as 0.5 g/cm^3 (ref. 6). It is believed that the carbonyl-iron microparticles, with a density of 7.8 g/m^3 and at velocities in the 2- to 20-km/sec regime, can produce impact craters with shapes characteristic of impact craters produced at higher velocities by lower-density particles. Evidence supporting this viewpoint has been presented by Morrison (ref. 7).

The two parts of the RADVS unpainted aluminum support tube, the E and B sections (fig. 3), were examined for evidence of hypervelocity impact by micrometeoroids or by secondary particles (lunar ejecta). The E section was examined by optical microscopy in two conditions:

(1) In the "undisturbed" state, as delivered to Ames via JPL from the Lunar Receiving Laboratory.

(2) After removal (by previous investigators) of most surface and embedded contaminants with replicating film.

In the undisturbed state, many micrometer-sized holes, or cavities, and foreign particles were observed. The holes were found with approximately the same distribution on all parts of the tube; the particles were concentrated primarily on the side facing the spacecraft and the lunar surface. Section E was again examined

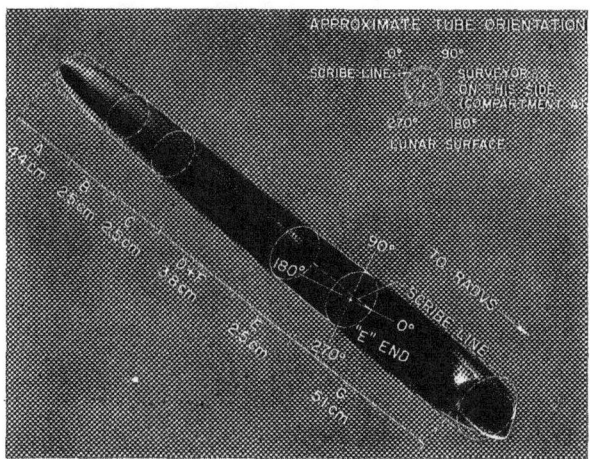

FIGURE 3.—Section of RADVS unpainted aluminum support tube, as cut from Surveyor 3 spacecraft by Apollo 12 astronauts. Locations of sections B and E are indicated.

after removal of the contaminants; the surface that had been covered with the particles appeared to have an eroded surface, where the "clean" surface appeared to have the same features as observed before contaminant removal. The B section was examined by both optical and scanning electron microscopy, but only after removal of contaminants by previous investigators. In general, this section appeared, by optical microscopy, to have the same surface characteristics observed on section E after the contaminants were removed.

Another approach, although somewhat qualitative, was used to characterize possible impact sites. Representative sites, including holes that were obviously voids and not mechanically produced, holes that were "rimmed" and, in one case, a particle embedded in the tube surface, were examined with the X-ray probe (fig. 4). Various smooth and irregular areas of the interior surfaces of the holes and smaller particles adhering to these surfaces were examined. The elements present in the aluminum tube (Al 2024–T3) were identified in appropriate proportions with significant amounts of aluminum and copper and lesser amounts of iron, silicon, manganese, and magnesium. Sodium and chlorine were found in most sites, in amounts that indicate the possibility that atmospheric salt contamination may have occurred during the

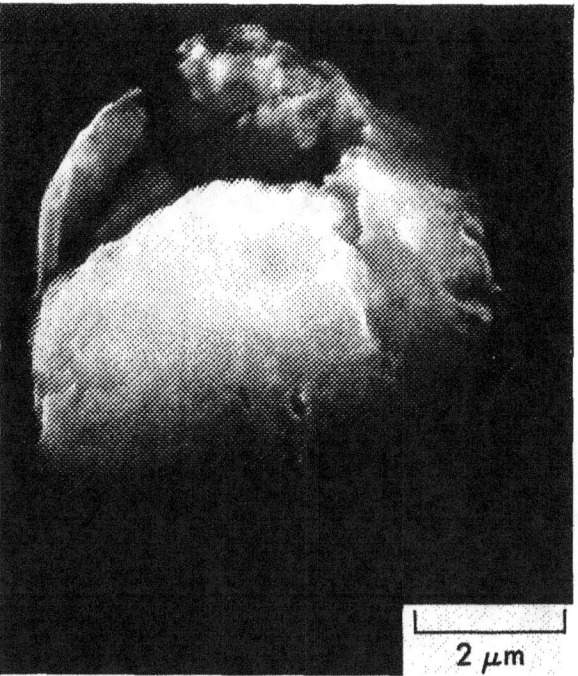

FIGURE 4.—Scanning electron micrograph of a particle embedded in the surface of section B of the RADVS support tube.

post-return exposure to laboratory environments in coastal areas. Of the principal elements known to be present in Apollo 12 lunar rocks and fines (Fe, Si, Mg, Ca, and O; see ref. 8), iron, silicon, and magnesium were identified in most sites. However, they were found in such small amounts that it was not possible to qualify them as being of lunar origin, because these elements were also present in the aluminum alloy. Calcium was identified in some sites. Although the X-ray probe lacks the resolution necessary to clearly resolve oxygen, this element was found in most sites examined. The results of the X-ray probe analysis indicate that lunar material containing calcium was deposited in unrimmed and rimmed holes. The embedded particle was composed only of elements present in the alloy or possibly of aluminum oxide. As an alumina polishing compound was used to polish the tube before assembly of the Surveyor 3 spacecraft, such particles could have been pressed into the tube to remain there until removed during replication of the tube surface.

A total of about 200 mm² of the surface of

section B was examined by the SEM; typical micrographs of the "dirty" and "clear" sides of this section are shown in figure 5. Note that the micrograph of the "dirty" side (fig. 5(a)) clearly shows the eroded surface (after contaminant removal) caused by impacting rocket fumes and/or lunar ejecta. General erosion such as shown in this micrograph has not been simulated

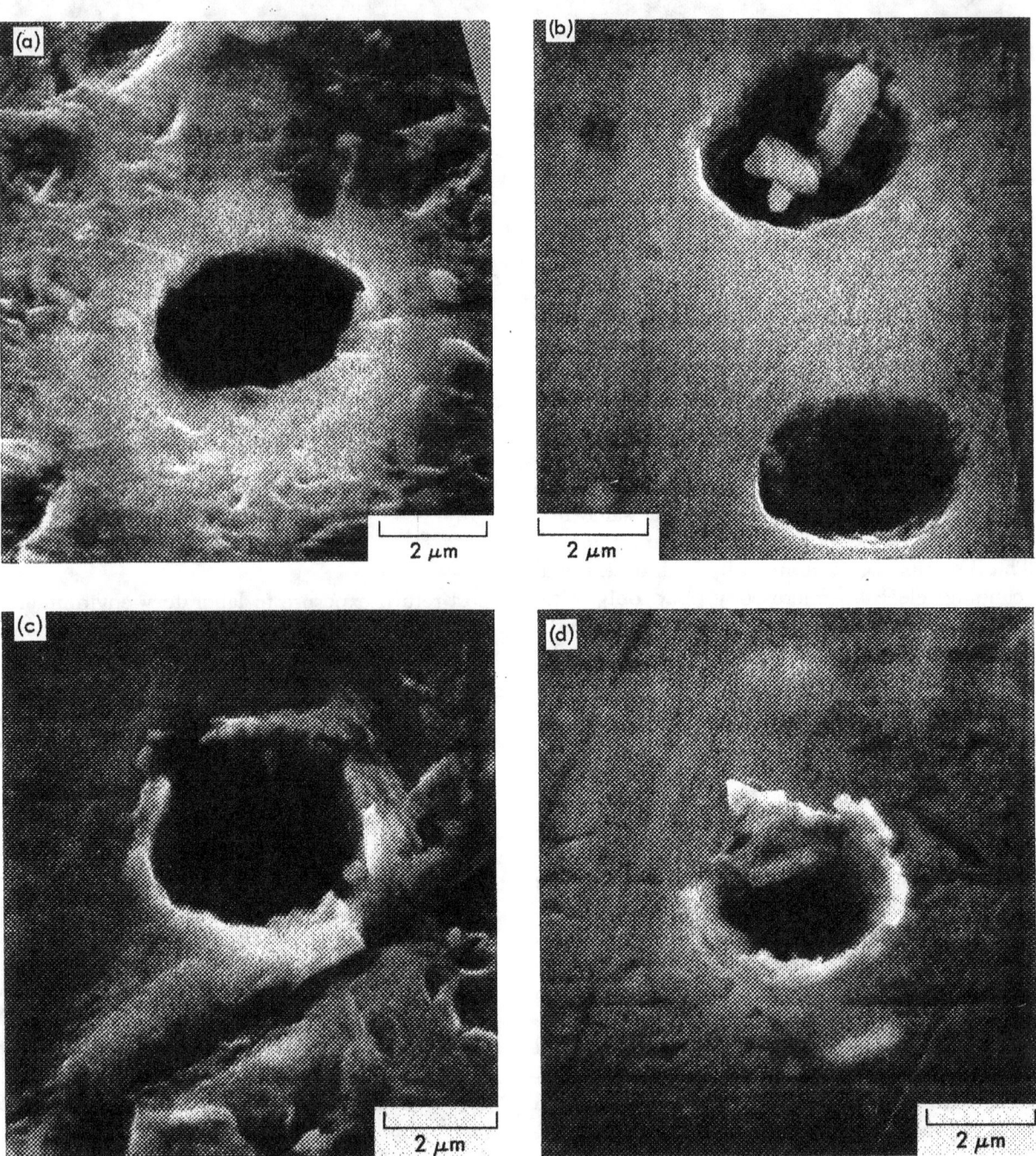

FIGURE 5.—Scanning electron micrographs of "dirty" and "clean" sides of RADVS section B. (a) "Unrimmed" hole on dirty side. (b) "Unrimmed" holes on clean side. (c) "Semi-rimmed" hole on clean side. (d) "Rimmed" hole on clean side.

in the ancillary laboratory program mentioned previously. Individual sites cannot be identified readily for study to determine whether or not the erosion was caused by impact or by chemical action. Therefore, no attempt has been made to characterize the cause of the erosion. Note that the "unrimmed" holes shown in figures 5(a) and (b) are representative of many holes found over the entire surface of the tube by both optical and scanning electron microscopy. These holes are from opposite sides of the tube. It has been judged by the authors and by other investigators (see ch. IV, pt. A; also see ref. 9) that the "dirty" side (fig. 5(a)) faced the lunar surface and the Surveyor 3 spacecraft. (See fig. 3.) As the tube was almost horizontal on the spacecraft, no direct micrometeoroid impacts should have occurred on this side. Therefore, holes such as that shown in figure 5(a) must, in some way, be characteristic of the tube manufacturing or flight preparation processes. Holes from the "clean" side, such as the one shown in figure 5(b), resemble the hole shown in figure 5(a) and, therefore, may have the same origin. Similar holes, although fewer in number, also were found on the surface of a "control" specimen (a section of aluminum tube, supplied by the Surveyor 3 manufacturer, of the same alloy and size as the RADVS support tube). Such holes may be voids introduced during tube manufacture. Rimmed holes were found on all sides of the tube. These holes seem to be caused by some kind of "mechanical" activity, although it is recognized that it is possible for such holes to be caused by impacting lunar ejecta or even by hypervelocity particles.

With this in mind, holes observed on the "clean" side were compared with micrographs of aluminum surfaces that had been bombarded with 2- to 20-km/sec microparticles in the space environment simulator at Ames. Shown in figure 6 are representative micrographs of these laboratory-produced impact sites; these sites have well-defined features:

(1) All have fully developed "flared-rim" craters.

(2) Oblique (nonperpendicular) entry produces "slanted," or elliptical, craters.

(3) Grazing entry produces long, elliptical "gouge" craters.

(4) Crater walls and floors are smooth and usually free of residue except for small, spherical particles.

A few of the holes, in section B, found by means of the SEM, have some of these features to a limited degree. However, such "impact-like" holes were found around the tube (as mentioned above in comments on features observed on the "dirty" side). It also was observed that not one of the holes examined resembled the holes made by "slanted" or "grazing" impact (fig. 6).

Evidence from the optical microscopy, SEM, and X-ray probe examinations indicate that none of the holes studied in this investigation were caused by hypervelocity micrometeoroid impact. This conclusion is based on somewhat meager data. Although both sections B and E were examined thoroughly by optical microscopy at magnifications sufficiently high to easily resolve 100-μm-diameter craters, none were found. On the other hand, only about 5 percent of the surface of section B was examined by SEM at magnifications sufficiently high to clearly resolve 1-μm-diameter impact sites. However, the results of this investigation were used to calculate an upper limit only for the lunar surface micrometeoroid environment, as Jaffe (ref. 10) did, by calculating an area-time product, A_t, for the surface area examined and the time of exposure. It was assumed that micrometeoroids in the 1-μm-diameter size range would be discrete, stony particles with a density of about 3.5 g/cm^3. It also was assumed that impacts by such micrometeoroids would make 3- to 6-μm-diameter craters. This is consistent with measurements obtained in the Ames space environment simulator experiments and by other investigators (ref. 11). For the 3.5-g/cm^3 density, the mass of a 1-μm particle would be about 1.8×10^{-12} g. For section B, the examined surface ($\sim$200 mm^2, 940-day exposure at a solid angle of about π-steradians) has an A_t of about 1.6×10^4 m^2 sec; therefore, the rate of impact was less than 6.2×10^{-5} particles/m^2 sec. This value indicates an upper limit to the lunar surface meteoroid environment for the 10^{-12}-g regime only slightly different from the upper-limit value of 4.5×10^{-5} particles/m^2 sec given in the 1969 NASA meteoroid design criteria (ref. 12).

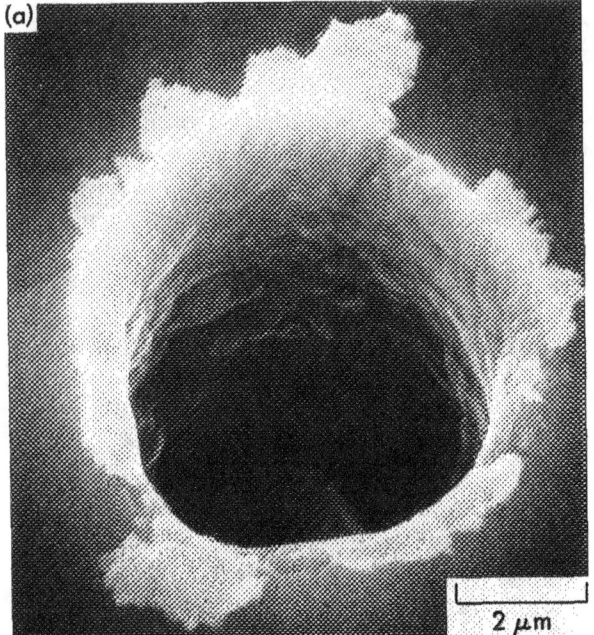
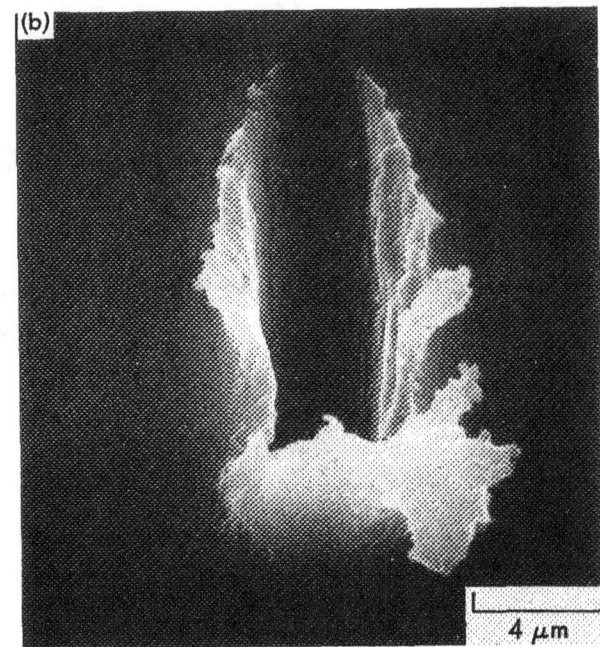
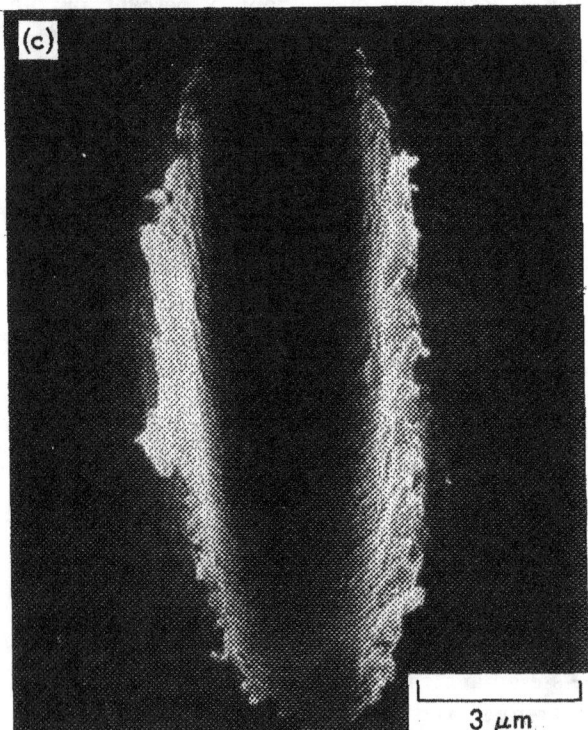

FIGURE 6.—Scanning electron micrographs of unpainted aluminum control tube with same dimensions as sections cut from Surveyor 3 RADVS support tube, showing hypervelocity impact sites produced by carbonyl-iron microparticles at about 7 km/sec. (a) Perpendicular impact. (b) Oblique impact. (c) Grazing impact.

Two sets of unpainted screws and washers from the lower shroud of the television camera were examined by optical and scanning electron microscopy and X-ray probe analysis. The orientation of the screws on the spacecraft was such that one was facing and the other shielded from the landing site of the LM. The complete front surfaces and the sides of the screws, as well as the exposed edges of the washers, were examined for possible micrometeoroid impact sites. No surface features found were similar to the hypervelocity impacts observed in the ancillary experimental program (fig. 6). It was concluded that none of the features observed could be definitely identified as hypervelocity impact sites.

Small particles of material up to 60 μm in size were distributed on the exposed surfaces of the screws and the washer. X-ray analysis of this material identified magnesium, aluminum, silicon, potassium, calcium, and titanium. The presence of these elements is consistent with the composition of the lunar fines, as determined by the Lunar Sample Preliminary Examination Team (ref. 8). The distribution of this material on the surface of the screws was statistically determined from SEM micrographs. Figure 7 shows a typical micrograph of particles of "lunar" material on the screw surface. The particle shapes vary from angular to spherical. Figure 8 shows the size distribution of all particles that could be identified in photographs of this magnification. In comparing the number of particles on the surface of the two screws, it was found that the bottom halves of the screws and washers, which were facing the lunar surface, had up to 45 percent more material on the surface than did the top halves. It was also determined that the screw shielded from the LM had three times as much material on its surface as did the exposed screw. These observations tend to indicate that:

(1) A considerable amount of lunar dust was ejected vertically around the Surveyor spacecraft during its own landing maneuver.

(2) Dust stirred by the exhaust gases from the LM descent rockets may have removed more lunar residue from the surface of the exposed screw than it deposited.

This difference in particle count between the shielded and exposed screws is consistent with the observation by Jaffe (ref. 13) in his analysis of the "blowing on" of lunar soil by Apollo 12 (see ch. IV, pt. I, of this document) and with the evidence gathered by other investigators (refs. 3 and 4).

The shapes of the material found on the surfaces of the screws were similar to those found on all other sections of the Surveyor 3

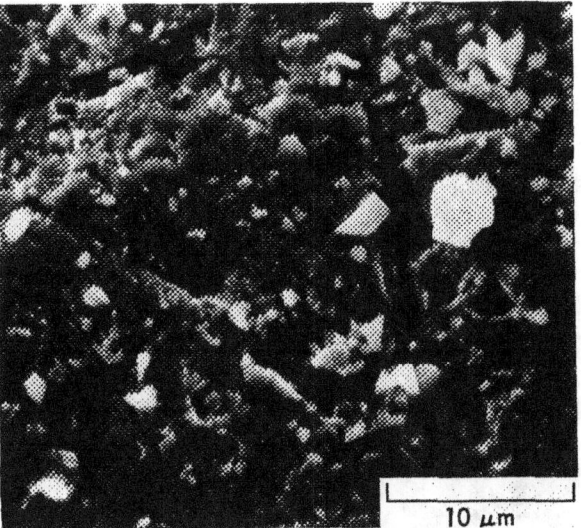

FIGURE 7.—Scanning electron micrograph of surface of unpainted screw from Surveyor 3 television camera, showing particles of "lunar" material.

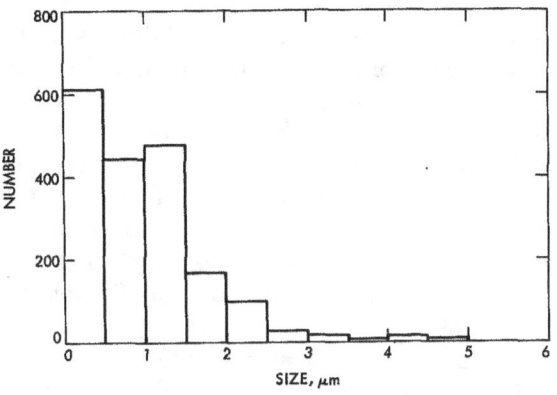

FIGURE 8.—Size distribution of particles found by scanning electron microscopy on the surfaces of the unpainted screws taken from the Surveyor 3 television camera. Particle shapes varied from very angular to spherical.

spacecraft surfaces studied. The only exceptions were the rod-like materials shown in figure 9(a). The rods, all 3 to 5 μm in diameter and up to 100 μm long, were discovered on the elevation-drive housing dust cover and on the screw shielded from the LM. Because similar rods, or fibers, were found in lunar material samples by investigators reporting at the Apollo 11 Lunar Science Conference (refs. 14 through 16), it is

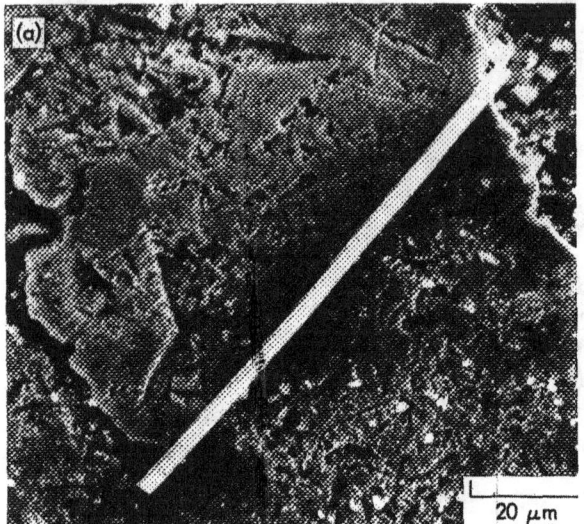 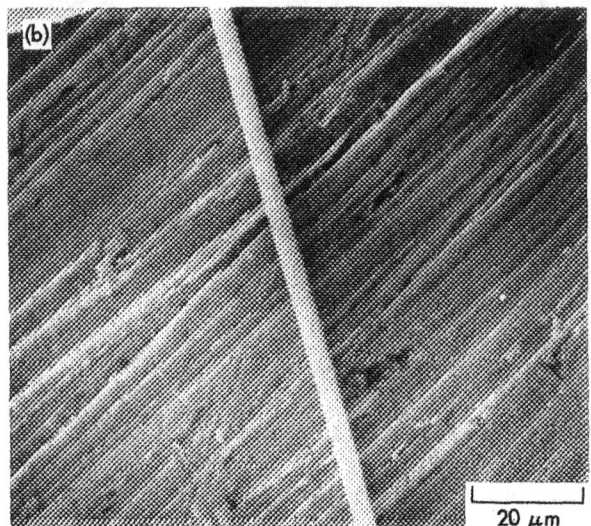

FIGURE 9.—Scanning electron micrographs of rod-like material found on unpainted screw of Surveyor 3 television camera and of fiber taken from beta-cloth space suit material of the Apollo astronauts. (a) Rod-like material on screw surface. (b) Beta-cloth fiber.

appropriate to comment here on the dispersive X-ray analysis of these fibers, conducted as part of this study. These rods were found by this analysis to be beta-cloth fibers from the Apollo 12 astronauts' gloves or back pack. Figure 9(b) is a micrograph of a beta-cloth fiber obtained for comparison. The spectrum obtained in the X-ray analysis of the fiber was identical to the spectra of the rods found on the Surveyor 3 components.

Identification of possible impact sites on the painted surfaces was more difficult than on the metallic surfaces. Very little was known previously about the physical characteristics of hypervelocity impacts in paints; therefore, paint samples prepared at Ames Research Center and standards prepared by the Surveyor 3 manufacturer and retained for control purposes were exposed to hypervelocity particle impacts in the ancillary test program discussed. Figure 10(a) is a micrograph that shows a typical laboratory-produced hypervelocity impact in a paint sample; figure 10(b) is a micrograph of one of the standards. It should be noted that there are several cracks in this unexposed paint surface. (This phenomenon has been referred to as "mud-cracking.")

Examination of the painted surfaces of the lower shroud of the television camera (fig. 10(c)) shows that a similar type of cracking has occurred. Although these cracks on the shroud are larger and more pronounced, the existence of similar cracks on an unexposed standard indicates the possibility that the cracks existed before flight and that they are not necessarily a result of exposure to the lunar environment. There are several holes on this standard that could easily be mistaken for impact sites. Figure 10(d) is a micrograph of the elevation-drive dust cover. All apparent impact sites were examined at higher magnifications, but none could be positively identified as formed by a micrometeoroid. Because of this similarity between the pores and impact sites, a comparison was made between the number and size of holes on the unexposed paint standard and the elevation-drive housing. Figure 11 shows the results of this statistical count. Although it is recognized that there may be some differences in porosity between several standards prepared at different times, these results indicate that at least the majority of apparent impact sites found on this part could be due to natural paint porosity.

The possibility was considered that some of the apparent impacts could have been caused by dust raised by the LM. The impacts from such

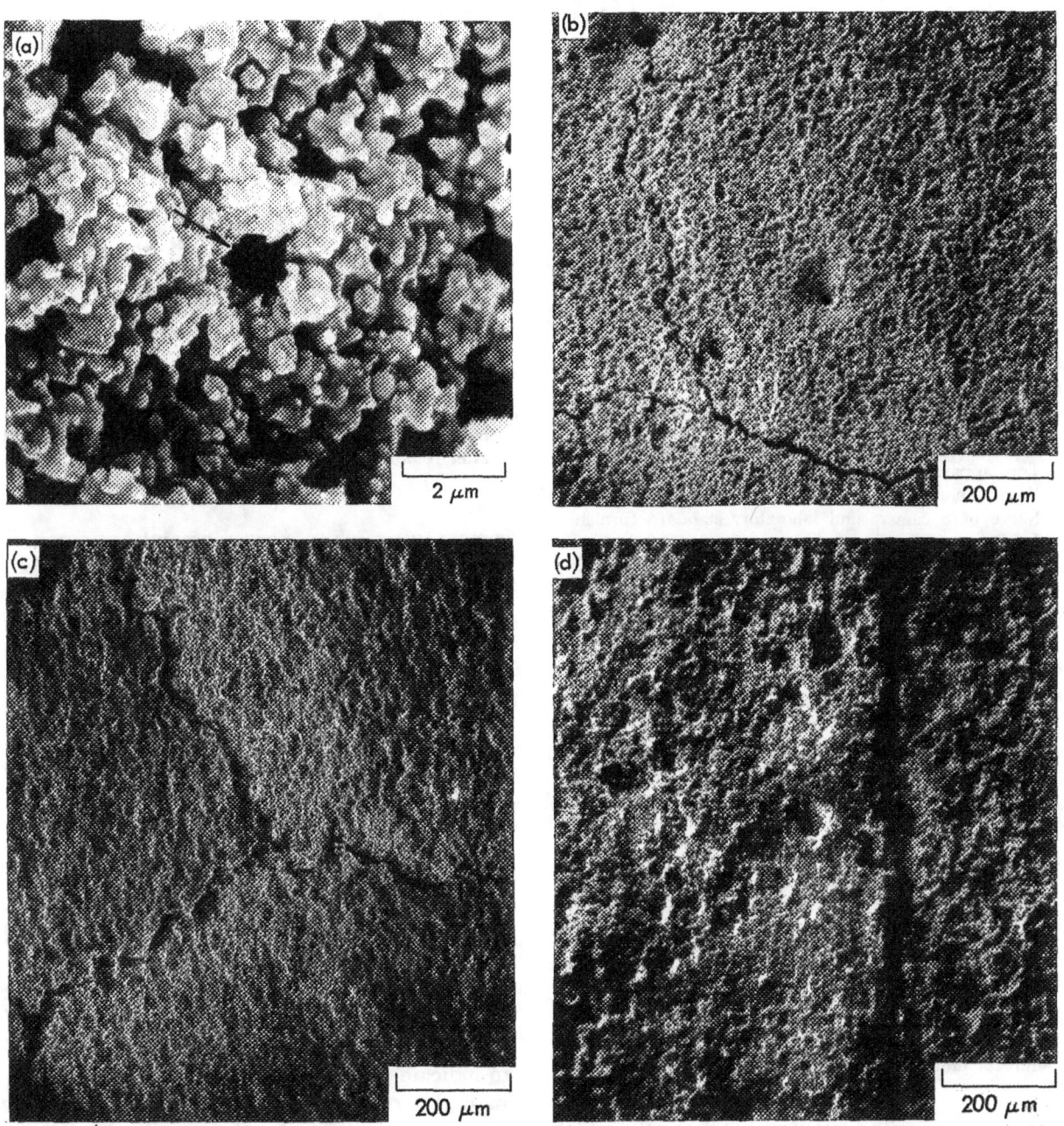

FIGURE 10.—Scanning electron micrographs of thermal-control paints from Surveyor 3 television camera and of laboratory standard thermal-control paint, showing hypervelocity impact sites and/or holes caused by natural porosity of paint. (a) Impact "crater" produced by carbonyl-iron microparticle at about 7 km/sec. (b) Natural-porosity holes in paint standard furnished by Surveyor 3 manufacturer. (c) Holes (or craters) in lower shroud. (d) Holes or craters in elevation-drive housing dust cover.

dust would occur at a much lower velocity than would impacts from micrometeoroids (ref. 13) and the newly exposed paint within the impact craters would differ from the "weathered" surface paint. Examination of small craters (30 to 100 μm in diameter) by optical microscopy showed

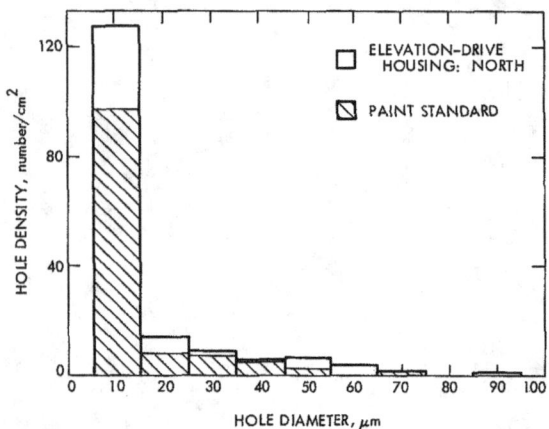

FIGURE 11.—Comparison of number and sizes of holes (or apparent impact sites) found in surfaces of thermal-control paint of elevation-drive housing of Surveyor 3 camera and laboratory standard furnished by Surveyor 3 manufacturer.

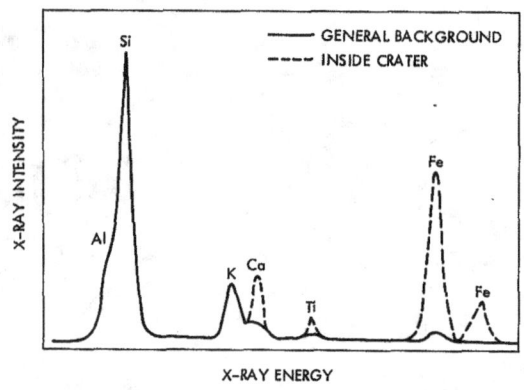

FIGURE 12.—Comparison of the X-ray spectra of the thermal-control paint of the Surveyor 3 television camera and the laboratory standard of this paint with the X-ray spectra of materials found with apparent impact craters.

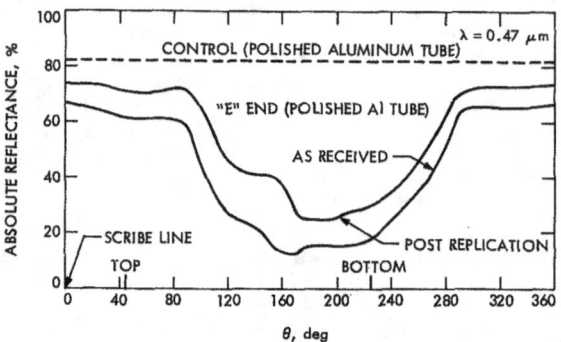

FIGURE 13.—Variation in absolute spectral reflectance around unpainted aluminum RADVS support tube compared with spectral reflectance of polished aluminum tube (from measurements taken before and after removal of contaminants, on section E).

many craters with very clean white walls, which would be expected if the crater was formed just before recovery with no time for additional contamination or degradation. In this size range, these white-walled craters account for about half of the observed difference in hole density between the standard and the exposed surface.

Identification of the chemical composition of residual debris in a hole was useful in determining the possible source of the impacting particle. The paint used on the Surveyor 3 television camera was a white inorganic paint composed of an aluminum silicate pigment and a potassium silicate binder. Figure 12 shows representative X-ray spectra typical of those obtained from several areas of the unexposed paint standard and from the degraded paint on the elevation-drive housing. As expected, the aluminum, silicon, and potassium peaks found in the unexposed paint also predominate the spectra for the degraded paint. Small amounts of calcium, titanium, and iron also are evident. This is consistent with the composition of the dust layer found on all parts studied. Analysis of lunar soil by other investigators (ref. 14) shows the presence of these elements. Several craters were found that contained residual material with greater relative amounts of the same three elements. No craters have yet been found that contain residual material of other chemical compositions. These results tend to indicate that the craters not accounted for by natural porosity probably are due to low-velocity impacts of lunar material from the landing of the LM.

Spectral reflectance characteristics of the two sections of the unpainted aluminum support tube were compared with similar measurements of a section of polished aluminum tube made of the same alloy. This tube was polished by the Surveyor manufacturer using the same techniques as those used on the flight hardware. For this "control" specimen, the solar absorptance, α_s, was about 0.15. The post-flight values ranged from an α_s of 0.26 on the "clean" side to 0.75 on the "dirty" side, with little variation along the axial length of each tube section. The variation in reflectance around tube section E is shown in figure 13 for a wavelength of 0.47 μm. The part of the tube with the lowest reflection (greatest contamination) was oriented toward the lunar surface and slightly toward the spacecraft descent engine 3. The reflectance was measured again after the surface was replicated, as described previously. As indicated in figure 13, this removal of loose material increased the reflectance on all sides of the tube. Figure 14 shows the distribution of total spectral reflectance around the tube. The contamination on the dirty side appears to be primarily of lunar origin or possibly from descent engine exhaust deposits. This contamination is not easily removed, however, as some traces of it remain even after repeated attempts by other investigators to remove it with ultrasonic cleaning (ref. 17) or with normal replication processes for transmission microscopy experiments (ref. 18).

As a basis for the analysis of the painted surfaces, reflectance measurements were made on several test samples of this paint coated at the same time as the flight spacecraft. The results indicated a pre-flight α_s of 0.20. The post-flight values depended upon the orientation of the surface relative to the Sun, with lunar surface, and the landing site of the LM. The final α_s values varied between 0.38 for a surface facing outer space to 0.74 for a surface facing directly toward the lunar surface. The distribution of the spectral reflectance is shown in figure 15. Note that the greatest change in reflectance occurred at the short wavelength end of the spectrum. Laboratory tests by other investigators (refs. 1 and 19) show that the reflectance of this inorganic paint could be degraded in the 0.25- to 1.5-μm range by exposure to ultraviolet radiation. The extent of this reflectance degradation is a function of the total Sun exposure; therefore, the different faces of the elevation-drive housing would exhibit different reflectances due to radiation damage only. In addition to radiation damage, a coating of lunar dust should modify the reflectance of the paint. For comparison, the spectral reflectance of lunar dust (ref. 20) is shown. This non-gray reflectance makes the influence of a dust-layer wavelength-dependent, with the greatest influence occurring at the short wavelengths. The decrease in reflectance, therefore, is due to a combination of contamination and degradation of the paint from ultraviolet radiation. A more detailed discussion of this subject is presented in reference 21.

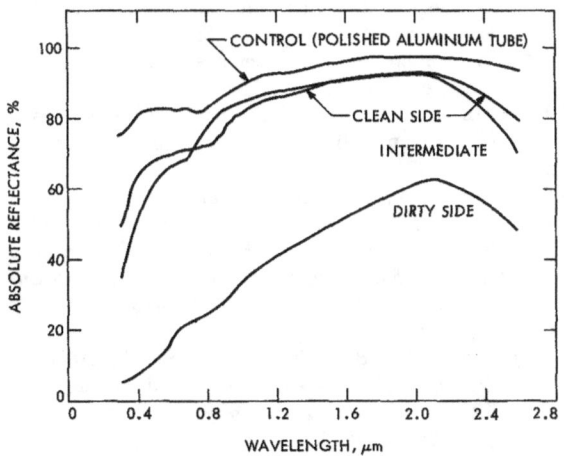

FIGURE 14.—Spectral distribution of reflectance (absolute reflectance as function of wavelength) on various portions of unpainted aluminum RADVS support tube (from measurements taken before removal of contaminants, on section E).

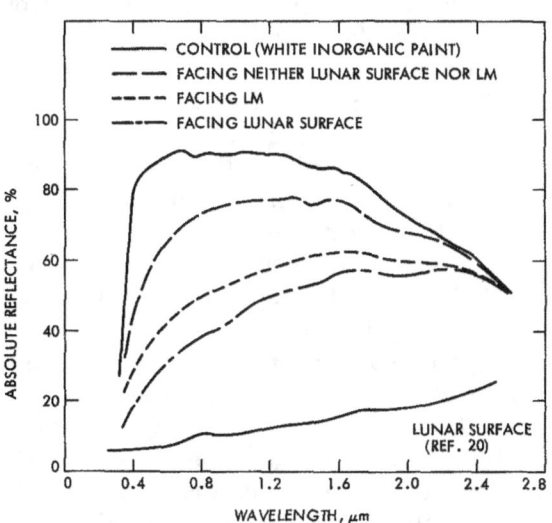

FIGURE 15.—Spectral distribution of reflectance (absolute reflectance as function of wavelength) for several areas of surface of thermal-control paint of elevation-drive housing dust cover of Surveyor 3 television camera.

Concluding Remarks

The results of this study indicate that the use of Surveyor materials as a means for definition of the lunar micrometeoroid environment on the lunar surface is difficult. Optical microscopy provides only limited identification of microparticles and surface defects; scanning electron microscopy provides excellent quantitative identification of surface characteristics. X-ray examination of residual material inside a hole can be a useful tool in determining the possible source of the impacting particle. Natural surface porosity and secondary impacts of lunar origin (landing of Surveyor and LM) not associated with the normal lunar environment account for most of the apparent impact craters studied. It was concluded that no sites were found that could definitely be characterized as micrometeoroid impact craters. The results of the study, based on somewhat meager data, indicate an upper-limit value for the micrometeoroid flux on the lunar surface for 1-μm-diameter particles (mass of about 10^{-12} g) of less than about 6×10^{-5} particles/m² sec. This rate of particle bombardment is far below that which would have been detrimental to the optical properties of the Surveyor 3 spacecraft surfaces within the time period of their exposure on the lunar surface. However, none of the surfaces studied retained their initial optical properties.

All of the surfaces examined were coated to some degree with lunar dust. In general, the surfaces exposed directly to the lunar surface had the greatest amount of dust. It was found that the spectral reflectance of both polished aluminum surfaces and thermal-control paints were affected by the 940-day exposure to the lunar environment. In the case of the polished surfaces, the most significant effect was erosion primarily of lunar origin or possibly from Surveyor 3 descent engine exhaust products. The post-flight values ranged from an α_s of 0.26 on the "clean" side to 0.75 on the "dirty" side, with little variation along the axial length of each tube section. The paints, however, were damaged significantly by solar radiation and surface contamination with a resulting change, for a surface facing outer space, in solar absorptance from the pre-flight value of 0.20 to a post-flight value of 0.38. For a surface facing directly toward the lunar surface, the post-flight solar absorptance was 0.74.

References

1. *Test and Evaluation of the Surveyor III Television Camera Returned from the Moon by Apollo XII*, vols. I and II, Report SSD 00545R, Hughes Aircraft Co., Culver City, Calif., 1971.
2. *Surveyor III Parts and Materials/Evaluation of Lunar Effects*, Report P70–54, Hughes Aircraft Co., Culver City, Calif., 1971.
3. *Surveyor III Mission Report. Part I: Mission Description and Performance*, Technical Report 32–1177, Jet Propulsion Laboratory, Pasadena, Calif., 1967.
4. *Apollo 12 Mission Report*, MSC 01855, Houston, Tex., 1970.
5. CUNNINGHAM, B. E.; AND EDDY, R. E.: "Space Environment Simulator for Materials Studies at NASA Ames." *J. Spacecraft and Rockets*, vol. 4, 1967, pp. 280–282.
6. COSBY, W. A.; AND LYLE, R. G.: *The Meteoroid Environment and Its Effects on Materials and Equipment*, NASA SP–78, Washington, D.C., 1965.
7. MORRISON, R. H.: *Simulation of Meteoroid-Velocity Impact by Use of Dense Projectiles*, NASA TN D–5734, 1970.
8. THE LUNAR SAMPLE PRELIMINARY EXAMINATION TEAM: "Preliminary Examination of Lunar Samples from Apollo 12." *Science*, vol. 167, 1970, pp. 1325–1339.
9. CARROLL, W. F.; BLAIR, P. M.; JACOBS, S.; AND LEGER, L.: *Discoloration and Lunar Dust Contamination of Surveyor III Surfaces*, presented at the Apollo 12 Lunar Science Conference, Houston, Tex., Jan. 11–14, 1971.
10. JAFFE, L. D.: Lunar Surface: "Changes in 31 Months and Micrometeoroid Flux." *Science*, vol. 170, 1971, pp. 1092–1094.
11. RUDOLPH, V.: *Investigation of Craters by Microparticles at a Velocity Range of 0.5 to 10 KM/SEC*. Ph.D. thesis, University of Heidelberg, 1967.
12. *Meteoroid Environment Model—1969 [Near Earth to Lunar Surface]*. NASA SP–8013, Washington, D.C., 1969.
13. JAFFE, L. D.: *Blowing of Lunar Soil by Apollo 12: Surveyor III Evidence*, presented at the Apollo 12 Lunar Science Conference, Houston, Tex., Jan. 11–14, 1971.
14. MORRISON, G. H.; GERARD, J. T.; KASHUBA, A. T.; GANGADHARAM, E. V.; ROTHENBERG, A. M.; POTTER, N. M.; AND MILLER, G. B.: "Multi-element Analysis of Lunar Soil and Rocks." *Science*, vol. 167, 1970, pp. 505–507.
15. BARGHOORN, E. S.; PHILLPOTT, D.; AND TURNBILL,

15. C.: "Micropaleontological Study of Lunar Material." Science, vol. 167, 1970, p. 775.
16. RAMDOHR, P.; AND EL GORESY, A.: Opaque Minerals of the Lunar Rocks and Dust From Mare Tranquillitatis. Science, vol. 167, 1970, pp. 615–618.
17. BÜHLER, F.; EBERHARDT, P.; GEISS, J.; AND SCHWARZMÜLLER, J.: Trapped Solar Wind Helium and Neon in Surveyor III Material, presented at the Apollo 12 Lunar Science Conference, Houston, Tex., Jan. 11–14, 1971.
18. BUVINGER, E. A.: Replication Electron Microscopy on Surveyor III Unpainted Aluminum Tubing, presented at the Apollo 12 Lunar Science Conference, Houston, Tex., Jan. 11–14, 1971.
19. BLAIR, P. M., JR.; PEZDIRTZ, G. F.; AND JEWELL, R. A.: Ultraviolet Stability of Some White Thermal Coatings Characterized in Vacuum, Paper 67-345, presented at AIAA Thermophysics Specialist Conference, New Orleans, La., Apr. 17–20, 1967.
20. NASH, D. B.; AND CONEL, J. E.: Luminescence and Reflectance of Apollo 12 Samples, presented at the Apollo 12 Lunar Science Conference, Houston, Tex., Jan. 11–14, 1971.
21. ANDERSON, D. L.; CUNNINGHAM, B. E.; AND DAHMS, R. G.: Thermal Radiation Degradation Analysis of Surveyor III Material, Paper 71-478, presented at AIAA Sixth Thermophysics Conference, Tullahoma, Tenn., Apr. 26–28, 1971.

PART G

EXAMINATION OF SURVEYOR 3 PARTS WITH THE SCANNING ELECTRON MICROSCOPE AND ELECTRON MICROPROBE

A. A. Chodos, J. R. Devaney, and K. C. Evans

Two screws and two washers, several small chips of tubing, and a fiber removed from a third screw were examined with the scanning electron microscope (SEM) and the electron microprobe. The purpose of the examination was to determine the nature of the material on the surface of these samples and to search for the presence of meteoritic material.

Examination of the screws consisted of detailed views of the shoulder portion at 60° intervals. Generally, low (22×), medium (550×), and high (2200×) magnification pictures were taken at each location. The washers were examined at 90° intervals, both on the face and the edge. The chips of tubing were examined at a minimum of four locations. As the chips were triangular in shape, areas near each corner and in the center were checked at 1000× magnification for presence of lunar dust.

The electron microprobe can be used only normal to the sample surface. Therefore, the top part of the washers and of the screws was examined. Each area that had been documented by the SEM was probed for the presence of selected elements. Although the search included Na, Mg, Al, Si, S, Ca, Ti, and Cr, the Fe and Ni were especially interesting. A similar search was made on the surface of the chips. Scanning photographs were made of selected areas that indicated the presence of interesting elements. A fiber, removed from a third screw, was examined in detail.

Lower Shroud, Screw 10

The surface was covered with fine lunar-dust particles ranging in size from 1 to 10 μm. Smaller particles may exist, but higher magnification was not performed. The distribution was relatively uniform around the screw. One 3-μm dumbbell-shaped piece was found. X-ray examination with the microprobe indicated nothing inconsistent with lunar soil except for a particle of impurity located in the letter "c" on the screw. This impurity gave a spectrum of Mg and Si and is probably talc or some similar material, possibly from handling during assembly.

Lower Shroud, Washer 10

SEM examination showed the exposed area of the washer to be uniformly covered with lunar dust. Microprobe examination indicated the presence of iron-rich particles, which provided an analysis similar to the steel of the screw.

Lower Shroud, Washer 22

SEM examination of these two parts showed very few dust particles. These particles could not be verified as lunar in origin. No microprobe examination was made.

Fiber From Head of Upper Shroud, Screw 125

Microprobe examination of the fiber indicated the presence of high Si and Ca, a small amount of Al, and minor-to-trace amounts of Mg, Ti, P, Fe, and Na. This fiber will be studied in more detail when the newer, high-resolution microprobe is operational.

Tubing Chips From Al Tube, Sections A–4 and G–2

As there were six chips of tubing on each mount, they were numbered A–4–1 through A–4–6 and G–2–1 through G–2–6. Samples A–4–1 through A–4–6 were generally clean. The surface had some scratches and small gouges in which small particles collected; these particles could not be verified as lunar by the SEM. Samples G–2–1, G–2–2, and G–2–6 were clean. The other chips are discussed in the subsequent paragraphs.

G–2–3

The entire surface was covered with lunar dust, except for a crisscross pattern of clean areas. The clean area was a rectangular pattern 0.4 by 0.5 mm. Examination of the clean area at 2400× magnification showed surface scratches in the same direction and spacing as the overall clean area pattern. The origin of the clean area and the scratches is not known. Various shapes and sizes of particles were found. Shapes varied from irregular particles to small spheres and dumbbells. The particles measured in size from 0.2 to 10 μm. A dumbbell 3.5 μm long had 0.1-μm particles on the surface. Microprobe examination indicated the presence of lunar material, but nothing of any unusual composition.

G–2–4

A uniform coating of dust covered the chip. Near the center, an interesting area was found that had an 18-μm chunk of material. On the chunk was a 2.8-μm dumbbell-shaped piece of material. Next to the chunk were 9.6- and 4.0-μm spheres. On the larger sphere were a 0.8-μm sphere and particles as small as 0.08 μm. Directly adjacent to the chunk was a clean area, 7.2 μm in diameter with a single 1.4-μm particle in the center.

Areas of this chip examined by the microprobe indicated a greater density of lunar material than G–2–3 but nothing of unusual composition. Specific lunar phases could be identified, but the size of the particles was near the resolution of the instrument.

G–2–5

This sample had a gradient of dust particles across the surface of the chip. The dust-covered side was similar to the other chips except that the dust was not as dense. The cleaner side had a few particles scattered about the surface.

No material found could be definitely stated to be meteoritic in origin. While there was a definite distribution of lunar material on the surface of some of the samples examined, none of this lunar material had an unusual composition. There were at least two cases of non-lunar material present.

PART H

SPUTTER-ION SOURCE MASS SPECTROMETER ANALYSIS OF SAMPLES CUT FROM THE SURVEYOR 3 CAMERA

F. G. Satkiewicz and F. F. Marmo

During the period that Surveyor 3 was on the Moon, the spacecraft was subject to lunar and solar interactions. As part of the effort to evaluate the component parts of the recovered material, the GCA Corp. received six samples for study; four of these have been studied. These samples were:

- 934: from lower shroud, bottom, polished, 1 by 1 cm (Al-Mg alloy).
- 935: No. 1, lower shroud toward Lunar Module (LM), 1 by 1 cm (Al-Mg alloy with a coating of paint consisting of Kaolin bonded with potassium silicate).
- 936: No. 2, lower shroud away from LM, 1 by 1 cm (same surface coating as 935).
- 933: ¼ of 906, hood (visor), 1 by 1 cm (same surface coating as 935 and 936).

Before analyzing these samples, GCA requested and received a calibration sample from Hughes Aircraft Co.

A thorough study was made of the surface from which the chemical composition was obtained. It was discovered that the potassium intensity increases at the surface when the paint is heated. An attempt was made to correlate this with a corresponding profile for another associated species; however, no other matrix species showed corresponding excursions in intensity. This suggests the possibility that the potassium is changing its bonding nature and thus the sputter-ion yield (or intensity); this behavior deserves attention because the presence of potassium in graded bonding may be related to the optical properties of the paint.

It also was observed that the craters produced in sputtering the paint in the calibration sample were discolored yellowish-brown. The intensity of discoloration was approximately proportional to the length of time the sample was sputtered. This discoloration was not related positively to any changes observed in the succession of spectra, although peaks associated with a silicon-rich oxide were observed in one crater after 4 hr of sputtering. The more logical explanation is related to the accumulation of solid-state defects arising from protracted ion bombardment. Blair [1] believes the discoloration may be similar to that produced by ultraviolet irradiation on the paint and suggests a thermal soak at 220°C for 24 hr to see whether the discoloration can be bleached.

The chemical makeup of lunar fines is very similar to the paint composition. Several lesser constituents with a high sputter-ion yield are more evident in the lunar fines. Thus, from a knowledge of the composition of both the paint and the lunar fines, it was possible to obtain calibration curves for relating total intensity of particular elements to the fraction of projected area taken up by the lunar materials. The elements in question are magnesium, calcium, titanium, chromium, and iron. (For sample 934, silicon and oxygen provided additional indicators.)

Not only was the coverage at the immediate surface determined, but the coverage in sputtering with time gave an approximate idea of particle size distribution.

The results obtained for the four samples are shown in tables 1 through 4; the average values of f are plotted in figure 1.

Intensity profiles for selected elements show that, within the first 200 through 400 Å, either another phase is present or the oxygen content of the lunar fines to this depth is less than stoichiometric. At this stage of the data interpretation, the former seems more likely.

---

[1] P. Blair, Jr., Hughes Aircraft Co., private communication.

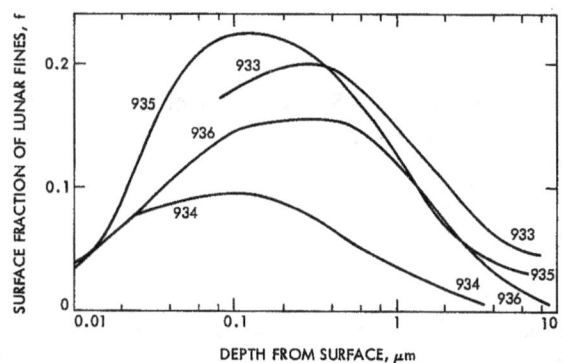

FIGURE 1.—Summary of lunar material coverage on Surveyor 3 camera samples.

Unlike most of the profiles, the Si^+ shows a virtually constant intensity from the immediate surface into the depth of the sample. Carbon is also present at the surface at higher concentration than in the "bulk." This could be interpreted in terms of the presence of some type of silicon-carbon compound. As the f values were derived by relating the individual intensities to silicon, the initial values of f are smaller. If aluminum were chosen, the initial portion would be a plateau. Accordingly, the maximum values of f can be taken as the maximum coverage on the surfaces.

An examination of the 935 and 936 profiles shows that the sample facing the LM has a higher coverage of very small particles, apparently originating from the LM landing. The profiles coincide for the interval between about 1 and 3 μm, and then diverge once more with a higher residue of lunar material for sample 935. Thus, in terms of mass, 933>935>936>934.

The smaller coverage on the Al-Mg alloy surface (934) may be due to the fact that particles are less well retained on a polished surface than on a rougher painted surface.

Chlorine was present on all of the sample surfaces. A study of the spectra shows that it is not associated with sodium (salt); the fluorine peak is too small to relate the chlorine to Teflon or Freon. Additional studies of the spectra will be made to explain this observation.

TABLE 1.—*Distribution of lunar fines on sample 933: painted surface from hood (visor)*

Area	Depth, μm	Surface fraction of lunar material, f_{x+}					
		Mg	Ca	Ti	Cr	Fe	Average
2	0.08	0.13	0.17	0.068	....	0.30	0.17
1	~.1	.17	.24	.17	0.18	....	.19
2	.2	.23	.18	.078	....	.27	.19
2	.32	.24	.19	.078	....	.27	.20
3	.56	.16	.15	.05	.16	.20	.14
1	.84	.18	.17	.12	.16	.33	.20
2	1.0	.15	.14	.076	.12	.23	.14
3	1.3	.11	.12	.06	.09	.21	.12
1	2.3	.09	.092	.062	.074	.20	.10
2	4.0	.062	.062	.033	.051	.12	.066
2	4.6	.057	.051	.032	.035	.11	.057
2	5.8	.045	.048	.026	.034	.10	.051
2	8	.045	.045	.024	.030	.09	.046

TABLE 2.—*Distribution of lunar fines on sample 934: polished Al-Mg alloy surface*

Area	Depth, μm	Surface fraction of lunar material, f_{x+}							
		Mg	Si	Ca	Ti	Cr	Fe	O	Average
2	0.01			0.053					0.053
3	.02		0.16						.16
3, 2	.04		.10	.079					.09
2, 3	.08		.15 .10	.083					.10
3	.12		.095						.095
3, 2	.16		.092	.069					.08
2, 3	.20	0.14	.092						.12
2	.24			.054	0.043		0.15	0.14	.097
2	.32		.11	.043					.076
2	.34	.069							.069
2	.36			.040					.04
2	.4		.081		.030		.11		.04
2	.44							.061	.061
1	.76	.045	.053	.025	.02	0.027	.095	.058	.046
1	1.6	.029	.024	.012	.0081		.042	.024	.023
1	3.6	.0048	.0029	.0006			.0049	.014	.0054

TABLE 3.—*Distribution of lunar fines on sample 935: painted surface toward LM*

Area	Depth, μm	Surface fraction of lunar material, f_{x+}					
		Mg	Ca	Ti	Cr	Fe	Average
2	0.01	0.038					
2	.02	.081					
2	.04	.18					
2	.08	.22					
2	.16	.22					
2	.28	.22					
2	.44	.19					
1	.51	.17	0.18	0.14	0.18	0.37	0.17
2	.64	.18					
2	.88	.14					
2	1.1	.12					
1	1.2	.10	.10	.058	.088	.18	.10
2	1.7	.084	.087	.050	.067	.16	.090
1	2.0	.068	.069	.043	.062	.13	.074
1	3.0	.048	.052	.032	.042	.10	.054
1	4.0	.038	.040	.024	.027	.07	.040
1	5.6	.029	.032	.017	.025	.063	.033
1	7.2	.029	.027	.017	.020	.054	.029

TABLE 4.—*Distribution of lunar fines in sample 936: painted surface away from LM*

Area	Depth, μm	Surface fraction of lunar material, f_x+					
		Mg	Ca	Ti	Cr	Fe	Average
2, 3, 4	0.01	0.018	0.021			0.13	0.056
2, 3, 4	.02	.045	.037			.17	.084
2, 3, 4	.04	.092	.075			.20	.12
2, 3, 4	.08	.15	.11			.18	.15
2, 3, 4	.12	.16					
		.14					
		.15					
2, 3	.16	.15	.12	0.069	0.11	.24	.15
			.11			.24	
2	.24	.15					
3	.28		.12				
2	.32	.16	.12	.075	.11	.22	.13
2	.38	.16					
1	.52	.13	.13	.11	.13	.29	.16
2	1.0	.11	.10	.06	.087	.19	.11
3	1.1	.11	.11	.094	.12	.24	.14
1	1.3	.098	.099	.072	.094	.19	.11
1	2.2	.064	.060	.039	.060	.11	.067
1	3.0	.041	.039	.024	.032	.08	.043
1	4.2	.027	.027	.020	.024	.056	.031
1	5.5	.018	.021	.017	.014	.046	.023
1	7.4	.0096	.011	.007	.0054	.022	.011
1	8.7	.0066	.0086	.0072	.0035	.021	.0094

PART 1

BLOWING OF LUNAR SOIL BY APOLLO 12: SURVEYOR 3 EVIDENCE

L. D. Jaffe

Surveyor 3 landed on the Moon on April 20, 1967. Thirty-one months later, on November 20, 1969, it was visited by the Apollo 12 astronauts, who noticed immediately that the color of the spacecraft had changed: white surfaces had become tan. This had been predicted before the Apollo 12 flight, as an effect of solar ultraviolet radiation on the Surveyor white paint (refs. 1 through 3; also see ch. IV, pt. A, of this report).

Astronauts Conrad and Bean removed the television camera from Surveyor 3 and brought it back to Earth. Ground examination showed that the surfaces which originally had been painted white showed patterns of discoloration. Some of the darker markings strongly resembled burnt-in permanent shadows of objects attached to the camera. For example, in figure 1, a dark marking on the painted surface looks very much like a shadow cast by the adjacent wire. The positions of these dark markings remain constant, however, independent of the lighting angle. The direction from which the "shadows" were thrown was approximately that of the Apollo LM. (See ref. 4.)

When I examined the camera, it seemed that the dark, upper portion of the cylindrical motor

FIGURE 1.—Housing of Surveyor 3 television camera (NASA photograph; contrast enhanced photographically). 1 indicates "permanent shadow" cast by adjacent wire (wire has been moved). 2 indicates "permanent shadow" cast by rectangular cover of the elevation drive train (upper left) onto cylindrical motor housing (at center). This "shadow" covers approximately the upper one-third of the cylindrical portion (arrows). "Shadow" of the lower left corner of the cover falls on the wire.

housing shown in the center of figure 1 corresponded in outline to a permanent shadow cast by the rectangular cover of the elevation drive train (top left in the figure). This was confirmed by viewing the camera from a distance, along the proper direction. The permanent shadow of the lower outboard corner of the housing falls on the wire. The distance from the corner to its permanent shadow is about 4 cm.

With a theodolite, it was possible visually to aline the outboard side edge and outboard bottom edge of the cover with the cast shadow and determine the direction of the "ray" casting the shadow. Two sets of measurements gave, for the direction in Surveyor camera coordinates:

Azimuth: $90.0° \pm 1.0°$
Elevation: $28.7° \pm 0.5°$

By using the appropriate coordinate transformation, the lunar directions are:

Bearing: $46.8° \pm 0.8°$ W of N
Zenith angle: $88.0° \pm 0.9°$

This transformation takes into account both the camera orientation during Surveyor operations (ref. 5) and subsequent rotation arising from sagging of the landing gear. (See ref. 6.)

Various reports (refs. 7 through 10) of the landed positions of Surveyor 3 and of the Apollo 12 LM, in the same coordinate system, are discrepant to the extent of about 10 m in their relative positions. By what seem to be the latest determinations (refs. 8 and 10), a line through the Surveyor camera at the bearing and zenith angles derived above passes 3 ± 3 m horizontally and 1 ± 2 m vertically from the point on the lunar surface directly under the center of the LM. (This point is 155 m away and 4.3 m higher than the Surveyor camera.) The agreement is well within the discrepancies mentioned. The discoloration pattern measured on the Surveyor camera apparently was produced by the Apollo LM, when the LM was very close to its surface position. In areas within line of sight of the LM, the Surveyor surface was whitened. Many shallow, white craters were noted on inspection of the Surveyor camera under a microscope by Cour-Palais (ref. 4; also see ch. VI, pt. E, of this document), predominantly on the side toward the LM; he attributed the surface whitening to these craters, and the craters to sandblasting by lunar particles ejected by LM exhaust during its landing.

To reach the Surveyor camera in a ballistic trajectory from the lunar surface directly below the LM, and arrive with the zenith angle mentioned, requires a particle velocity of 70 m/sec or greater and an emission angle at or slightly below the horizontal.

It is true that particles entrained by LM exhaust would not follow a ballistic trajectory initially, but this is probably a good approximation away from the LM. The sharpness of the "permanent shadows" on the Surveyor camera shows that the incident particles were well collimated.

Thus, the discoloration pattern on the Surveyor 3 camera not only provides excellent evidence that the camera surface was whitened by the impact of particles blown from the lunar surface by the exhaust of LM as it landed, but also indicates the velocity and direction at which these particles were ejected. Many of the lunar particles moved at very low angles to the horizontal.

References

1. BLAIR, P.; AND BLAIR, G. R.: *Summary Report on White Paint Development for Surveyor Spacecraft*, TM–800, Hughes Aircraft Corp., Culver City, 1964.
2. HAGEMEYER, W.: "Surveyor White Paint Degradation." *J. Spacecraft and Rockets*, vol. 4, 1967, pp. 828–829.
3. ZERLAUT, G. A.; AND GILLIGAN, J. E.: *In Situ Degradation of Thermal Control Surfaces*, Illinois Inst. Tech. Res. Inst. Report U6061–17 to NASA Marshall Space Flight Center, Chicago, Ill., 1969.
4. COUR-PALAIS, B. G.: NASA News Release 70–23, Washington, D.C., 1970.
5. RENNILSON, J. J.; AND BATSON, R. M.: *Surveyor Program Results*, NASA SP–184, Washington, D.C., 1969, pp. 38–42.
6. SCOTT, R. F.; LU, T.-D.; AND ZUCKERMAN, K. A.: "Movement of Surveyor 3 Spacecraft." *J. Geophys. Res.*, vol. 76, 1971, pp. 3414–3423.
7. ARMY MAP SERVICE: *Lunar Map 1:500; Surveyor III Site*, Army Map Service, Washington, D.C., 1968.
8. BATSON, R. M.: *Surveyor Program Results*, NASA SP–184, Washington, D.C., 1969, pp. 45–54.
9. MISSION EVALUATION TEAM: *Apollo 12 Mission Report*, MSC–01855, Houston, Tex., 1970.
10. LUNAR SAMPLE PRELIMINARY EXAMINATION TEAM: "Preliminary Examination of Lunar Samples from Apollo 12." *Science*, vol. 167, 1970, pp. 1325–1339.

ACKNOWLEDGMENTS

I thank J. J. Rennilson for providing the transformation matrix relating camera coordinates during Surveyor 3 operations to lunar coordinates and for other assistance, W. F. Carroll for helpful discussions, and R. F. Scott for information on landing gear condition.

PART J

LOW-TEMPERATURE OXYGEN-PLASMA EFFECTS ON SURVEYOR PLASMO-CLAY COATING

R. B. Gillette

The objective of this study was to determine whether the reflectance of the degraded plasmo-clay thermal-control coating could be restored by exposing it to an oxygen plasma. Previous experiments showed that the reflectance of similar coatings, irradiated in the laboratory, could be increased (restored) by exposure to a low-temperature oxygen plasma. Therefore, it was of interest to demonstrate whether this process could be used successfully to restore the reflectance of a coating degraded in the lunar environment. Results of previous experiments suggest that oxygen-plasma treatment may be a technique of prolonging coating lifetime in space.

It is believed that the primary causes of degradation of the Surveyor coating are the deposition of lunar soil and bulk radiation damage in the metal-oxide pigment crystals. The possibility also exists that organic compounds from the rocket plume or outgassing materials may have deposited on some surfaces. Exposure of the coating to an oxygen plasma could restore reflectance either by removing the organic contaminant film (if one is present) or by eliminating bulk radiation damage. Organic contami-

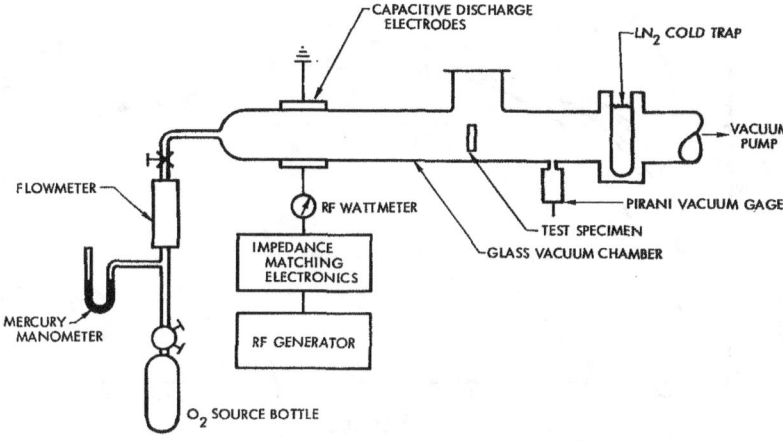

Figure 1.—Schematic of plasma generator.

nant films would be removed by their conversion to gaseous constituents. It is believed that bulk radiation damage would be removed by a reabsorption of oxygen ions into the pigment crystal lattice. This apparently eliminates color centers that have formed at either oxygen ion vacancies or interstitial metal ions.

This article contains a discussion of this experiment. Some observations regarding a pit, observed in the exposed paint surface, are also presented.

The apparatus used for the oxygen-plasma exposure is shown in figure 1. Oxygen (99.5 percent label purity) was bled into a flow meter at 1-atmosphere pressure, with reduction to 0.5 torr occurring in the glass reaction chamber containing the coated Surveyor specimen. Before entering the reaction chamber, the gas was excited in a capacitive radio-frequency (RF) discharge. Power from a 300-W RF generator was matched to the gas load impedance by inductive and capacitive tuning coils. An oxygen flow of about 250 std cm³/min was maintained in the experiment. The plasma streamed over the test specimen surface, was deactivated, and exhausted via a cryogenic trap using a mechanical vacuum pump. It has been estimated from nitric oxide titration data that the incident flux of oxygen atoms on the specimen during plasma treatment was on the order of 2×10^{19} to 4×10^{19} atoms/cm² sec.

The plasmo-clay-coated specimen used in the experiment was removed from the clamp ring assembly that supported the Surveyor camera. A photograph of the specimen is shown in figure 2. The exposed paint surface was oriented approximately vertical and did not view the Apollo 12 landing point. It is evident in the photograph that a portion of this surface had been sanded before launch (scratch marks). It also can be noted in the figure that a relatively large pit is present (on the bend line), and some touchup

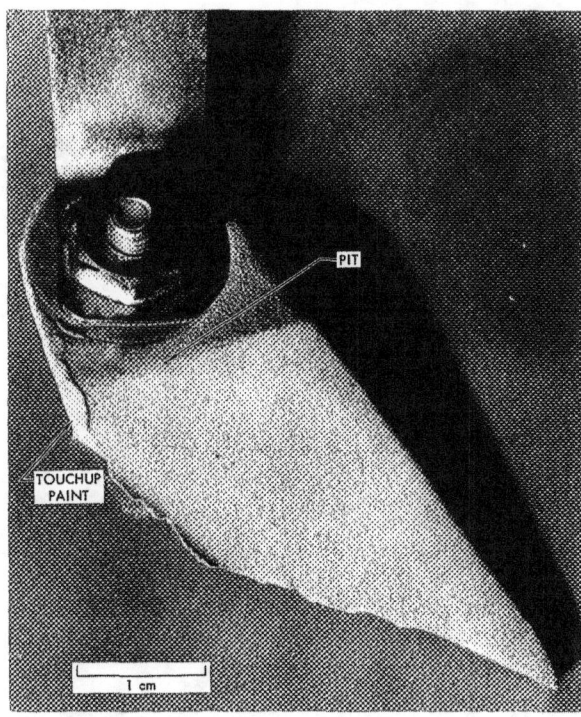

Figure 2.—Section of camera clamping bracket.

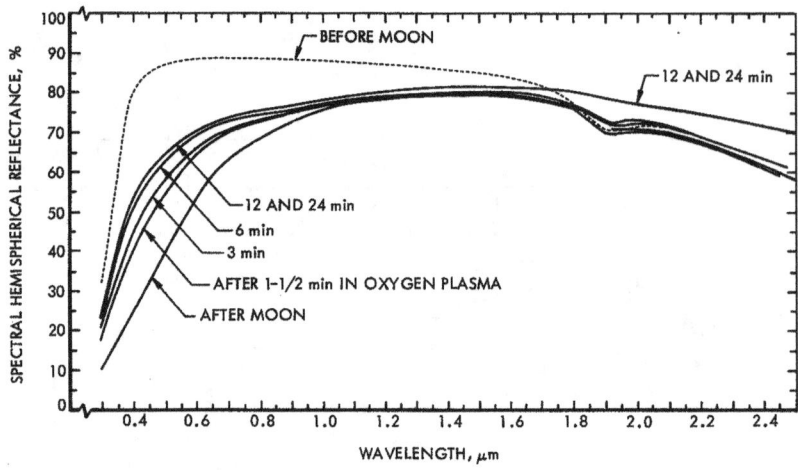

FIGURE 3.—Effect of lunar exposure and subsequent oxygen-plasma treatment on Surveyor plasmo-clay coating.

paint had been applied along one edge before launch. Reports from the spacecraft manufacturer state that the coating should be 5 to 8 mil thick; however, measurements of thickness along edges indicate a thickness of only about 2 to 3 mil on this specimen. Observations of the surface along broken edges and in the pit revealed that—

(1) Coating was discolored only in a thin surface layer.

(2) Lunar soil was present only on the discolored surface.

Results of the oxygen-plasma treatment experiment are shown in figures 3 and 4. Spectral hemispherical reflectance is shown in figure 3 and the change in spectral reflectance (ΔR) is shown in figure 4. In figure 4, a negative ΔR indicates a decrease in reflectance with respect to a control specimen, and a positive ΔR indicates an in-

FIGURE 4.—Effect of lunar exposure and subsequent oxygen-plasma treatment on Surveyor plasmo-clay coating.

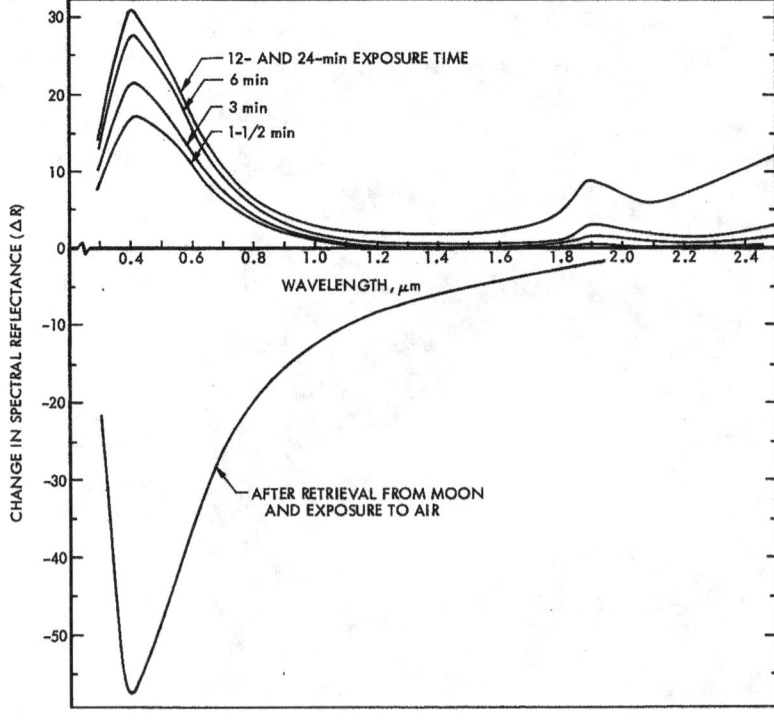

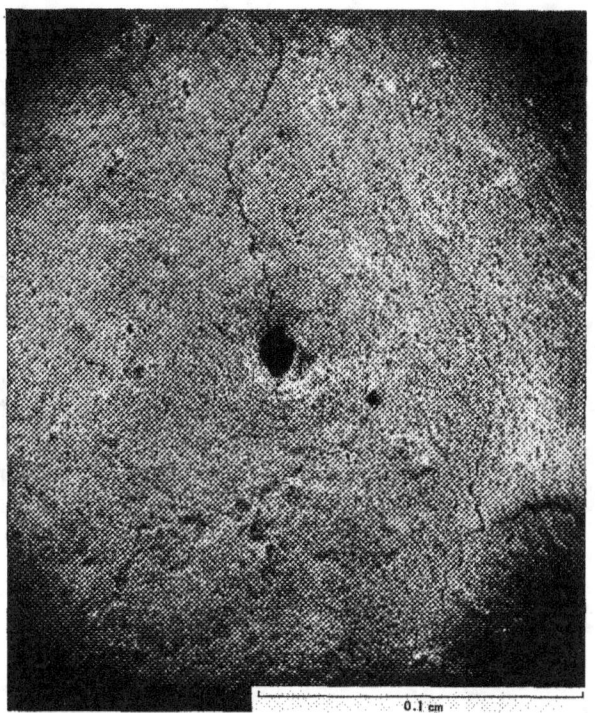

FIGURE 5.—Scanning electron microscope photograph of pit in plasmo-clay coating.

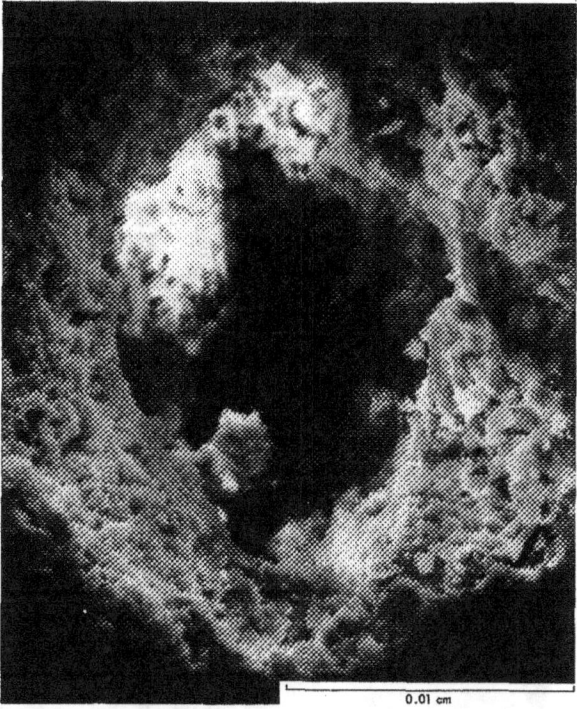

FIGURE 6.—Scanning electron microscope photograph of pit in plasmo-clay coating.

crease in reflectance with respect to the "as received" specimen. Reflectance and $\triangle R$ are plotted in figure 3 for a typical control specimen, the as-received Surveyor specimen, and after plasma exposure times of 1½, 3, 6, 12, and 24 min. The data show that—

(1) The degradation occurred in a wavelength band extending from about 0.2 to 1.1 μm, with a peak degradation ($\triangle R$) of about —58 percent at 0.4 μm.

(2) An increase in reflectance was induced by plasma treatment in the same wavelength band (0.2 to 1.1 μm) and in the infrared wavelengths beyond about 1.7 μm.

(3) The reflectance ceased to increase after about a 12-min plasma exposure.

(4) A maximum $\triangle R$ of about 31 percent occurred after 12 min at a wavelength of 0.4 μm.

Solar absorptance values calculated from the data were 0.17 for the control specimen, 0.44 after lunar exposure, and 0.31 after oxygen-plasma exposure.

These results show that the oxygen-plasma treatment can eliminate some of the lunar-environment-induced degradation on the plasmo-clay coating. Insufficient data were obtained to enable any conclusions to be made regarding the mechanism of degradation or plasma restoration. However, it was noted that the behavior of the Surveyor coating was similar to other metal-oxide pigmented coatings tested at Boeing Co. in unreported research. Most white coatings that have been irradiated in vacuum develop strong optical absorption adjacent to the short-wavelength cutoff (the wavelength equivalent to the electronic conduction band gap), and in the near-infrared wavelength region. It is generally observed that degradation in infrared wavelengths quickly disappears upon exposure to air following irradiation; however, the degradation band near the short-wavelength cutoff disappears slowly or not at all in air. Exposure to atomic oxygen (plasma treatment) vs. molecular oxygen (air) causes this latter absorption band to disappear partially or completely. Such was the case with the short-wavelength absorption band in the Surveyor plasmo-clay coating. The results of these experiments indicate that the short-wavelength absorption band observed in the Surveyor coating is related to the depletion of oxygen from pigment crystals during irradiation in vacuum.

In regard to the reflectance increase experienced during plasma treatment at wavelengths longer than about 1.7 μm, it is speculated that this is the result of dehydration.

Visual observations of the lunar soil on the surface after plasma treatment indicated that it was still highly absorbing to light. Therefore, part of the residual discoloration after plasma treatment could have been the result of lunar soil.

Scanning electron microscope photographs of the pit noted in figure 2 are shown in figures 5 and 6. The dimensions of the pit are about 80 by 125 μm (at the smallest cross section) and about 275 μm (11 mil) deep. Visual observations with an optical microscope indicated that the pit may extend into the aluminum substrate. Considering the relatively deep penetration, it is possible that the pit resulted from a high-velocity particle impact. To confirm this hypothesis, it is recommended that an elemental analysis, using a scanning electron microscope or other suitable technique, be performed on material at the pit bottom.

In conclusion, it was shown that the plasmoclay coating, degraded in the lunar environment, can be partially restored by oxygen-plasma treatment. This result and similar experiments on other white coatings confirm the concept of using oxygen-plasma generators for prolonging spacecraft coating lifetime in space.

PART K

EXAMINATION OF THE SURVEYOR 3 SURFACE SAMPLER SCOOP

R. F. Scott and K. A. Zuckerman

The Surveyor 3 spacecraft, launched from Cape Kennedy to the Moon on April 17, 1967, carried the surface sampler (fig. 1) for the purpose of performing mechanical tests of the lunar surface. Three days later, the spacecraft landed on the Moon's surface in Oceanus Procellarum and became operational. After many checks of the spacecraft subsystems, the surface sampler was turned on and, after calibration tests above the lunar surface, was used on April 21, 1967, to conduct the first controlled tests of the physical and mechanical properties of the lunar surface material (ref. 1).

During the initial calibration sequence, it was apparent that the surface sampler was operating normally except in the extension and retraction mode. In this mode, the commanded movements were about one-third of those recorded in the pre-flight calibrations. The anomaly persisted throughout the entire period of operation of the sampler on the Moon, and no changes in it were observed. It was concluded, on the basis of an evaluation of the possible failure modes of the surface sampler, that the problem lay in the

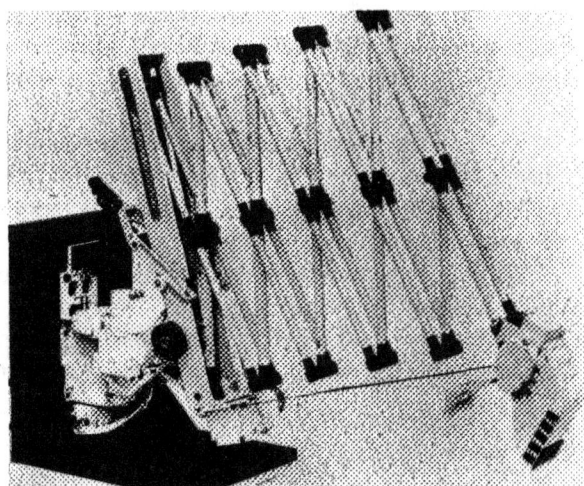

FIGURE 1.—Pre-flight photograph of Surveyor 3 surface sampler.

electrical circuit of the retraction motor rather than in any frictional characteristics developed in the joints during the landing of the Surveyor spacecraft.

The surface sampler was used for a period of about 18½ hr and responded to a total of 1900 commands. Contact was made with the lunar surface in 25 bearing and impact tests. In trenching tests, it is estimated that the scoop of the surface sampler traveled a distance of 6 m through or in the lunar soil. In impact tests, the base of the scoop door came in relatively violent contact with the lunar surface material 13 times, as it was dropped from a height range of between 30 and 60 cm above the lunar surface. Figure 2 is an enlargement of a Surveyor 3 television picture and shows the scoop on May 1, 1967.

After the end of lunar operations on the first lunar day (May 3, 1967), the surface sampler and the spacecraft remained inactive for the lunar night. At this time, the surface sampler had been positioned to the extreme right of its operational area and elevated almost to its maximum extent so that the scoop was at a height of about 75 cm above the lunar surface. The spacecraft evidently did not respond to commands sent at the beginning of the second lunar day. No other responses were received from the spacecraft.

Following the success of the first lunar manned mission, Apollo 11, in July 1969, plans were made for a second spacecraft, Apollo 12, to land as close as possible to the Surveyor 3 landing site in order that the astronauts could visit the spacecraft and its vicinity and possibly remove parts of the spacecraft for return to Earth. The Lunar Module (LM) landed about 155 m to the northwest of the Surveyor 3 spacecraft, which was visible to the astronauts on their emergence from the LM.

The astronauts made two excursions outside their spacecraft; on the second of these, the Surveyor spacecraft was visited, photographed, and examined. Figure 3 is an enlarged picture of the right [1] side of the surface sampler taken on the Apollo 12 mission; figure 4 shows the left side. The pictures were originally taken by the astronauts in black and white.

[1] "Right" and "left" are used from the point of view of the Surveyor 3 television camera.

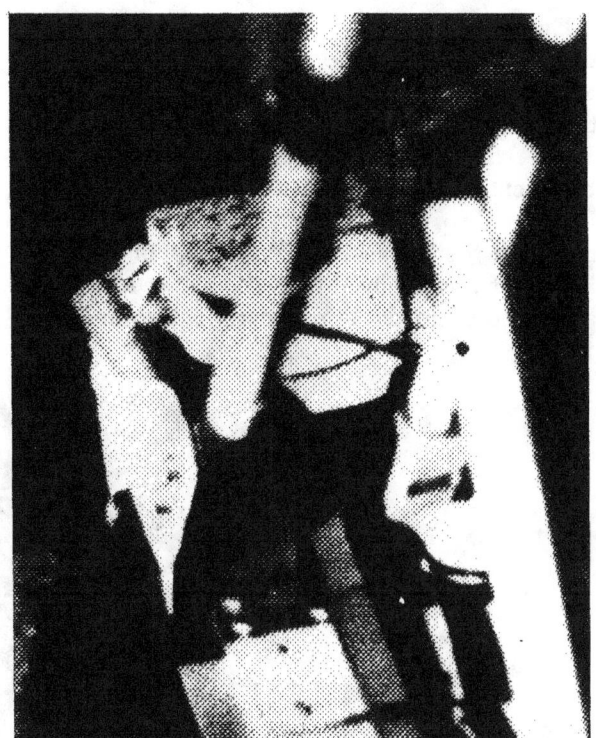

FIGURE 2.—Surveyor 3 picture showing soil on top of surface sampler scoop (GMT Day 121, 16:14:14).

During the operations around Surveyor 3, the astronauts were successful in recovering parts of the spacecraft, including a portion of the scoop and the first joint of the surface sampler. Before the flight, it was not thought possible that the scoop could be brought back because the cable-cutting tool supplied to the astronauts for removal of the other components had not proved suitable in pre-flight tests for cutting the retraction tape of the surface sampler. However, astronaut Conrad reported that when he applied the tool to the Surveyor 3 retraction tape and twisted it, the extension tape broke away from the surface sampler. As no part of the tape was returned, it is likely that the tape broke at a point where the tape was welded to itself near the scoop. Conrad then severed three arms of the scoop behind the first joint. The scoop and the attached portions of the arms up to the first joint were put in a bag and returned to the LM. The retrieval of the scoop was facilitated by the fortuitous positioning of the sampler at its maximum elevation in 1967.

Eventually, the part of the surface sampler in the bag was transferred to the Command Module, returned to Earth, and stored in quarantine in the Lunar Receiving Laboratory (LRL) at the NASA Manned Spacecraft Center, Houston, Tex., until its first release on January 7, 1970. During this time, no attempt was made to maintain the surface sampler scoop in vacuum; in fact, it was removed from the bag at least once and exposed to the atmosphere inside the quarantine facility. It was not, therefore, to be expected that the lunar soil accompanying the scoop would exhibit the same properties as lunar soil in the high vacuum conditions existing in the lunar surface.

When the Surveyor 3 operations ended in 1967, the door of the surface sampler scoop was closed, and an unknown amount of lunar soil was contained inside the scoop. Because the scoop, while inside the plastic bag, was subjected to a great deal of handling during the various stages of its journey back to Earth and in the LRL, the soil inside emerged through openings between the scoop door and the body and had free access to all other parts of the mechanism. Consequently, although only the scoop base had direct contact with the lunar surface during the Surveyor 3 mission, it was found that the entire outer surface of the scoop, the motor mechanism and housing, and parts of the arms that were returned were coated with lunar soil when the plastic bag was opened in the LRL for preliminary examination of the scoop.

FIGURE 3.—Apollo 12 photograph of soil on right-hand side of scoop (AS12–48–7107).

The soil adhered in varying degrees to the different parts of the surface sampler, although it is not known whether the mechanism of adhesion is the same as that which existed on the lunar surface. For example, in the conditions of atmospheric humidity in the LRL, the soil may have acquired enough moisture so that it adhered to the scoop by virtue of its dampness, as fine-grained terrestrial soil sticks to some surfaces. Adhesion of the soil to the scoop had been observed during the Surveyor 3 operations. During Surveyor operations, some estimates had been made of the magnitude of the adhesion of the lunar soil to Surveyor spacecraft components, but it was not possible to make measurements of this property.

In the preliminary examination in the LRL, it was noticed that there was a concentration of lunar soil on the right-hand side of the scoop in the area shown in figure 3 to be covered with lunar soil on the undisturbed scoop on the lunar surface. Possibly some of this material still represented pristine lunar surface material adhering to the scoop. Elsewhere on the scoop surface, it was not possible to identify on the astronauts' pictures areas of definite soil cover that could be correlated with the scoop appearance and soil coating at the time of the initial examinations.

After preliminary examination in January at the LRL, the scoop was transferred to its designers and manufacturers, the Hughes Aircraft Co. (HAC), Culver City, Calif. In the following 2 months, plans were established for the examination, handling, and testing of the scoop and the material accompanying it. The surface sampler scoop remained in the Surveyor test facility at HAC, and detailed examinations, which are described in the subsequent paragraphs, were performed.

Detailed Examination of the Scoop Surface

In the period before the detailed scoop examinations took place, a study was made along similar lines of a surface sampler in the Soil Mechanics Laboratory at the California Institute of Technology. This sampler, No. SN 44107, is

FIGURE 4.—Apollo 12 photograph of left-hand side of scoop (AS12-48-7128).

a flight model conforming in all essential details to the device mounted on Surveyor 3. It contains the same materials and is painted with the same original paint. It differs from the Surveyor 3 scoop in only a few essentially minor exterior details. They are:

(1) The Surveyor 3 surface sampler had short black sleeves painted on the arms adjacent to the joints (see fig. 1); this was not done on any of the other surface samplers.

(2) The "laboratory" scoop possesses two screws inserted in its top surface; these are not present on the Surveyor 3 scoop.

(3) Some of the screws on the laboratory scoop have a different head size and shape.

(4) No epoxy has been applied to the screws and electrical connections of the laboratory scoop.

In terms of geometry, design, and dimensions, as will be seen in subsequent pictures, the scoops are identical. In the following discussion, the laboratory and the returned scoops will be compared, with reference to pictures taken of the laboratory scoop alone, the returned scoop alone, and pictures of the two scoops side by side.

Plans for the examination of the returned Surveyor parts were completed by the end of March, and a detailed study of the scoop began on April 1, 1970.

External Appearance

Before removal of any of the lunar soil coating the exterior of the scoop, the surface was examined and photographed in detail at various levels of magnification and in electronic flash, 3200° K tungsten (standard artificial light for type B color film), infrared, and ultraviolet illumination conditions.[2]

[2] All illustrations have, however, been reproduced here in black and white, with the exception of fig. 11.

Many changes in appearance are evident in the returned sampler. The blue paint that covers most of the surface has faded in color from the original light blue color to a whitish blue in the relatively protected or concealed areas of the arms and scoop. The original color of the paint is 5.0 PB 7/6 on the Munsell scale; the paint on the returned sampler is now 10.0 B 8/2 on the Munsell scale in the cleaner (not soil covered) areas and 10.0 B 7/2 on less clean parts. However, on the upper surfaces of the arms and on the upper and side surfaces of the scoop itself,

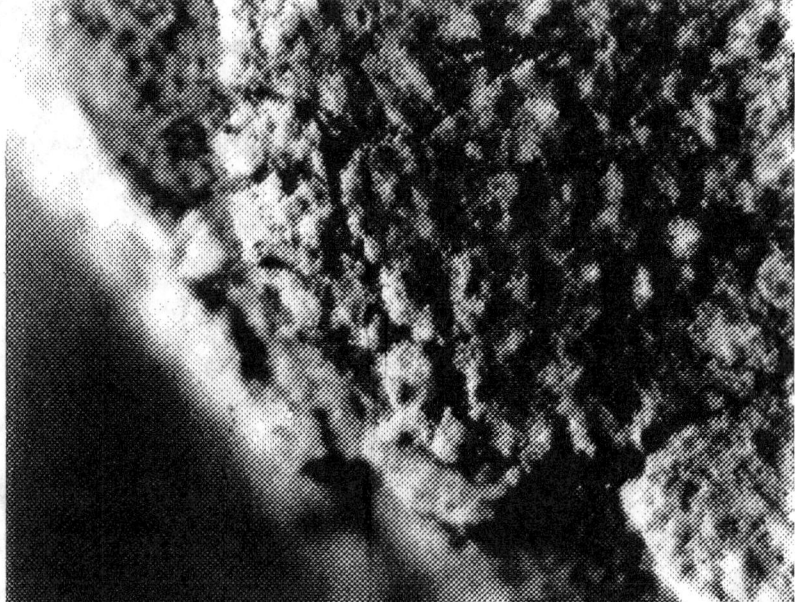

FIGURE 5. — Microphotograph of cracked and chipped paint on left-hand side of surface sampler door (width of field: 3 mm).

the color of the paint has been changed to a light tan. This tan is most pronounced on the upper surfaces and shades into the whitish blue on the underside of, for example, the arms. A microscopic examination of the paint surface at a magnification of 80× (figs. 5 and 6) indicates that the tan is a change in the painted surface rather than a light coating of surface particles.

Figure 5 is an enlarged photograph of the painted surface on the left-hand side of the base of the scoop door. It is thought that during transit from the Moon, and subsequent handling in the LRL and elsewhere, some of the paint around the edge of the scoop door may have been abraded and removed. Some of the paint probably also was removed during operations on the

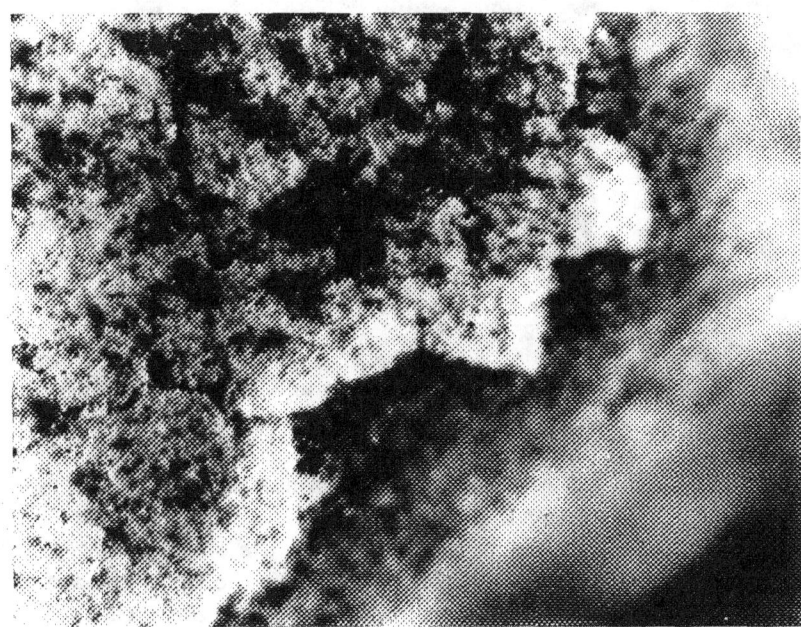

FIGURE 6. — Microphotograph of cracked and chipped paint on right-hand side of surface sampler door (width of field: 3 mm).

lunar surface. In figure 5, a gradation is observable from the light blue color of the paint, which is very close to its original color, near the edge of the scoop door to the tan color, which is more characteristic of the major portion of the scoop surface. It also can be seen that many lunar soil particles, including a substantial proportion of small glassy spheres, are present. The irregular, bumpy texture of the painted surface is characteristic of the original painted coating. The color change is not uniform, as can be observed by a comparison of figures 5 and 6, and it seems to depend on the degree to which the surface was exposed to solar radiation. In figure 6, which is the right-hand side of the scoop door base, the color change is less than on the portion of the scoop door shown in figure 5.

Coloration patterns on both the right- and left-hand sides of the scoop are shown in figures 7 through 9 in black and white and natural color; the pattern is also apparent in varying degrees under different lighting conditions in figures 10 (infrared), 11 (ultraviolet), 12 (ultraviolet), 13 ultraviolet, and 14 (ultraviolet). (The conditions under which the pictures were taken are described below.) To some extent, the patterns of color change can be correlated with the extent to which the scoop was covered with lunar soil before it was touched by the astronauts. On the right-hand side of the scoop, a comparison of figures 3 (taken by the astronauts) and 7 clearly indicates that the bottom part of the scoop side, which was covered with lunar soil, has not changed in color to the same extent as the rest of the scoop. This would indicate that the color change process is related to the irradiation of the

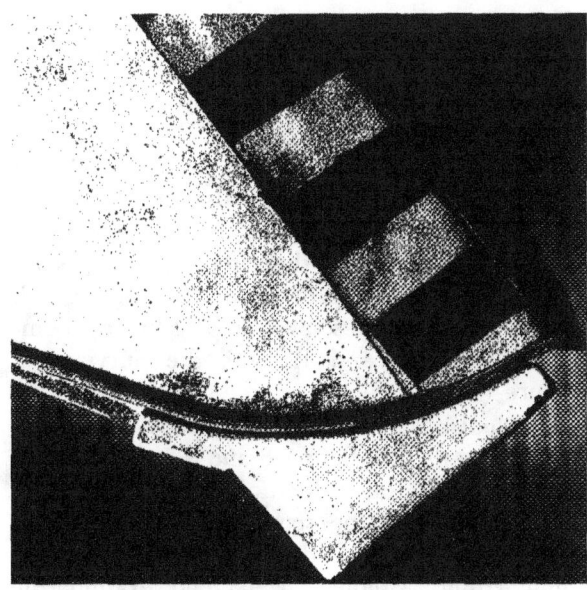

FIGURE 7.—Right-hand side of surface sampler door showing reference stripes (width of field: 5 cm).

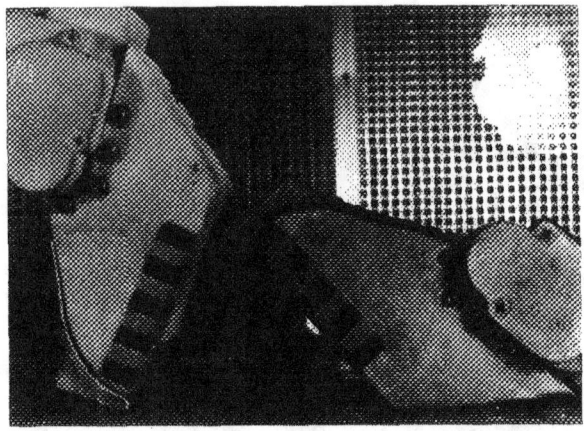

FIGURE 8.—Comparison of Surveyor 3 surface sampler with laboratory surface sampler. White-light photograph of right-hand side.

FIGURE 9.—Comparison of right-hand side of Surveyor 3 surface sampler with left-hand side of laboratory surface sampler. White-light photograph of top of scoop and motor box.

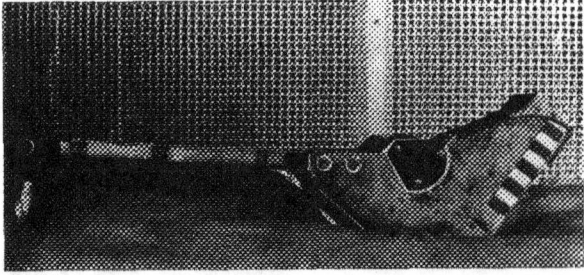

FIGURE 10.—Surveyor 3 surface sampler (compare with fig. 25).

painted surface. It can be seen in figures 9 and 12 that the top of the scoop also has a blotchy appearance; the tan color is lighter inside the blotches. In this area, the effect again appears to be related to a protective covering of soil clumps or aggregates, as can be seen by comparing figures 9 and 12 with the Surveyor 3 picture (fig. 2).

A visual examination of the scoop, as shown by a comparison of figures 8 and 9, indicates that the intensity of tan coloration is greatest on the upper surface of the scoop, less on the sides, and still less on areas that have been shaded to some extent. Although a detailed examination of this point has not been made, it appears that the degree of alteration of the painted surface is related to the duration and angle of surface exposure to the Sun on the lunar surface. Even the base of the scoop, which was exposed to some solar radiation in the lunar morning, has been changed somewhat in color, as seen in figure 15. On the left-hand side of the scoop (fig. 4), a pattern of color is apparent; the tip of the scoop appears lighter than the rest of the area on this side. This effect is still observable on the returned scoop, but is less clear than shown in figure 4. Possibly the illumination condition of figure 4, as well as the soil-coated condition of the returned scoop, made the contrast between the tanned and less tanned zones not so obvious. It seems likely that the blotchy appearance of the grooves or dents on the upper surface of the scoop (as seen in fig. 13, for example) developed from an accumulation of some lunar soil in the bottom of the grooves, with a resulting protective action. It is not known why general gradational differences in the degree of color change exist on apparently uniformly exposed plane sides of the scoop. These may arise from local changes in the thickness or composition of the scoop paint, or may be due to the presence on the Moon of differing thicknesses of dust coatings resulting from lunar surface operations. It has been shown (ref. 2) that, at some point between the end of Surveyor 3 operations in 1967 and the visit of the Apollo 12 astronauts, two of the spacecraft's legs had collapsed. It is possible that some soil was shaken from the scoop at this time. This may have contributed to variations in the degree of color change in the paint in areas where no soil covering can be seen in the Apollo photographs. Because the left side of the scoop was more exposed to the sandblasting of the Apollo 12 descent engine (ref. 3), soil removal and addi-

Please See Figure 11 (which has been moved to the end of the book)

FIGURE 12.—Ultraviolet stimulation, visible recorded on film. Comparison with figure 9 focus sharp on top of motor box.

tional color changes may have been effected on this side during the Apollo 12 landing. Another possibility is that the abrasion of the paint that took place during the lunar surface testing resulted in different sensitivities of the paint to the possible irradiation in different areas. As the color change is more visible in the ultraviolet photographs and less so in white light, it may be inferred that the change resulted primarily from the exposure of the paint to radiation of ultraviolet wavelengths.

A second item of interest concerning the painted surface is the crazing or cracking of the paint on the sides and base of the scoop door. Polygonal fracture patterns are apparent in figures 5 through 7, 14, and 15. This part of the scoop was made of a glass-fiber-impregnated resin coated with the standard paint. The fracture pattern does not appear on the painted metallic surfaces of the rest of the scoop, and may therefore be related to the different thermal conduction and expansion characteristics of the paint, the resin, and the metal. It is also possible that radiation damage to the paint could have resulted in volume changes. In this case, the appearance of fracture patterns on the scoop door would be related to either the different thickness of the paint or different nature of bonding of the paint to that surface as compared with the other metallic surfaces. The chipping of the paint from the tips of the scoop door indicate that the bonding between the paint and the resin was weaker there than elsewhere, since paint at the edges of the scoop body was equally subjected to contact with the lunar surface.

A careful study of the Surveyor 3 television

FIGURE 13.—Comparison of Surveyor 3 surface sampler with laboratory surface sampler. Ultraviolet stimulation, visible recorded on film. Top of scoop shows grooves.

FIGURE 14.—Ultraviolet stimulation, visible recorded on film (compare with fig. 15).

FIGURE 15.—Comparison of Surveyor 3 surface sampler with laboratory surface sampler. White-light photograph of base of bearing plate.

pictures was inconclusive as to the presence of chipping or flaking at these points during the lunar surface operations in 1967. Observations

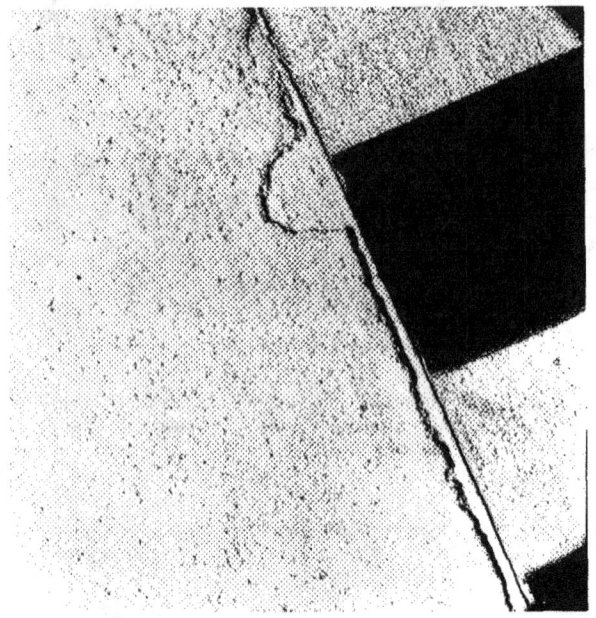

FIGURE 16.—Microphotograph of right-hand side of laboratory surface sampler (width of field: 1.5 cm).

during handling of the returned sampler indicated that the paint at the corners of the scoop base chips quite easily. Fragments of paint were observed in the lunar soil collected from the inside and outside of the scoop.

Even a cursory examination of the returned scoop shows that it has been subjected to a considerable amount of scratching and abrasion. Some of the typical larger scratches are apparent in figure 7, and photographs made at higher magnification show them clearly. For example, two photographs, magnified originally 11.5 ×, shows the condition of the laboratory scoop (fig. 16), in comparison with that of the returned scoop (fig. 17) in the same area of the surface. The terrestrial scoop has been used in a variety of soil-testing operations in various soils on Earth. The general effect of this soil contact has been to smooth down the irregularities in the painted surface without the development of scratches. (See fig. 16.) Considerably less soil contact took place with the Surveyor 3 scoop, but it is evident, as shown in figure 17, that its surface has been abraded. A general smoothing of the surface of the paint is also evident in figure 17. An undisturbed painted surface close to its original condition is shown in figure 18 (the laboratory scoop gear box), which demonstrates the rough nature of the surface developed by the spray painting process.

It was thought initially that the scratches on the Surveyor 3 scoop were formed during lunar surface operations, but it has since been learned that the painted surface of the scoop may have been lightly sandpapered (and in places repainted) before launch to remove defects.

Some months after the initial examination of the Surveyor 3 scoop, it was disassembled for study of the individual components. When this was done, it was found that the inside surface of the scoop presented an appearance essentially identical to that of the laboratory scoop in figure 18. Because the inside had been subjected to almost as much sliding contact with the lunar soil as the outside, it must be concluded that the lunar material has not substantially abraded the painted surface and that the scratches visible in figures 7 and 17 result from pre-flight surface treatment.

Another comparison of the two surface samplers is shown in figures 19 (laboratory) and 20 (Surveyor 3 scoop). It can be seen in these figures that terrestrial operations have also re-

FIGURE 17.—Surveyor 3 surface sampler (width of field: 1.5 cm; compare with fig. 16).

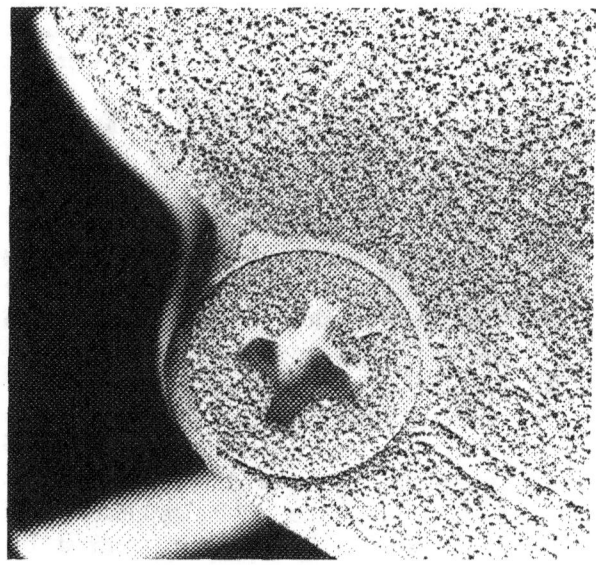

FIGURE 18.—Microphotograph of laboratory surface sampler's painted screw head on side of gear box (width of field: 1.5 cm).

FIGURE 20.—Surveyor 3 surface sampler (width of field: 5 cm; compare with fig. 19).

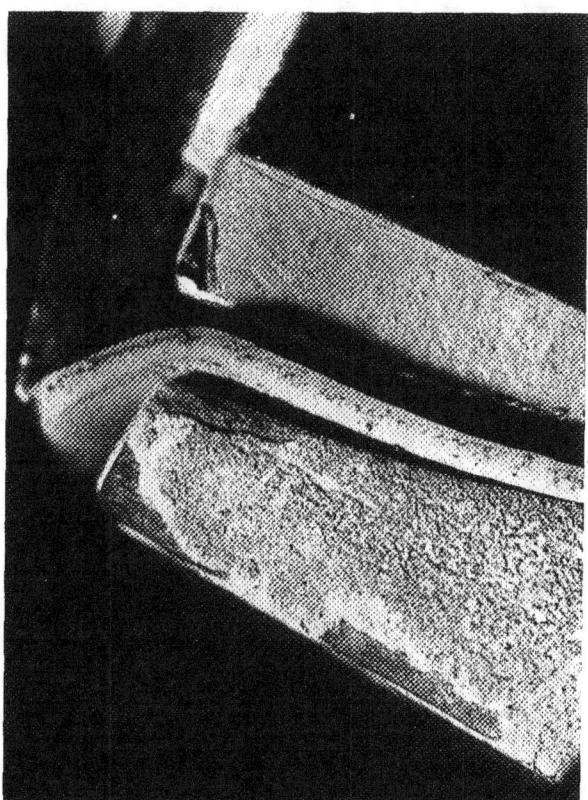

FIGURE 19.—Microphotograph of left-hand side of laboratory surface sampler showing Teflon and chipped paint (width of field: 5 cm).

sulted in the removal of paint chips from the side of the scoop tip, and that some crazing of the paint in this area has also occurred.

Adhesion of the lunar soil to all surfaces of the returned scoop is readily apparent in figures 17 and 20. In figure 20, even the Teflon seal of the scoop door is heavily coated with lunar soil particles. The lunar soil scattered about the surface sampler during and after its return to Earth seems to adhere differentially to the different surfaces of the sampler. The most obvious observation is that the lunar material adheres more readily, in order, to (1) painted, (2) Teflon, and (3) metallic surfaces. Figures 21 and 22 show, for comparison, the operating mechanism of the scoop door of the terrestrial sampler, and the same area of the Surveyor 3 sampler door. It can be seen that lunar soil is adhering to the painted surface of the door in considerable quantities, and that the metallic surface, the screw heads, and the door axle are relatively free from lunar soil. It should be noticed that the metallic surfaces are not absolutely clean. It was not possible to tell in a superficial examination if there

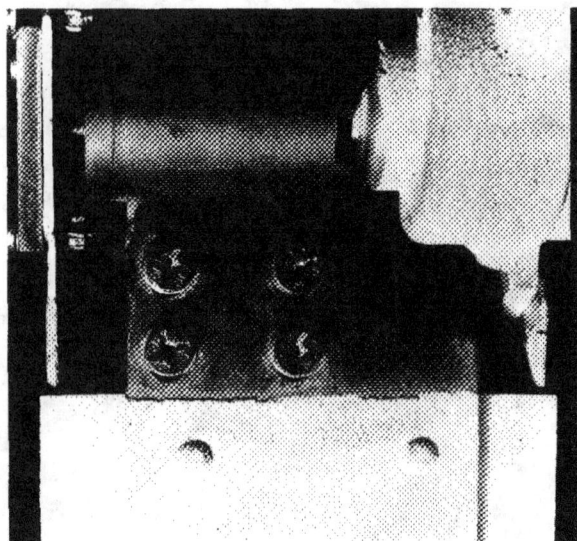

FIGURE 21.—Laboratory surface sampler. Microphotograph of door and hinge (width of field: 5 cm).

was selective adhesion of various components of the lunar soil, expect in the case of glassy spheres, as noted later. In figure 23, adhesion of the soil to the Teflon also can be clearly seen, as well as a slight color change of the Teflon itself. The Teflon appears slightly brown on its outer edges shading to the original milky white next to the metal part of the scoop door. It is apparent that this change took place rather quickly on the lunar surface by referring to figure 2, which clearly indicates the same shading on the visible portion of the Teflon door after only 10 days on the lunar surface. The discoloration is also clearly apparent in figure 23, and to a lesser degree in figures 14 and 15. As with the color change of the paint, the discoloration of the Teflon probably resulted from its exposure to solar radiation.

In spite of the considerable amount of contact with a variety of soils in laboratory bearing tests and trenching work, the surface of the gear housing (fig. 18) of the terrestrial surface sampler exhibits almost the original appearance of the painted surface. The strong contrast between this and the lunar sampler is evident in figure 24, where it is evident that the gear box had been repainted a number of times before launch.

To examine in more detail the changes in the surface sampler, photographs were taken, under different lighting conditions, of both the laboratory device and the returned Surveyor 3 sampler. Using Ektachrome infrared color film with a medium yellow filter, the appearance of the

FIGURE 22.—Surveyor 3 surface sampler (width of field: 5 cm; compare with fig. 21).

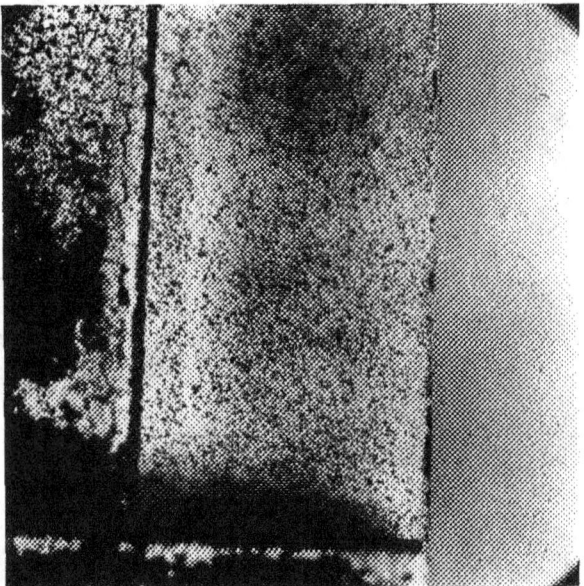

FIGURE 23.—Surveyor 3 surface sampler showing discolored Teflon (from color slide, used as black and white; width of field: 1.5 cm).

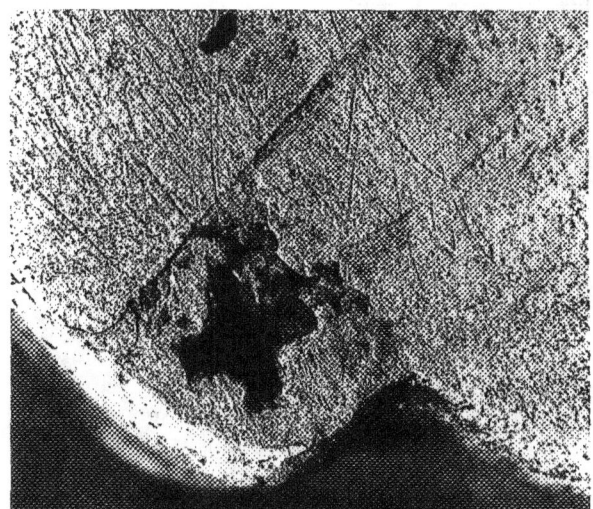

FIGURE 24.—Surveyor 3 surface sampler (width of field: 1.5 cm; compare with fig. 18).

ture was taken, is obvious. For comparison, the appearance of the two samplers under ultraviolet light is shown in figure 11. This technique enhances details of the painted surface that are not obvious under ordinary illumination. In the second black marking from the bottom of the picture (figs. 8 and 11) in the striped area of the returned surface sampler, a light streak can be seen. This streak was a defect in the anodized aluminum surface and existed before launch of the Surveyor spacecraft. It can just be seen in figure 1, for example.

In figure 11, various stages in the painting or repainting of the terrestrial surface sampler can be seen by the different shading of the paint. The wiggly, light-colored line halfway down the laboratory scoop in that figure (the same mark appears in darker blue in fig. 8) is the result of conducting tests with the terrestrial sampler in

terrestrial sampler is shown in figure 25. The pinkish appearance of most of the sampler, in contrast to its light-blue color under normal lighting and film conditions, indicates its reflective characteristics in the infrared portion of the spectrum. The different appearance of the returned sampler is obvious in figure 10, which was made under identical lighting, film, and filter conditions. The metallic parts of the surface appear to be least changed, and the painted surface itself no longer exhibits the pink appearance of the terrestrial sampler. This indicates that the sampler has become more highly absorbing to infrared radiation. Such a change in the painted surface is of interest from the point of view of thermal control of various spacecraft compartments in extended missions in space.

The changes in the surface condition of the returned sampler are most strongly evident in pictures in which the surface samplers were illuminated by soft ultraviolet light and photographed on color film through a No. 2A filter which excluded ultraviolet light. The film, therefore, records the emission of visible light stimulated by the ultraviolet light source. Figure 8 shows a comparison of the terrestrial and returned lunar samplers under normal lighting and film conditions. The marked change in the appearance of the returned surface sampler, which had been cleaned of lunar soil before this pic-

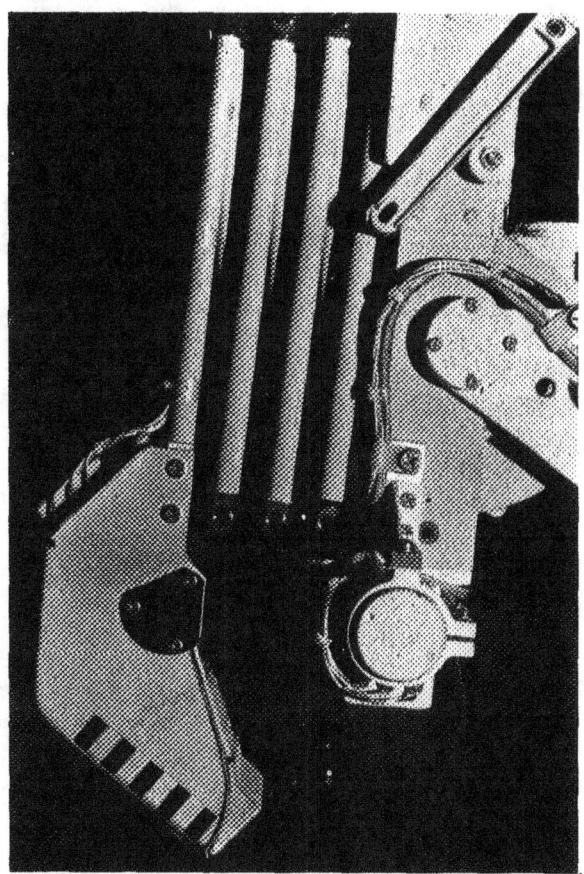

FIGURE 25.—Left-hand side of laboratory surface sampler. Infrared photograph; K–2 medium yellow filter.

a soil saturated with water in order to simulate bearing capacity tests at lower g-levels. The mark at the side of the terrestrial surface sampler at the tip appears to be a fingerprint; other fingerprints are also apparent on the painted surface. However, the most striking change under these lighting conditions is the completely different color of the returned surface sampler. It is obvious that under lunar conditions the surface properties of the painted surface have been substantially altered.

In figure 11, the brown color change which is apparent on earlier pictures, such as figure 8, appears as a dustier pink, contrasting to the lighter bluish pink, for example, at the scoop tip where the sampler was protected by lunar soil. It is not known why the shading pattern on the side of the returned scoop is apparent; it may be related to the abrasion of the surface during bearing and trenching tests. In the picture, the light blue-green flecks that appear on both scoops and on the table on which they are resting are fluorescent pieces of organic material which were present in the laboratory. They probably were derived from a variety of fabrics that were present.

The most striking change in the appearance of the samplers can be observed by comparing the tops of the two scoops in figures 9 and 12. The brown of the returned scoop is deeper on the upper surface; this is made even more apparent by the photograph (fig. 12) in ultraviolet illumination. Figure 12 shows a yellow region at the bottom of the housing that covers the scoop door motor. This probably is due to irradiation of a spill of the epoxy coating which was applied to the terminals of the wires for protection, as it does not appear on the laboratory surface sampler to which no epoxy was applied. Alternatively, it may be the result of the irradiation of this part of the scoop which was altered by heating when the wires were soldered in place. The upper surface of both scoops is shown in figure 13 in which it is seen, as remarked earlier, that some protection to the paint was probably afforded by patches of soil at the bottom corners of the grooves. It is likely that the lighter appearance of both scoops around the edges is due to abrasion during transport and handling.

The origin of the dark splotch (which is real) at the bottom of the left-hand groove of the returned scoop is not known. It is also not clear why the protection, which it is surmised was offered by lunar soil collecting in the grooves, is so obvious. If the brown coloration is a result of solar radiation, it might be expected that it would shade gradually from the color in the completely shielded area into the appearance of the unprotected surface.

A comparison between the bases of the scoop doors is shown in figures 14 and 15. In the case of the laboratory scoop, much of the paint was removed from the scoop door during the test in water-saturated soil; at one stage of testing, some of it was pulled off by stripping a piece of adhesive tape that was attached to the scoop base. It is evident that the bonding of the paint to the resin of the scoop base is not very strong. In the returned scoop, on the right in figure 14, the pattern of crazing on the base is apparent, as is also the browning of the edges of the Teflon sealing the door. Once again, even though the base of the scoop was relatively protected from solar illumination, it has also undergone the color change apparent in the previous photographs. In the portions of the glass-impregnated resins, which are revealed where the paint has chipped away from the scoops, little or no color change is obvious in either figure 14 or 15. This may be an indication that the paint was removed in these areas only during and following the return of the surface sampler to Earth. The change in appearance of the Teflon surface is shown in figure 23 for a comparison with the view of the same area, as seen by the television camera of Surveyor 3, in figure 2.

The wires to the scoop door motor were attached to terminals on the scoop (see figs. 1 and 12). These connections were covered with a clear epoxy plastic to protect them. The present appearance of the plastic covering one of these terminals is shown in figure 26. This photograph indicates that the epoxy material has changed from its originally water-clear state to a yellow-amber color. The bubbles, which are apparent in the photograph, probably were included when the epoxy was cast originally. A crack reaching to the surface runs through the epoxy in the center of the picture; it is not known whether it was originally present or not. It appears to bear some relation to the large bubble in the middle

FIGURE 26.—Microphotograph of epoxied screw head on Surveyor 3 surface sampler (width of field: 3 mm).

of the picture. The adhesion of lunar soil particles to the epoxy surface is evident in figure 26.

In October 1970, after the returned scoop had been disassembled, an examination was made of the individual parts of the surface sampler. The inside of the scoop, which was painted with the standard blue paint, was heavily coated with lunar soil. The soil, which was 1 to 2 mm thick in the corners, still exhibited its cohesive properties and was not dislodged from the painted surface when the scoop was turned over. An area, inside the cutting edge, which had made frequent contact with the lunar soil, was cleaned of its soil covering. The paint appeared unmarked even under 20× magnification. In fact, the appearance of the painted surface was similar to that shown on the laboratory sampler in figure 18. The paint also lacked the tan coloration characteristic of the outside surfaces and retained the pale blue color of the laboratory sampler. It is concluded, therefore, that the scratches observed on the scoop exterior were caused by the pre-flight sandpapering process.

The Teflon seal on the sampler door also was examined. Although the color of the outer edges had changed, as reported previously, the larger area on the inside, which was protected from direct solar radiation, had the same milky-white color as the Teflon on the laboratory sampler. Two of the nylon ties, which had secured electrical wires to one of the extension arms, also had been removed. The areas beneath the ties were pale blue in color. These protected areas were subjected to essentially the same thermal and vacuum conditions on the Moon as the discolored areas, yet retained their original appearance. It is concluded, therefore, that solar radiation was the cause of the discoloration.

Measurement of Adhesion of Lunar Soil to Surface of Returned Scoop

An attempt was made to measure the magnitude of the existing adhesion (whatever its nature) between the lunar soil and the various surfaces of the scoop by the following technique. A small vacuum-cleaning apparatus was built in order to remove the soil from the surface sampler surface. It consisted of a small pump, plastic hose, and two Lucite chambers containing different sizes of filter papers. At the input end, a pen holder was supplied to retain a nozzle through which air and the lunar soil were drawn in. Four different nozzle sizes were tested.

In practice, the experiment and cleaning operation consisted of starting the vacuum pump and bringing the nozzle closer to the surface of interest while holding it at right angles to the surface. It was generally observed that, at some particular distance from the surface, a circular area under the nozzle tip would suddenly become clean leaving, in most cases, a very abrupt discontinuity between the clean surface and the adjacent soil-covered area. This result was interpreted to mean that the adhesion of the lunar soil to itself was somewhat greater than its adhesion to the scoop surface. Thus, when a critical surface shearing stress was reached because of the air flow over the surface, the soil detached itself from the surface and passed into the nozzle and thus into the collection chambers. In a formerly well coated painted area, the clear demarcation line between the clean and dirty surfaces is shown in figure 27. By carefully measuring the distance of the nozzle from the surface of the scoop and the radius of the area which was made clean at the critical distance of approach, an estimate could be made of the surface shearing stress required to remove the soil. To make this estimate, the nozzle was calibrated by measuring the mass rate of flux of air into the nozzle at different distances of approach from various flat plates. From these tests, it was estimated that the adhesive strength of the lunar soil to the painted surface was on the order of 10^4 dynes/cm^2 (0.1 psi). The adhesion of soil to the metallic surfaces of the sampler seemed to be somewhat less and was in the range of 10^3 to 10^4 dynes/cm^2 (0.01 to 0.1 psi).

It was observed that, in an area of painted surface that had been cleaned by this technique, the remaining particles consisted almost entirely of glassy spheres. This can be seen in a careful examination of figure 27. It would appear that the adhesion of the spheres to the paint, at least, was considerably greater than that of granular fragments of other shapes, as one might expect that angular grains would exhibit a greater degree of mechanical interlocking with a rough surface than spherical particles.

FIGURE 27.—Microphotograph showing cleaned area next to dirty area on painted surface of Surveyor 3 surface sampler (width of field: 3 mm).

References

1. SCOTT, R. F.; AND ROBERSON, F. I.: "Soil Mechanics Surface Sampler: Lunar Surface Tests, Results and Analysis." *J. Geophys. Res.*, vol. 73, 1968, pp. 4045–4080.
2. SCOTT, R. F.; LU, T.-D.; AND ZUCKERMAN, K. A.: "Movement of Surveyor 3 Spacecraft." *J. Geophys. Res.*, vol. 76, 1971, pp. 3414–3423.
3. JAFFE, L. D.: "Blowing of Lunar Soil by Apollo 12: Surveyor 3 Evidence." *Proceedings of Apollo 12 Lunar Science Conference*. To be published.

PART L

MOVEMENT OF THE SURVEYOR 3 SPACECRAFT

R. F. Scott, T.-D. Lu, and K. A. Zuckerman

The scientific and engineering results from the Surveyor 3 lunar mission have been reported (ref. 1); the results pertinent to this discussion are repeated briefly here.

No communication was returned from the spacecraft after the first lunar night. At the end of its multiple-impact touchdown, the spacecraft came to rest on the inner eastern slope of a 200-m-diameter crater. The ground slope was about 10° to 12°; the inclination of the spacecraft's vertical axis from the lunar vertical was determined to be 12.4°. Footpad 2 was within the

FIGURE 1.—Surveyor 3 television picture of footpad 2 taken on April 21, 1967. The image is hazy because a film of lunar dust was deposited on the camera mirror during the landing (GMT Day 111, 07:43:38).

field of view of the television camera, footpad 3 was partially visible, and footpad 1 was obscured by spacecraft components. In the last stages of landing, footpad 2 left an impression on the lunar surface some distance from its final location. This apparently penultimate contact and the footpad itself from the point of view of the Surveyor television camera are shown in figure 1.

Only the right side of footpad 3 could be observed; it had plowed downhill through the soil; in its final position, the visible part of its top surface was about 10 cm above the soil level.

On November 19, 1969, the LM landed near Surveyor 3. On November 20, astronauts Conrad and Bean took many photographs of Surveyor and removed several spacecraft components for return to Earth. Some of the photographs, when compared with the original Surveyor pictures, exhibit some features of interest which will be discussed here. It is tentatively concluded that, at some time between Surveyor 3 shutdown on May 3, 1967, and the time the photographs were taken by Conrad and Bean, the Surveyor spacecraft moved 7 or 8 cm.

Movement

Figure 2 is a photograph taken by Conrad and Bean on their way toward Surveyor 3. If the mast angle in this picture is measured with respect to the visible lunar horizon, it is found to be about 15°, in the plane of the picture. The maximum downslope angle of tilt would be somewhat greater. If the lunar horizon differs from the true horizontal in this picture by less than 2.5°, then it would appear that the spacecraft has increased its inclination downslope since 1967. More positive evidence for this is apparent in figures 3 and 4, respectively, also taken by the astronauts. In these photographs, the shock absorbers of legs 1 and 3 are collapsed. Their normally extended position can be seen from the position of the leg 2 shock absorber in figure 2. Here the extended shock absorber and its supporting strut form a straight line, in contrast with the angle that the leg 1 members make in figure 3. Study of the position of the leg 2 shock absorber in figure 2 and comparison with the shock absorber of leg 3 in figure 4 show that the leg 3 shock absorber is also collapsed.

All the shock absorbers were extended during the landing and communication life of Surveyor 3 in 1967. The shock absorbers contained helium gas at high pressure; the gas was retained by

FIGURE 2.—Surveyor 3 photograph taken by astronaut. Leg 2 and the surface sampler are at the right of the picture. The upper member of the leg is the extended shock absorber, which lies almost in a straight line with the fixed support running from the upper end of the shock absorber to the spacecraft structure. To the left and pointing almost toward the camera is leg 1 with the footpad embedded in the soil. The shock absorber on this leg is at an angle to the supporting member (AS12–48–7121).

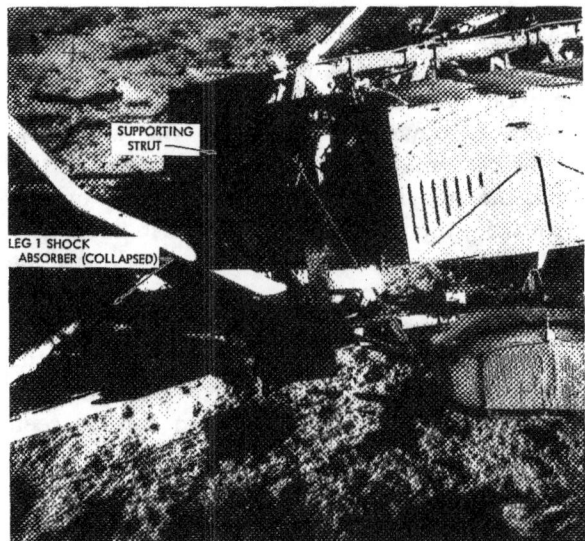

FIGURE 3.—Astronaut photograph showing detail of collapsed shock absorber connection on leg 1, to left, partially in shadow (AS12-48-7118).

FIGURE 4.—Astronaut photograph of footpad 3 and part of leg 3. The collapsed shock absorber is the upper tubular member. The footpad shows some soil, with adjacent lighter colored areas (AS12-48-7124).

FIGURE 5.—Enlargement of part of Apollo 12 photograph showing footpad 2 of Surveyor 3 and lunar surface imprints (AS12-48-7110).

seals, which can fail. It is concluded that the shock absorbers on legs 1 and 3 of Surveyor 3 collapsed at some time after the termination of communication with the spacecraft.

An indication that the failure of the shock absorber on leg 3 may have been sudden is seen in figure 4, which shows footpad 3. This picture indicates that the edge of footpad 3, not visible to the Surveyor 3 camera, dug into the lunar soil so that its upper surface became covered with soil. However, a lighter shading appears around the edge of the pile of soil on the footpad. This was at first interpreted (ref. 2), it is now thought erroneously, as lunar soil of a lighter color.

However, it was reported by the astronauts that the exposed spacecraft parts which were originally white were a light-tan color at the time of their visit. This observation was subsequently confirmed by examination of the returned spacecraft parts (ref. 3). It is conjectured, therefore, that the footpad received a partial covering of soil during the landing in April 1967; this soil protected the underlying footpad surface from a process that either coated or, more probably, altered the white painted surface in an unknown length of time to a tan color. When the footpad was jerked by the hypothetical shock absorber collapse, the soil on the pad moved, and the protected white footpad surface was revealed in contrast to the tanned surface, as shown in figure 4. An argument against this explanation is that the lunar soil has repeatedly demonstrated the property of adhering to spacecraft surfaces. Thus, it is not clear that the soil on the footpad could have slid sideways to reveal a relatively white, rather than a soil-covered, surface. However, the appearance of the footpad in

FIGURE 6.—Laboratory photograph simulating position of Surveyor 3 footpad 2, and lunar soil imprints. Picture taken from Surveyor 3 camera position for comparison with figure 1. In this picture, *both imprints* visible in figure 5 are present in their correct positions relative to each other, but the second imprint is concealed from the camera by the footpad.

figure 4 is difficult to account for any other way. The explanation would have to be that, since the lunar soil probably adheres to itself more strongly than to the spacecraft under lunar conditions, an impulse such as that of the postulated sudden shock absorber collapse generated footpad accelerations high enough to cause shearing at the soil/footpad interface rather than through the soil. The soil deposited by the Surveyor 7 surface sampler on the upper surface of the alpha-scattering instrument slid over the surface in the manner conjectured above, when the instrument was subsequently moved (ref. 4).

In figure 1, the spacecraft's view of footpad 2 showed an impact mark some distance uphill of the footpad's final resting place. The same footpad, as viewed by the astronauts' camera, is shown in figure 5, in which a second imprint can be seen between the previously observed mark and the footpad. The clarity of this second imprint was somewhat surprising, as it is not apparent in figure 1, although its presence was suggested in the Surveyor 3 report. From this unexpected result and the consideration discussed, the following question was raised. Was footpad 2, as observed by Conrad and Bean (fig. 5), in the same position as it had been 31 months earlier (fig. 1)?

It was decided to attempt an answer by simulating the geometrical arrangement of footpad 2, lunar soil imprints, and both Surveyor and Apollo 12 cameras. It was not difficult to arrange a Surveyor footpad and the Surveyor 3 camera position correctly because the location and orientation of the spacecraft parts were known. To obtain the first imprint position, a slide projector was set at the correct angle at the Surveyor camera location. A slide of figure 1 was inserted in the projector; the full-scale footpad and imprint were adjusted until the projected image overlay them correctly. The result of this operation is shown in figures 6 and 7. Figure 6 is a photograph of the final arrangement taken by a camera in the Surveyor 3 television camera position. It may be compared with figure 1. For figure 7, the projector was set up at the Surveyor 3 camera position and projected an image of figure 1 on the footpad and soil. A camera, positioned as closely as possible in the line of sight of the projector, took the photograph shown as figure 7 using the illumination of the projected image. It can be seen that the overlap of the projected Surveyor 3 image on the laboratory model is reasonably good except at the left edge of the pad. The slight mismatch there does not affect the

FIGURE 7.—Laboratory photograph of projection of figure 1 on footpad and soil arrangement of figure 6.

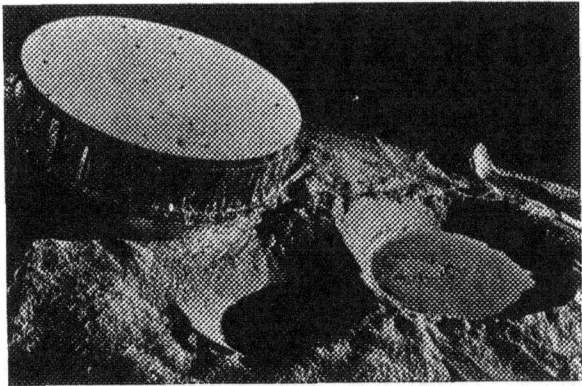

FIGURE 8.—Laboratory photograph of footpad 2 in original Surveyor 3 location. Compare with figure 5.

FIGURE 9.—Laboratory photograph of footpad 2 and imprints in position best matching figure 5.

FIGURE 10.—Laboratory photograph from Surveyor camera position of footpad 2 and imprints in position best matching astronaut photograph. The second imprint is clearly visible. Compare with figures 1 and 6.

conclusion. It was found that the appearance of the footpad, as viewed from the Surveyor 3 camera position, was extremely sensitive to the angle of footpad tilt. It is considered that the angle in the simulation is within ±1° of the angle in the Surveyor pictures.

A collimated light source was directed to light the scene at the Sun angle of the Apollo 12 photograph (fig. 5), and the position and orientation of the footpad *imprints* in that photograph were duplicated. The footpad was maintained at the position and orientation of the Surveyor 3 pictures (figs. 1, 6, and 7). With this arrangement, it was impossible to obtain a photograph that matched figure 5 with respect to *footpad* position and orientation. The closest reproduction is shown in figure 8. The footpad then was adjusted until a photograph was obtained that was a close duplication of figure 5. This required a footpad translation of about 7 cm, in effect obtained by a lateral rotation of the spacecraft about footpad 1, and a footpad tilt of about 5° in the counterclockwise direction when viewed from the astronaut position of figure 5. The resulting photograph is figure 9, which should be compared with figures 5 and 8.

A view of this arrangement, from the Surveyor 3 camera position, is seen in figure 10, in which the second imprint is clearly observable, in contrast with figures 1 and 6. In figure 6, the second imprint was present in the correct position with respect to the first imprint according to the Apollo 12 photograph of figure 5. The appearance of the footpad, because of its change of tilt, is entirely different in figure 10 from that in figure 6 or 1.

From this simulation study, it seems that a television picture of footpad 2 on a Surveyor 3 spacecraft in the same position as observed by the astronauts would have shown clearly the second imprint. It also would have shown a footpad tilt angle different from that in the original Surveyor 3 picture (fig. 1). Alternatively, an astronaut picture of the Surveyor 3 footpad 2 in its April 1967 position would have shown a less obvious second imprint, and a footpad at a different angle.

Another minor piece of evidence for spacecraft rotation is that the original Surveyor pictures show the inside edge of footpad 2 resting on an essentially level soil surface. Pictures taken by the astronauts show a ridge of soil along this edge almost to the top of the conical portion of the pad. However, the viewing angles are so different in the Surveyor 3 and Apollo 12 pictures that it is difficult to be sure that the same area is being observed.

Conclusions

It is tentatively concluded that the Surveyor 3 spacecraft moved, probably as a result of a sudden failure of the leg 3 shock absorber, between May 1967 and November 1969. The movement at footpad 2 was in the amount of 5° of tilt and 7 to 8 cm of lateral translation in the form of a rotation about footpad 1, which is embedded in the lunar soil.

Because a number of fairly close views of the Surveyor spacecraft and surface sampler appear on the Apollo 12 roll of film before the photographs presented as figures 6 and 7, the possibility arose that the spacecraft may have been moved by the astronauts. Post-mission questioning of Conrad and Bean indicated that this was not the case.

The time at which the movement occurred can be estimated only from the comparison of the shielded and unshielded portions of footpad 3 and a knowledge of the mechanism and rate of the process that tans the painted surface. The nature and magnitude of the spacecraft movement are pertinent to studies of the possible movement of lunar surface particles adjacent to Surveyor 3 (ref. 5). They also have significance for any spacecraft examinations in which its orientation is important.

References

1. *Surveyor III Mission Report. Part II: Scientific Results*, TR 32–1177, Jet Propulsion Laboratory, Pasadena, Calif., June 1967.
2. SCOTT, R. F.; CARRIER, W. D.; COSTES, N. C.; AND MITCHELL, J. K.: "Mechanical Properties of the Lunar Regolith." *Apollo 12 Preliminary Science Report*, NASA SP–235, Washington, D.C., 1970, pp. 161–182.
3. BENSON, R. E.; COUR-PALAIS, B. G.; GIDDINGS, L. E., JR.; JACOBS, S.; JOHNSON, P. H.; MARTIN, J. R.; MITCHELL, F. J.; AND RICHARDSON, K. A.: "Preliminary Results from Surveyor 3 Analysis." *Apollo 12 Preliminary Science Report*, NASA SP–235, Washington, D.C., 1970, pp. 217–223.
4. SCOTT, R. F.; AND ROBERSON, F. I.: "Soil Mechanics Surface Sampler." *Surveyor VII Mission Report. Part II: Science Results*, TR 32–1264, Jet Propulsion Laboratory, Pasadena, Calif., March 1968, pp. 135–185.
5. JAFFE, L. D.: "Lunar Surface: Changes in 31 Months and Micrometeoroid Flux." *Science*, vol. 170, 1970, pp. 1092–1094.

PART M

ANALYSIS OF SURVEYOR 3 TELEVISION CABLE

F. C. Gross and J. J. Park

The Apollo 12 astronauts Charles Conrad, Jr., Richard F. Gordon, and Alan L. Bean returned the Surveyor 3 television camera to Earth in November 1969. The camera was delivered to the Lunar Receiving Laboratory in Houston, Tex., where it remained in quarantine until January 7, 1970. Following various tests on the camera and its components at the Hughes Aircraft Co. facilities in Culver City, Calif., the component parts were distributed to selected investigators for additional testing and evaluation.

A sample of cable described as "4 inches of TV cable, fabric wrapped," which had been exposed to the atmosphere for an unknown period of time, was received by Goddard Space Flight Center (GSFC) for extensive chemical analyses of the various components. The cable was a combination of 19 insulated wires covered by a

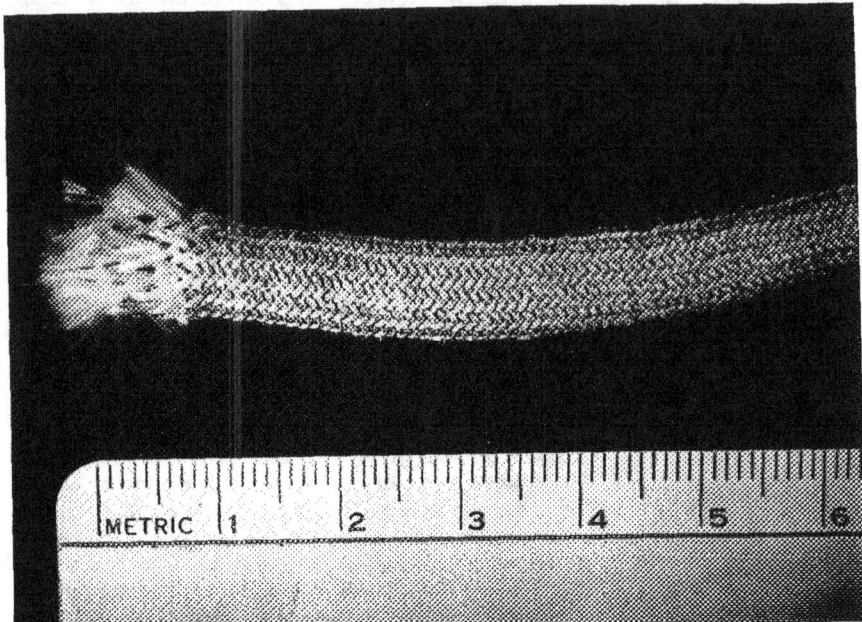

FIGURE 1.—Piece of Surveyor television cable showing glass fabric that covers more than 19 insulated wires (magnified 1.5 ×).

sleeve of woven fabric of braided glass yarn with a diameter of 0.5 mm for the strand and a woven thickness of about 0.8 mm. A similar sample, from the type approval test (TAT) equipment vehicle, was also received. This TAT sample was identical except that it had not undergone the exposure on the Moon. Thus, it was possible to compare these samples and their analyses. It should be re-emphasized that the Surveyor sample had not been kept in vacuum, but had been exposed to the atmosphere for a period of time before delivery to GSFC.

Procedure

Because of the value of the Surveyor cable, the cable was divided into three parts in order

FIGURE 2.—Enlarged view (magnified 20 ×) of woven glass fabric; black particles are believed to be Moon dust.

to perform a minimum of duplicate tests and to have the third part available for removing any doubtful results. The TAT sample was used liberally to assist in determining procedures and to work out potential problems before beginning tests on the Surveyor cable.

Each sample was considered to consist of two parts: (1) the glass fabric outer covering and (2) the wires with their insulation. It was possible, in some instances, to use a sample for more than one test. For the glass fabric, it could be examined in its "as received" state by attenuated total reflectance-infrared spectroscopy (ATR–IR). One portion was extracted with chloroform and filtered. The chloroform extract was evaporated and weighed, with the extract also undergoing gas chromatography and infrared analysis. The same fibers were extracted again with hot water and filtered; the filtrate was dried and weighed, with the residue also undergoing gas chromatography and infrared analysis. A second portion was selected for emission spectroscopy and X-ray diffraction. A third portion was used for pyrolysis gas chromatography, and a fourth portion for pyrolysis infrared. Neither of the two pyrolysis samples could be used again.

The wire insulation could be stripped from the wires and examined separately. Separated and individual portions were needed for pyrolysis gas chromatography, for differential thermal analysis, for differential scanning calorimetry, for pyrolysis infrared, and also for the tensile tests.

Specimen

The Surveyor cable, as mentioned previously, consisted of a glass fabric sleeve that covered more than 19 insulated wires. (See fig. 1). The fabric appeared to be a dirty gray, darker than the TAT sample. Some small particles, presumably Moon dust, were noted on the fabric (fig. 2). The TAT sample had loose pieces of metal or dirt in its fabric sleeve; however, it is certain that the Surveyor sample had more particles and also was darker.

Each wire consisted of 19 braided copper strands with a thin silver plating. Each copper strand was about 0.8 mm in diameter; the silver plating was approximately 0.003 mm thick. The insulation on these strands was a yellow polymer. A cross section revealed that there were two insulation layers; the outer one was thinner than the insulation around the cable.

Fabric

Many tests were conducted on the glass fabric. In comparing the results, some tests showed a definite difference between the Surveyor and the TAT; most of the tests were negative, i.e., showed no apparent difference between the two samples.

The emission spectroscopy of the washed fabric revealed the presence of its elemental constituents and permitted an estimate of the percentages of the constituents. The constituents were—

Constituent	Amount, percent
Silicon	>10
Aluminum, magnesium, boron	1 to 10
Iron	0.1 to 1
Calcium, titanium, sodium	0.01 to 0.1
Zirconium	0.001 to 0.01
Manganese	0.0001 to 0.001

The fabric is high in silicon and has much aluminum, magnesium, and boron, similar to a borosilicate glass. X-ray diffraction showed that it was amorphous, as a glass. The results were negative, with no detectable difference between the Surveyor and the TAT samples.

Attenuated Total Reflectance

The examination of the fabric, by reflectance in the infrared region, was scanned in the 2.5- to 25-μm range using a Perkin-Elmer Model 621 spectrophotometer. The spectrum obtained was that of a noncrystalline inorganic silicate. Again, the results were negative.

Chloroform Extract

Samples were extracted with boiling chloroform. The chloroform was evaporated and the weight of extract was determined. These results were positive. The TAT sample had a 0.32-percent residue; the Surveyor sample had a 0.21-percent residue. The residue was again dis-

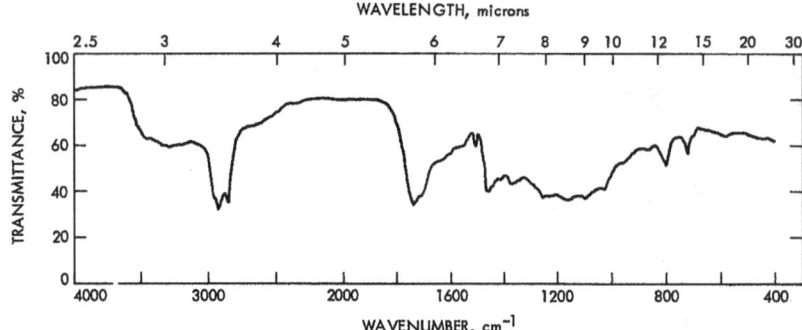

FIGURE 3.—Spectrum of chloroform extract from glass fabric on TAT sample.

solved and its infrared spectrum obtained. The spectrum (fig. 3) of the TAT sample showed primarily aliphatic esters and other carbonyl-containing compounds such as fatty acids. The spectrum of the Surveyor sample (fig. 4) was a more clearly defined pattern of aliphatic esters, indicating that some volatilization of the lower boiling constituents may have occurred in the space vacuum.

The extract was examined by means of gas chromatography, a method that separates the constituents of a vaporized sample into distinct fractions. This confirmed that the extract from the Surveyor had fewer volatile components than did the TAT sample.

The extract was further examined by gas chromatography/mass spectrometry. This confirmed the evidence that volatilization of some constituents had occurred from the Surveyor fabric. The relatively low-boiling chlorinated hydrocarbons and other compounds found in the TAT extracts were absent in the Surveyor samples. The higher boiling constituents of both samples remained about the same.

Aqueous Extraction

After the chloroform extraction, the samples were subjected to boiling water extraction; the extract was used for infrared analysis, emission spectroscopy, and X-ray diffraction. The extraction also gave a positive result. The amount of the residue was 1.75 percent for the TAT sample and 1.58 percent for the Surveyor sample. The infrared pattern indicated that the extract was an inorganic silicate. X-ray diffraction gave only three broad, weak diffraction lines, insufficient for positive identification. The emission spectrographic analysis also gave a negative result, showing the presence of primarily silicon, sodium, and magnesium as follows:

Constituent	Amount
Silicon	Major
Sodium, magnesium	Major
Boron, aluminum, calcium	Minor
Iron, copper, titanium	Not detected

The extracted material probably is a form of water glass, or sodium silicate.

Pyrolysis Infrared

The infrared spectrum was obtained of a pyrolyzed sample of the fabric. Pyrolysis consists of burning the sample and collecting the condensable gaseous products. The spectrum indicated the presence of a small amount of organic material, probably hydrocarbons, though the results were negative.

Reflectance Spectroscopy

The discoloration of the glass fiber was apparent upon visual examination. However, repeated attempts to obtain transmission and reflectance patterns in the range from the near infrared to the ultraviolet, from 25 000 to 1900 Å, in samples of fibers or as ground particles, did not show a difference between the Surveyor and TAT samples.

Wire Insulation

The insulation was a yellow color, consisting of two layers.

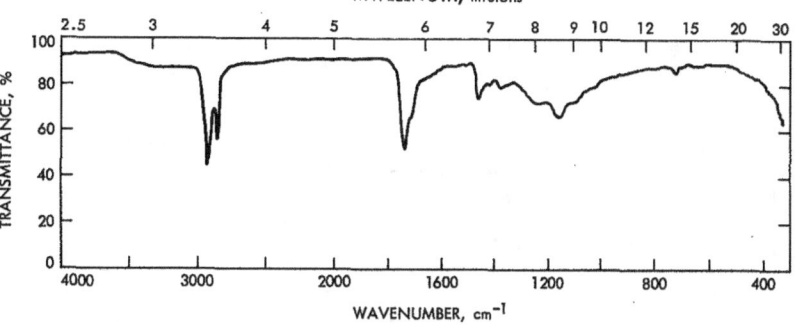

FIGURE 4.—Spectrum of chloroform extract from glass fabric on Surveyor sample.

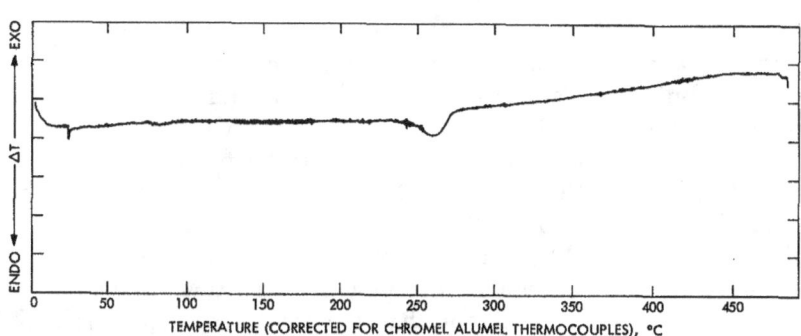

FIGURE 5.—Differential thermal analysis recording of TAT wire insulation.

Attenuated Total Reflectance of Outer Surface

The infrared spectrum of the outer surface of the insulation indicated that it was a polyimide resin, similar to Kapton film (H film), but probably applied as a liquid and called "liquid H." The results of the comparison were negative; there was no detectable difference in the patterns of the Surveyor insulation and the TAT insulation.

Pyrolysis Infrared

The spectra of the pyrolyzate indicated the presence of a fluorocarbon resin, similar to FEP Teflon, but modified slightly. This test was negative. It is interesting to note that pyrolysis of a polymide gives no infrared pattern, so that the FEP Teflon was the only noticeable constituent.

Pyrolysis Gas Chromatography

The separation of the pyrolyzed constituents was conducted by heating the insulation to about 1000°C in a helium gas carrier. The separation gave negative results.

Differential Thermal Analysis

Small amounts of the two-layered insulation were heated to over 450°C in air and also in nitrogen. The results were negative, although both samples showed an endothermic reaction starting at about 240°C with the peak at about 260°C. (See figs. 5 and 6.) The differential scanning calorimeter results also were negative.

Attenuated Total Reflectance Subsurface

Dissolution of the outer polyimide layer by a 15-percent hot potassium hydroxide solution was performed. The underlying Teflon layer of the Surveyor sample showed some small areas of slight discoloration. The ATR–IR patterns showed some slight differences (figs. 7 and 8), though it would be difficult to assign positive significance to this due to the normal variation of the test.

Tensile Tests

The most obvious change in the Surveyor insulation was apparent in its tensile strength and elongation. The tensile tests were conducted on

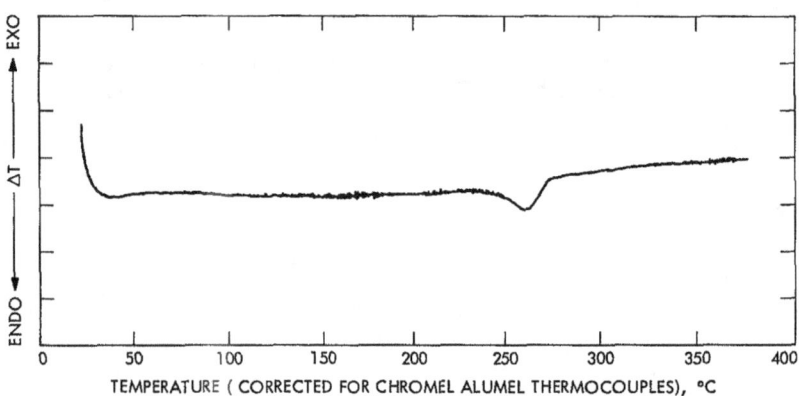

FIGURE 6.—Differential thermal analysis recording of Surveyor wire insulation.

the dual-layered insulation; the wires were pulled out, leaving the insulation in the tubular shape. During the tensile tests, it was observed that the outer polyimide layer separated early in the test, but that the Teflon inner layer remained intact for the continuation of the test. The results are given in table 1.

The third Surveyor sample compares well with the TAT sample in load and elongation, except for the elongation of the polyimide layer. The two Surveyor samples show considerable changes in the elongation and in the ultimate load for both the polyimide and the Teflon layer.

The one Surveyor sample which is apparently anomalous is believed to be from a wire within the bundle, rather than at the surface of the bundle. The presence of the change, however slight, in the infrared pattern is a clue to a change in the polymer structure. It had been pointed out that the tensile strength probably would be more obviously changed by exposure to a hostile environment. This apparently is the

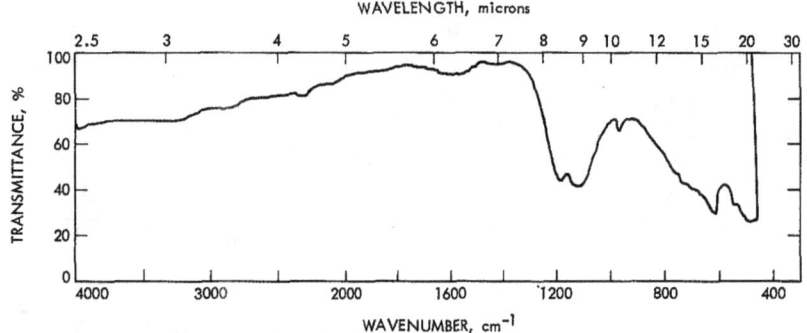

FIGURE 7.—Attenuated total reflectance-infrared reflectance chart of TAT Teflon insulation.

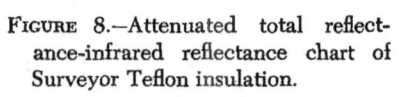

FIGURE 8.—Attenuated total reflectance-infrared reflectance chart of Surveyor Teflon insulation.

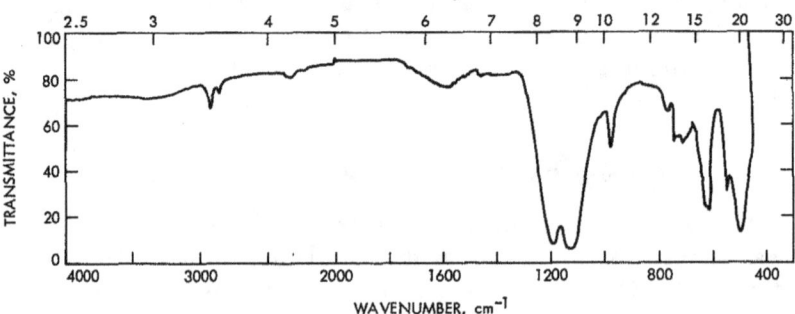

TABLE 1.—*Tensile tests*

Sample	Polyimide		Teflon	
	Ultimate load, g	Elongation, percent	Ultimate load, g	Elongation, percent
TAT	1010	35	960	990
	930	33	920	800
Surveyor	790	72	760	340
	770	60	740	400
	1000	62	920	1180

case. However, the increase in elongation of the polyimide layer implies an additional curing of the polymer, even though the ultimate load did decrease.

Other Observations: Corrosion

Certain parts of at least two Surveyor wires showed black areas (fig. 9). It was determined that these areas were under the insulation, rather than on the surface. These areas were examined in an electron microprobe analyzer. The results showed primarily silver and copper, but also sulfur and iron; the sulfur was associated with the silver, rather than the iron. The X-ray diffraction pattern identified the black areas as silver sulfide. There were also occasional areas on the wires determined to be high in copper and sulfur, possibly copper sulfate. It must be assumed that the sulfide corrosion was on the wires before the trip to the Moon, though it should be pointed out that no such areas were observed on any of the TAT wires.

Summary

A detailed physical and chemical analysis was conducted of the Surveyor television cable. In comparing the analysis of the Surveyor cable to that of the TAT cable, only a few notable changes were apparent. These changes included some loss of volatile constituents from the glass fabric outer covering and the discoloration of the glass. The insulation on the wires appears to have developed a slight discoloration and possibly slight changes in the infrared spectrum of the Teflon layer. A more noticeable change occurred in the tensile strength and the elongation of the outer polyimide layer and the inner Teflon layer.

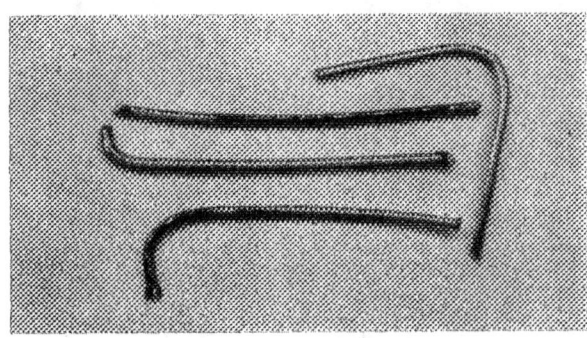

FIGURE 9.—Surveyor wires showing corrosion on three separate wires. Fourth wire with large bend is from TAT sample.

ACKNOWLEDGMENTS

We thank the following GSFC personnel of the Materials Engineering Branch for their help in the tests described: Frank Briden for differential scanning calorimetry, William Campbell for thermal analysis, Joe Colony for gas chromatography, Ronald Hunkeler for transmittance and reflectance spectroscopy, Jane Jellison for scanning electron microscope, Carl Johnson for tensile tests, Larry Kobren for electron microprobe analysis, William Latham for metallography, Edward Nelson for infrared analysis, and Pedro Sarmiento for emission spectroscopy and X-ray diffraction.

V. Organic Contamination Analysis

HIGH-RESOLUTION MASS SPECTROMETRIC ANALYSIS OF SURFACE ORGANICS ON SELECTED AREAS OF SURVEYOR 3

B. R. Simoneit and A. L. Burlingame

The Apollo 12 astronauts, C. Conrad and A. Bean, on November 20, 1969, recovered the television camera, the surface sampler scoop, and a few smaller items from Surveyor 3, which had landed on the Moon 31 months earlier (refs. 1 and 2). It was immediately noticed that the white surfaces of the spacecraft had become tan; on subsequent examination, optical interference patterns were found on the camera mirror. The effects of lunar particles on this mirror had been observed earlier by Jaffe and Rennilson (ref. 3). The decision was made to investigate the possible organic contamination of the mirror surface and camera exterior (shroud) due to spacecraft outgassing, Lunar Module (LM) descent engine blasting, possible Surveyor 3 engine exhaust products (although thought to be unlikely because of the configuration of the spacecraft components), and unknown sources.

Chips from the lower shroud, one facing the LM, one away from the LM, and one unpainted and shielded also were analyzed by an ion microprobe analyzer (IMA).

During this investigation, a thin film of possible polymeric organic matter was found under the surface dust of the mirror. It was not removed by the acetate replication; thus, it was decided to scrape selected areas for analysis.

Experimental

The mirror and middle shroud were extracted for organics by washing the surface with solvent applied by a syringe and collected in a beaker. The area of the mirror that was washed with benzene is labeled in figure 1. The whole operation was carried out in a class 100 organic clean room. The wash solvents used were benzene and a 3:1 mixture of benzene and methanol. These redistilled nanograde solvents were analyzed for background contaminants by high-resolution mass spectrometry only. The sample washes were concentrated on a rotary evap-

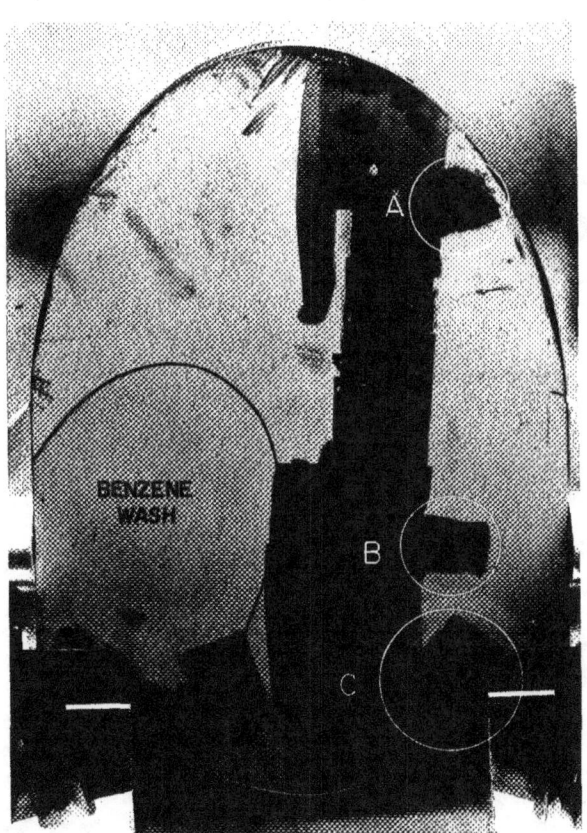

FIGURE 1.—Surveyor 3 mirror surface with the areas of organic sampling indicated (benzene wash; scrapings A, B, and C).

orator and the total residue subjected to high-resolution mass spectrometry. The mass spectrometer system consisted of a modified G.E.C.-A.E.I. MS–902 high-resolution mass spectrometer online to an XDS Sigma 7 computer (refs. 4 through 7). Multiple spectra of each sample were recorded online under the following spectrometer conditions: resolution of 10 000, ionizing current of 500 µA, ionizing voltage of 55 eV, ion source temperature of 200° to 220°C, and mass range of 12 to 400 atomic mass units at a scan rate of 16 sec/decade with a 20-kHz clock

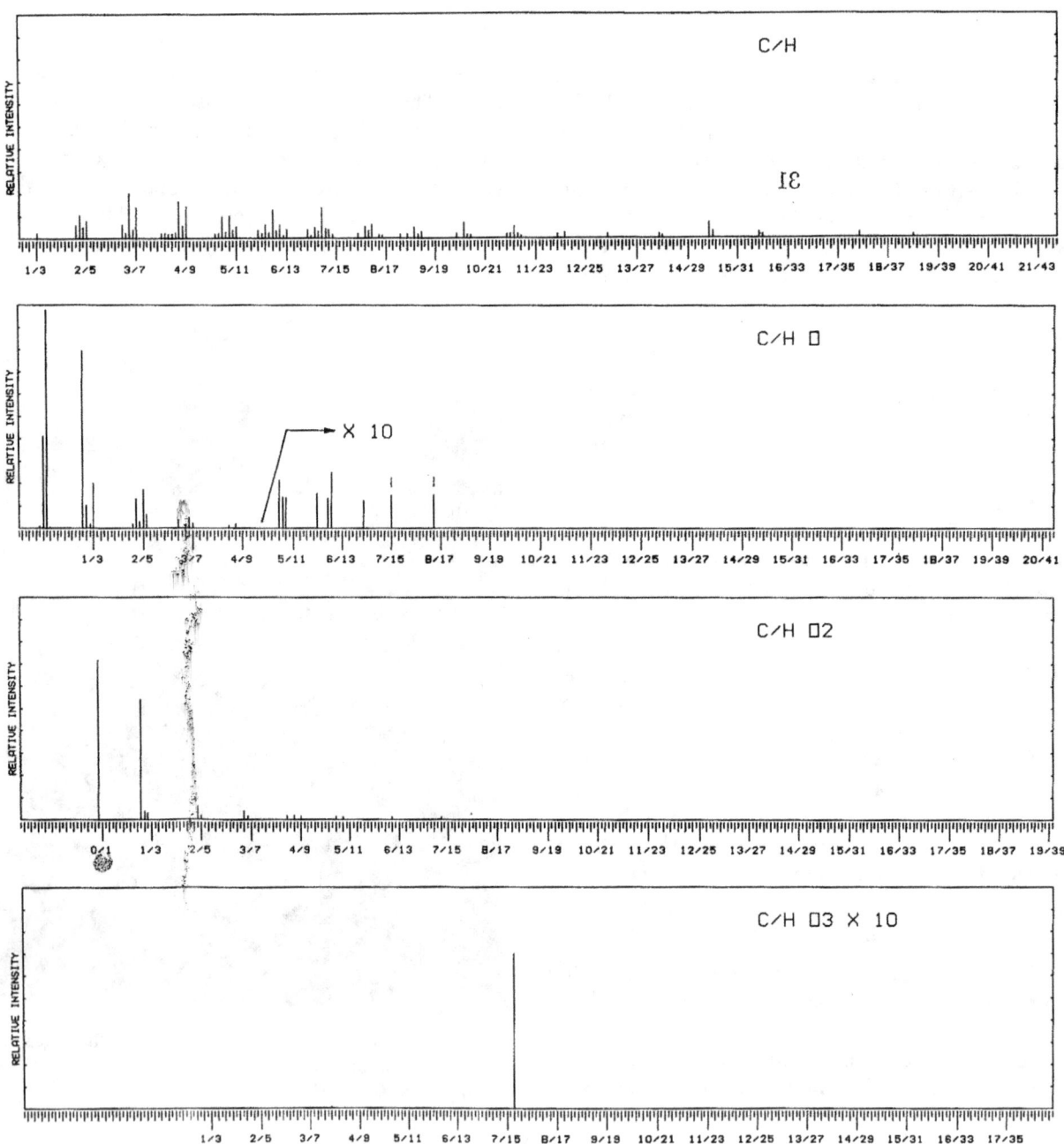

FIGURE 2.—Partial high-resolution mass spectral data for the background residue of the benzene wash solvent.

rate. The samples were inserted into the mass spectrometer on a ceramic direct introduction probe.

The lower shroud chips were analyzed by an IMA. The sample, introduced through a vacuum lock, is bombarded by an inert gas ion beam (Ar^+) at an energy of 7 keV. Secondary ions, sputtered off the sample surface, are separated and measured in a double focusing mass spectrometer. The primary ions, generated in a conventional electron bombardment source, are accelerated and focused to a 200-μm spot on the sample. The sample and the surrounding target area are held at a potential of +3 kV, which serves as the accelerating potential of the mass spectrometer. The secondary ions are expelled through a grounded port (0 V) positioned at a 45° angle to the bombarding beam. The ions are energy focused in a 90° electric sector and separated in a 90° magnetic sector. The detector system consists of a beam defining slit, electron multiplier, and a unity gain amplifier. All spectra were obtained by decreasing the magnetic field exponentially at a rate of 6 min/decade in mass and recording the detector output on a strip chart recorder.

As mentioned previously, the sample and surrounding target area serve as an accelerating potential for the mass spectrometer. This presented a problem with these samples because the painted surfaces were electrically nonconducting, resulting in a loss of the accelerating potential over most of the target area. The space charge on this surface distorted the remaining fields so that no ions were detected. To overcome this problem, each sample was wrapped in tantalum foil such that a 1-mm strip of the painted surface was left exposed. The bombarding beam was focused within this strip, and the tantalum foil supplied the necessary potential. There was no noticeable difference in the intensity of the secondary ion beam using this method and the normal method in which the sample is a conductor.

The mirror surface was scraped with a steel blade to remove the particulate matter and the film of possible organic matter underneath. Three areas were sampled this way; they are indicated in figure 1 by A (top), B (middle), and C (bottom) of the mirror. The samples then were analyzed by high-resolution mass spectrometry using the same instrumentation cited earlier. The samples were introduced into the ion source in a sidewell of a ceramic direct probe. The ion source temperature was 350°C, and the operating conditions were the same as discussed earlier.

Results

Only analyses by high-resolution mass spectrometry were possible on the trace amounts of organic material isolated from the various washes. The summed high-resolution mass spectral data for the benzene residue is shown in figure 2 and serves as the background example for all solvent washes. These data are presented as heteroatomic plots (ref. 8) where the relative ion intensity is plotted vs. the carbon-to-hydrogen ratios of the respective heteroatomic compositions.[1] The hydrocarbons (C/H plot of fig. 2) are relatively low in concentration; the major series has the composition C_nH_{2n-6} for $n=6$ to 19,

---

[1] In these heteroatomic plots (ref. 8), the masses are plotted in methylene units. On the abscissa, each principal division marker corresponds to the saturated alkyl fragment (even-electron ion), for example, C_nH_{2n+1}, with the number of carbon and hydrogen atoms given subsequently. Each principal division of the abscissa is further divided into 14 units. The number of hydrogen atoms of an unsaturated or cyclic-fragment ion is obtained by subtracting the number of units (hydrogen) from the $2n+1$ hydrogen atoms of the respective saturated principal division, C_nH_{2n+1}. A peak with a tick mark above it has more than seven degrees of unsaturation. Fragments of this kind are plotted below the next lower major saturated division, i.e., below the $C_{n-1}H_{2n-1}$ division. To convert the composition of these ions as they appear on the plot to their actual composition, add one carbon and subtract 12 hydrogen atoms. The origin of the abscissas is the same mass to charge (m/e) ratio for each plot; thus, the nominal masses from plot to plot lie directly above one another, and a superposition of the plots would yield a "low" resolution mass spectrum of the sample. All plots are normalized to a base peak (usually the base peak of the entire spectrum, unless otherwise specified) on the relative intensity scale. In order to make high-mass, low-intensity features of the spectrum observable, the whole spectrum or any region thereof can be multiplied by a scale factor. This factor is indicated by $\times \underline{000}$ at the point of scale expansion. In all high-resolution mass spectrometric data cited in this article, no peaks due to the ^{13}C isotope contributions are present; these peaks have been deleted by computer sorting (refs. 4 and 5).

and not every homolog is present. The prior series is further substantiated by the fragment ion series C_nH_{2n-7} (i.e., $C_nH_{2n-6}—CH_3\cdot$) ranging from $n = 5$ to 18. There are other lower molecular weight hydrocarbons present, but in less significant amounts. There are few oxygenated species present. Carboxylic acid fragments of the series $C_nH_{2n-1}O_2$ for $n = 3$ to 7 are indicated. The peak of composition $C_8H_5O_3$ (C/HO$_3$ plot of fig. 2) is derived from phthalate esters and the ion has Structure I.

A more detailed discussion of the common organic contaminants encountered, for example, in the Apollo lunar sample program, has been presented by Simoneit and Flory (ref. 9).

Mirror Wash

The summed high-resolution mass spectral data for the total benzene wash residue from the mirror are shown in figures 3 and 4. This residue consisted primarily of phthalate esters, hydrocarbons, carboxylic acids, and (in minor amounts) silicones and LM engine exhaust products. The hydrocarbon series found are C_nH_{2n+2} for $n = 3$ to 15, C_nH_{2n} for $n = 3$ to 14, C_nH_{2n-2} for $n = 3$ to 15, C_nH_{2n-4} for $n = 3$ to 16, and C_nH_{2n-6} to C_nH_{2n-10} for $n = 5$ to 12 (C/H plot of fig. 3). Not every member of each series is present and the more saturated series are the most abundant. A peak of composition $C_{19}H_{32}$ (possible androstane) is also found with a peak, $C_{18}H_{29}$, indicating loss of a methyl radical. One of the major components of the extract residue is dioctyl phthalate as substantiated by the peaks of compositions $C_8H_5O_3$ (Structure I) in the C/HO$_3$ plot of figure 3, $C_8H_7O_4$ (Structure II) and $C_{16}H_{23}O_4$ (Structure III), both in the C/HO$_4$ plot of figure 3. Dioctyl phthalate does not exhibit a molecular ion. Other phthalate esters were not detected. In the C/HO$_3$ data (fig. 3), there is evidence for two fragment ion series, $C_nH_{2n-3}O_3$ for $n = 5$ to 10 (probably derived from dicarboxylic acids), i.e., molecular ion (M$^+$) minus a hydroxyl radical (OH$\cdot$), and $C_nH_{2n-1}O_3$ for $n = 15$ and 17 to 20 of unknown derivation. There is a significant quantity of palmitic acid, $C_{16}H_{32}O_2$, present (C/HO$_2$ plot of fig. 3), as well as minor amounts of other carboxylic acids. These are discerned from the fragment ion series $C_nH_{2n-1}O_2$[M$^+$ minus a methyl radical (CH$_3\cdot$)] for $n = 1$ to 13, 15, and 18 (the lower homologs are probably rearrangement ions) and $C_nH_{2n-1}O$ (M$^+$—OH$\cdot$) for $n = 2$ to 16 and 18.

Silicone oil is indicated present by a series of peaks. The ions of compositions C_3H_9Si (Structure IV) in the C/HSi plot of figure 3, $C_5H_{15}OSi_2$ (Structure V) in the C/HOSi$_2$ plot of figure 3, and $C_7H_{21}O_2Si_3$ (Structure VI) in the C/HO$_2$Si$_3$ plot of figure 4 are derived from straight chain silicones. In the C/HO$_3$Si$_3$ plot of figure 4 are found two peaks of compositions $C_4H_{13}O_3Si_3$ (Structure VII) and $C_5H_{15}O_3Si_3$ (Structure VIII). These structures are derived from the cyclic silicones (e.g., Structure IX).

LM descent engine exhaust products are evi-

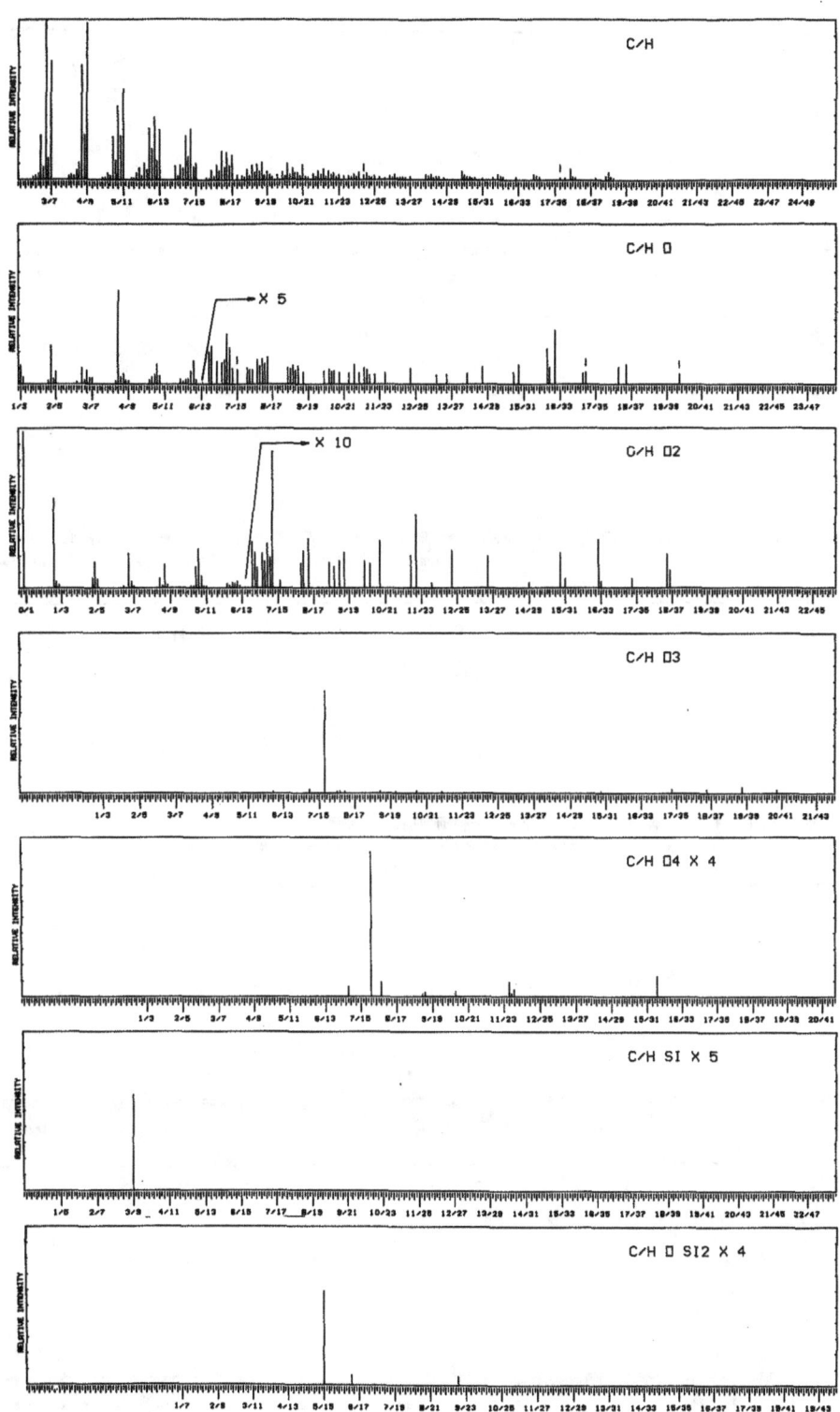

FIGURE 3.—Partial high-resolution mass spectral data for the benzene wash residue from the Surveyor 3 mirror.

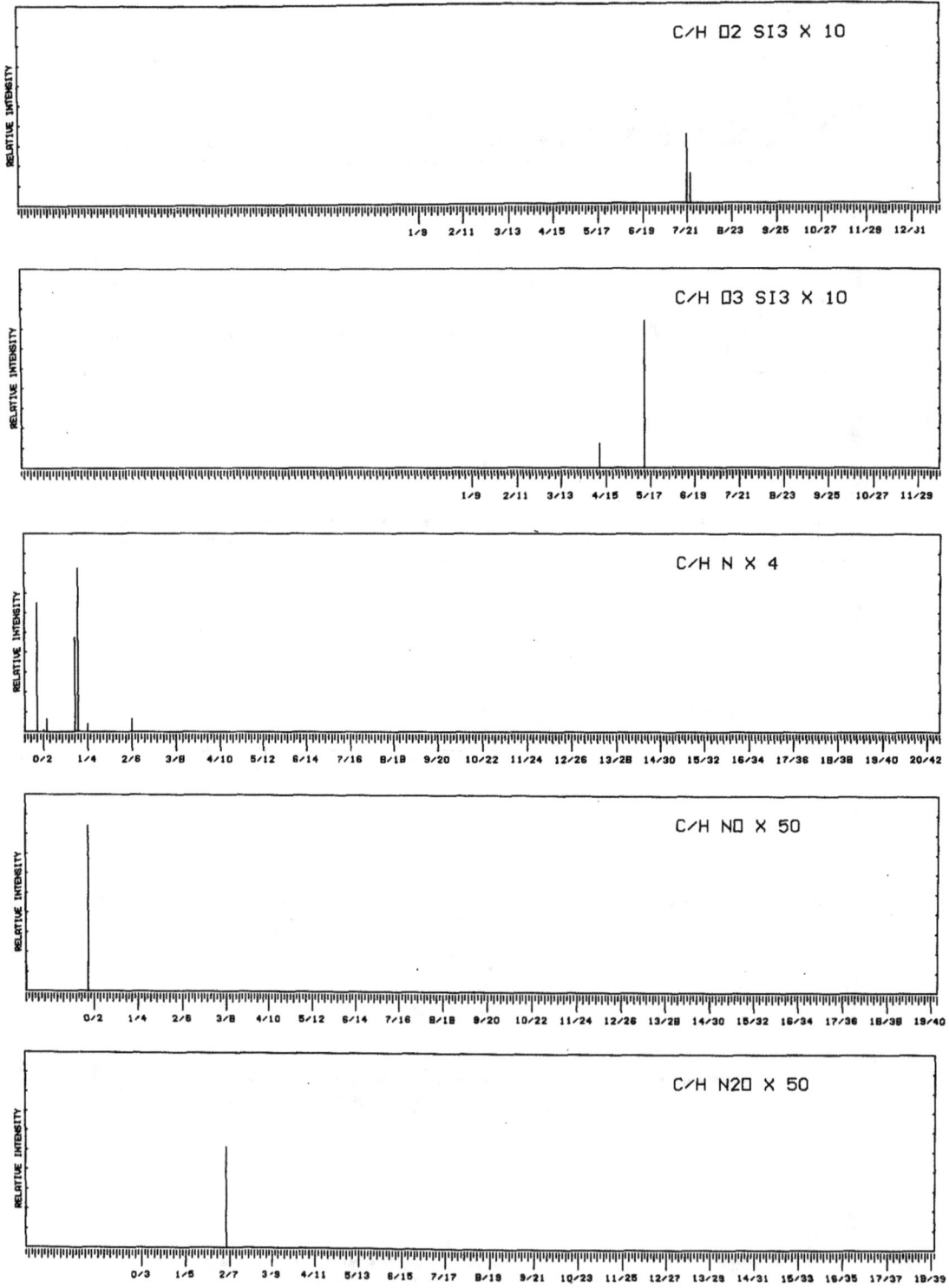

FIGURE 4.—Partial high-resolution mass spectral data for the benzene wash residue from the Surveyor 3 mirror.

IX

dent only in trace amounts probably because of the limited solubility of the salts and polar products in benzene. The C/HN plot of figure 4 exhibits ions of compositions NH_3 for ammonia; CN and HCN for hydrogen cyanide (the composition N is from N_2 in air); CH_4N (Structure X), a fragment from either methyl or dimethyl amine; and C_2H_6N (Structure XI), a fragment

X   XI

from either dimethyl or trimethyl amine, or possibly from dimethyl formamide, or from Structure XII. The C/HNO plot of figure 4 indicates a peak of composition NO (nitric oxide); in the C/HN_2O plot, a peak of composition $C_2H_6N_2O$ (Structure XII) is found. The latter is a partial oxidation product of unsymmetrical dimethyl hydrazine (UDMH, Structure XIII), the LM

XII   XIII

fuel. The nitric oxide is the major reduction product of the LM oxidizer, N_2O_4.

Middle Shroud Washes

The middle shroud was washed with benzene and methanol on the side toward the LM and on the side away from the LM. The summed high-resolution mass spectral data for the two wash residues are shown in figures 5 and 6 (toward LM) and 7 and 8 (away from LM). The data for the wash on the shroud toward the LM will be discussed in detail and the qualitative differences of the leeward sample will be covered. The major constituents found are hydrocarbons, dioctyl phthalate, and silicones.

The hydrocarbon series are C_nH_{2n+2} for $n = 2$ to 19, 21, and 23; C_nH_{2n} for $n = 2$ to 13; C_nH_{2n-2} for $n = 2$ to 14; C_nH_{2n-4} for $n = 2$ to 16; and in minor amounts C_nH_{2n-6} to C_nH_{2n-10} for $n = 6$ to 12. The presence of these series was deduced from the respective stronger fragment series due to $M^+-CH_3\cdot$ ions. The dioctyl phthalate is a significant constituent of the sample. The peaks of compositions $C_8H_5O_3$ (C/HO_3 plot of figure 5, Structure I), $C_8H_7O_4$ (C/HO_4 plot of figure 5, Structure II), and $C_{16}H_{23}O_4$ (C/HO_4 plot of figure 5, Structure IV) confirm this compound. Silicone oil is again found in considerable amount. The ions of compositions C_3H_9Si (Structure IV), C_2H_7Si (Structure XIV) in the

XIV

C/HSi plot of figure 5, $C_5H_{15}OSi$ (Structure V) in the C/$HOSi_2$ plot of figure 6, $C_7H_{21}O_2Si_3$ (Structure VI) in the C/HO_2Si_3 plot of figure 6 indicate the straight chain silicones. The peaks of compositions $C_4H_{13}O_3Si_3$ (Structure VII) and $C_5H_{15}O_3Si_3$ (Structure VIII) in the C/HO_3Si_3 plot of figure 6 are derived from the cyclic silicones. In this sample, there appears to be a larger quantity of the cyclic silicones than was the case for the mirror wash.

The minor components of this sample are nitrogenous and oxygenated compounds. A group of oxygenated peaks indicates a partially depolymerized vinyl alcohol and styrene copolymer. The peaks of compositions $C_{13}H_{19}O_3$ (Structure XV) in the C/HO_3 plot of figure 5, $C_{11}H_{15}O_2$ (Structure XVI) in the C/HO_2 plot of figure 5, $C_9H_{11}O$ (Structure XVII) in the C/HO plot of figure 5, and C_7H_7 (tropylium ion) in the C/H plot of figure 5 fit the following fragmentation pattern for the above copolymer (Structure XVIII; see refs. 10 and 11).

There are small amounts of free carboxylic acids present as the peak $(C_2H_4O_2)$ from the McLafferty rearrangement is rather strong, and the series $C_nH_{2n}O_2$ ranges from $n = 1$ to 6, with the $M^+-CH_3\cdot$ series, $C_nH_{2n-1}O_2$, ranging from $n = 1$ to 8.

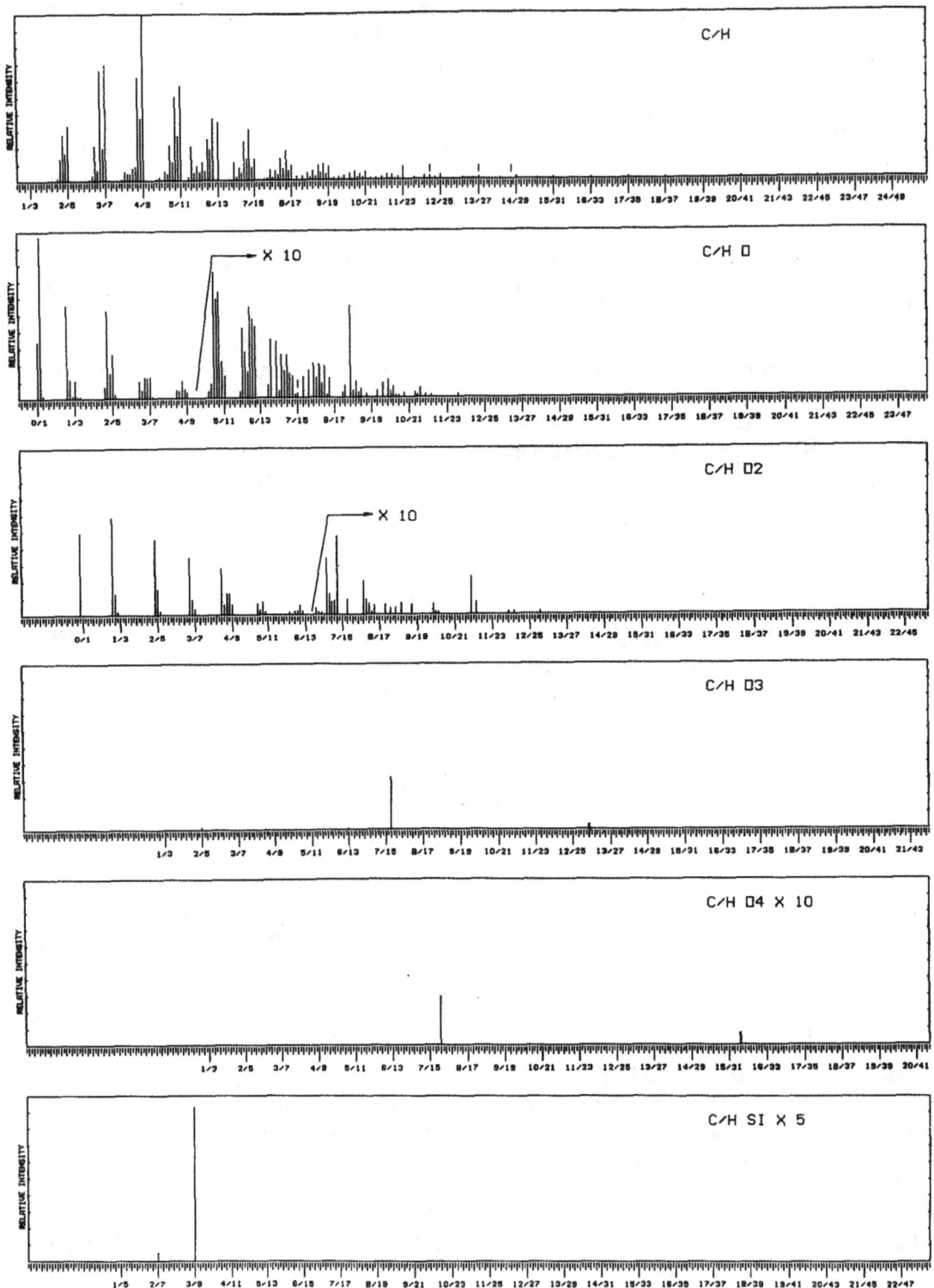

FIGURE 5.—Partial high-resolution mass spectral data for the benzene/methanol wash residue from the Surveyor 3 middle shroud facing toward the LM.

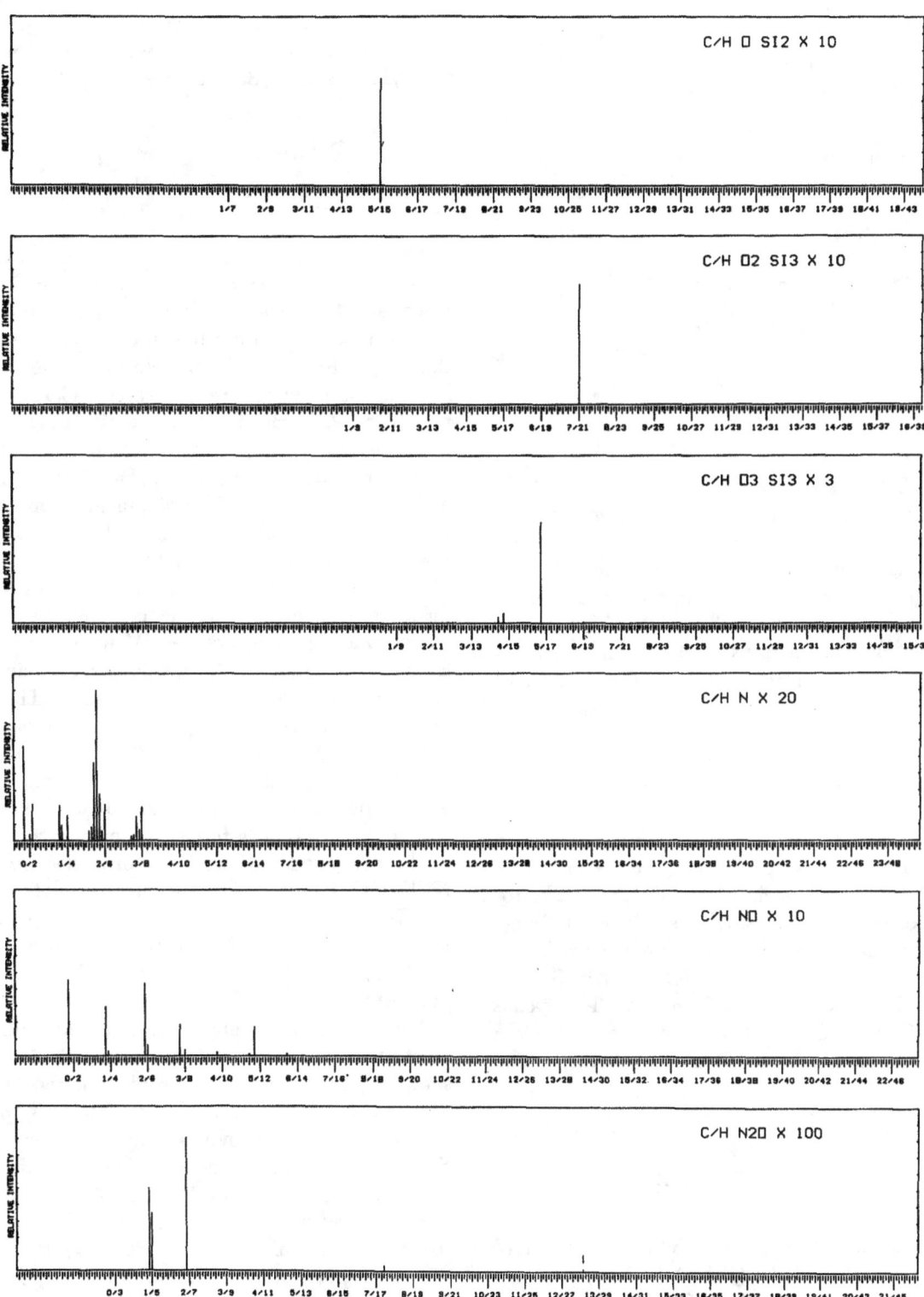

FIGURE 6.—Partial high-resolution mass spectral data for the benzene/methanol wash residue from the Surveyor 3 middle shroud facing toward the LM.

The LM descent engine products are a bit more varied in this sample as compared to the mirror wash, but still very minor. The C/HN plot of figure 6 exhibits ions of ammonia, hydrogen cyanide, acetonitrile, propionitrile, and fragments such as Structures X and XI. The groups of peaks of compositions $C_2H_{0-4}N$ and $C_3H_{4-6}N$ are various ions of acetonitrile (Structure XIX) and propionitrile (Structure XX), respectively.

$$CH_3-CN \qquad CH_3CH_2-CN$$
$$\text{XIX} \qquad\qquad \text{XX}$$

These two compound groups also were observed in the LM exhaust products (refs. 9 and 10). The C/HNO plot of figure 6 indicates peaks for nitric oxide, nitrosomethylene, nitrosomethane, N-hydroxyaziridine (Structure XXI; see ref. 10), and fragment series $C_nH_{2n}NO$ for $n = 3$ to 5 and $C_nH_{2n-2}NO$ for $n = 5$ and 6. The peaks C_3H_6NO and C_3H_8NO may be derived from dimethyl formamide (Structure XXII). The

C/HN$_2$O plot of figure 6 exhibits the peak of composition $C_2H_6N_2O$ (Structure XII) and the new peaks of compositions CH_4N_2O (Structure XXIII) and CH_5N_2O (a protonated species of Structure XXIII). These new peaks possibly

can be derived from the Surveyor 3 engine exhaust, since the fuel used was mainly monomethylhydrazine (Structure XXIV).

The sample away from the LM consists again mainly of hydrocarbons, dioctyl phthalate, and silicones. The overall total ionization for this sample is about double that for the sample from the opposite side of the shroud. The major hydrocarbon series are saturated: C_nH_{2n+2} for $n = 2$ to 29, C_nH_{2n} for $n = 2$ to 19, C_nH_{2n-2} for $n = 2$ to 16, C_nH_{2n-4} for $n = 2$ to 14, and (in minor amounts) C_nH_{2n-6} to C_nH_{2n-10} for approximately $n = 6$ to 13. The significant concentration of dioctyl phthalate is indicated by the peaks due to Structures I to III (C/HO$_3$ and C/HO$_4$ plot of fig. 7). The silicone oil is present in large amounts, as evidenced by the peaks corresponding to Structures IV to VI and XIV from the straight chain species and Structures VII and VIII from the cyclic species. The following compounds are present as minor constituents of the mixture, but relative to the other sample they are about double in concentration. The copolymer of vinyl alcohol and styrene (Structure XVIII) is found present by the same peaks discussed earlier, Structures XV through XVIII (C/HO, C/HO$_2$, and C/HO$_3$ plots of fig. 7). The carboxylic acids, $C_nH_{2n}O_2$, range from $n = 1$ to 9. In the C/HO plot of figure 7, the series $C_nH_{2n-11}O$ is evident for $n = 7, 11, 13, 15, 16,$ and 22.

The LM descent engine products are, for the most part, twice as abundant in this leeward sample, and essentially the same compounds are present. In the C/HN plot of figure 8, peaks are found for ammonia, hydrogen cyanide, Structures X and XI, acetonitrile and other C$_2$ species, propionitrile and other C$_3$ species, and the composition C_4H_8N. The C/HNO plot indicates a strong nitric oxide peak and the same group of peaks discussed earlier. The peaks of compositions $C_5H_{10}NO$ and $C_6H_{10}NO$ are rather intense and are thought to be reaction products of hydrocarbons on the shroud with LM exhaust

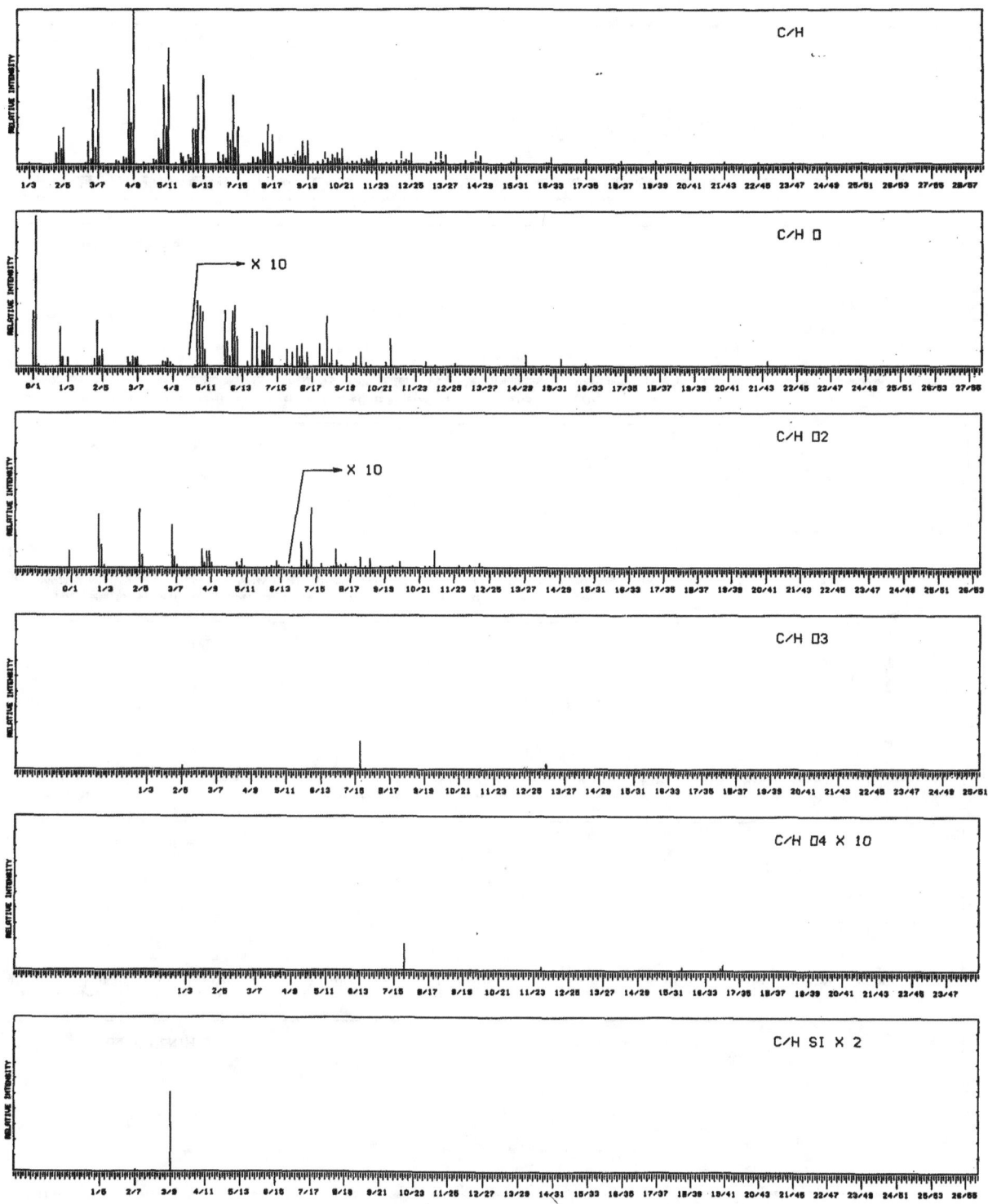

FIGURE 7.—Partial high-resolution mass spectral data for the benzene/methanol wash residue from the Surveyor 3 middle shroud facing away from the LM.

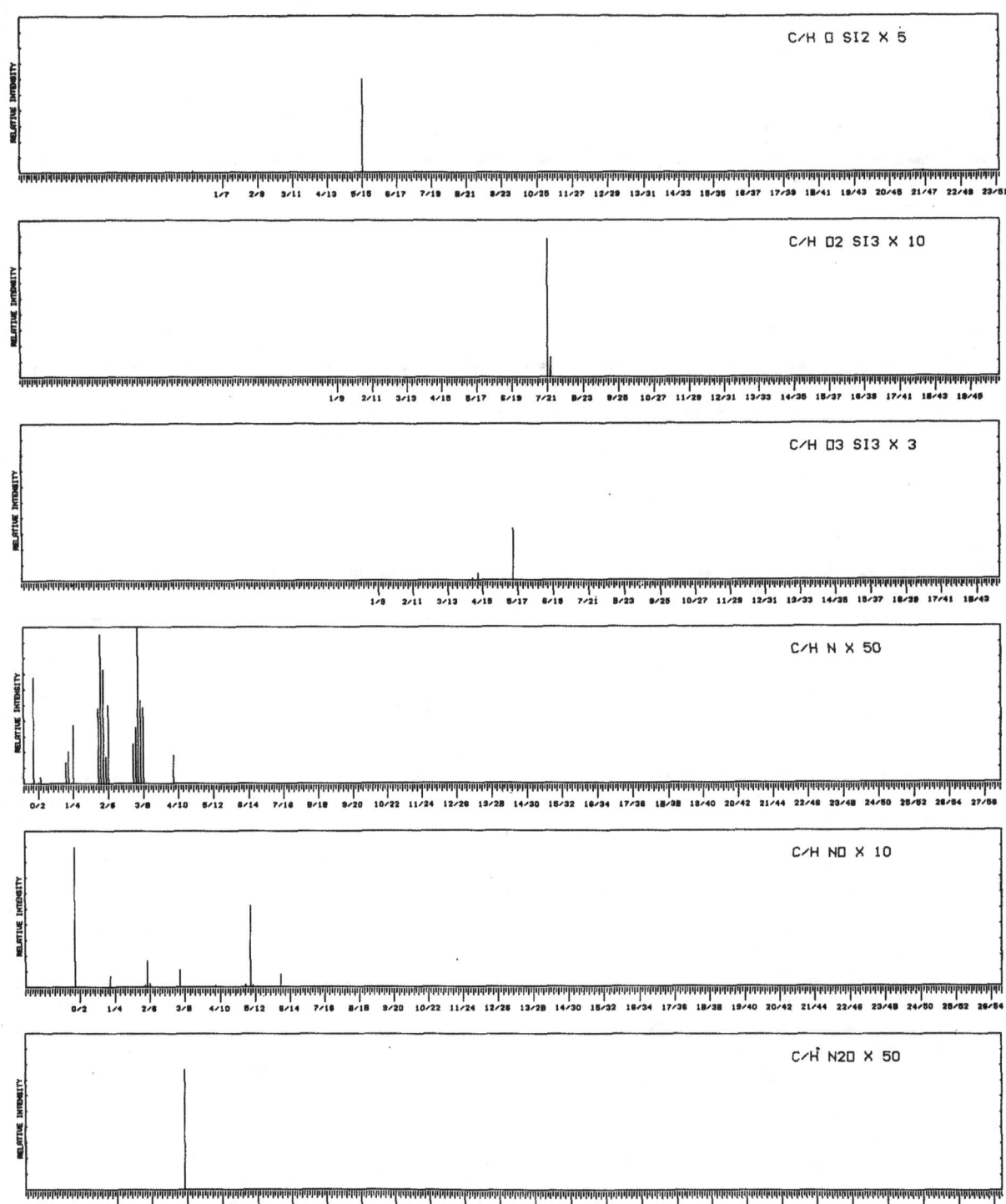

FIGURE 8.—Partial high-resolution mass spectral data for the benzene/methanol wash residue from the Surveyor 3 middle shroud facing away from the LM.

products (ref. 10). The C/HN_2O plot of figure 8 exhibits only a single peak fitting Structure XII and no trace of Structure XXIII.

Ion Microprobe Analyses

The IMA results are limited to the observation of primarily atomic species. The data for two runs on each sample are shown in figure 9. Sample 1008 is from the lower shroud on the side toward the LM and sample 1010 is from the same shroud on the side away from the LM. Both these chips have paint on them (kaolin and potassium silicate binder). Sample 1012 is an unpainted piece also from the lower shroud, but from a shielded area on the bottom. The main overall observation is the difference in the spectra of the unpainted chips vs. the two

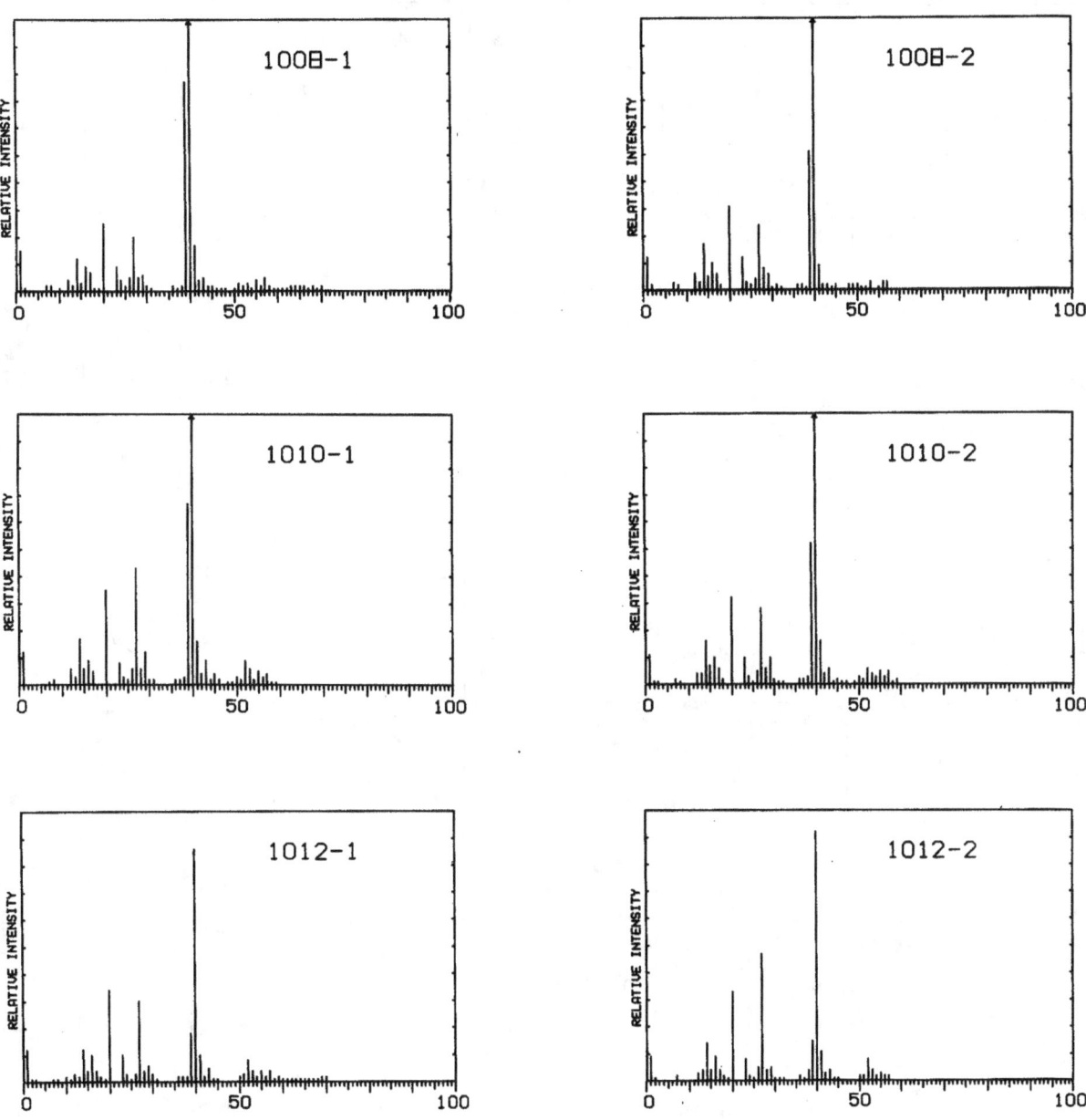

FIGURE 9.—Low-resolution ion microprobe analyzer spectra for chips 1008, 1010, and 1012 from the Surveyor 3 lower shroud.

painted chips. The peak at m/e 40 is argon, from the bombarding beam, and m/e 20 is mainly Ar^{2+}. To facilitate comparison of intensities, all ions other than Ar^+ and Ar^{2+} were summed and are reported as the percentage of the total ionization, excluding the bombardment ions. These data are also listed in table 1 with the terrestrial elemental abundances. There are

TABLE 1.—*Ion microprobe analysis of Surveyor 3 lower shroud chips*

Nominal mass	Element	Natural abundance	Element	Natural abundance	Total ionization, percent					
					1008-1	1008-2	1010-1	1010-2	1012-1	1012-2
1	H	99.985			5.55	5.43	5.83	4.62	4.08	4.26
2	H	.015			.37	.90	.49	.51		.39
3	H						.49			.39
4	He	100.00								
5										
6	Li	7.42								
7	Li	92.58			.74	1.36	.49	1.03	.34	.78
8					.74	.90	.49		.68	.39
9	Be	100.00								
10	B	19.78			.37		.97			
11	B	80.22					.49			.39
12	C	98.89			1.48	2.71	1.46	1.54	2.04	1.55
13	C	1.11			.74	1.36	.97	2.05	1.02	1.55
14	N	99.63			4.44	7.69	5.83	7.18	5.78	6.20
15	N	.37			1.11	2.26	1.94	2.05	2.04	2.71
16	O	99.76			3.33	4.52	4.85	4.62	2.72	3.88
17	O	.037			2.59	2.71	1.94	2.05	1.70	2.33
18	O	.209			.37	.90	.97	1.03		.78
19	F	100.00			.37		.49	.51		
20	Ne	90.92			(a)	(a)	(a)	(a)	(a)	(a)
21	Ne	.257								
22	Ne	8.82								
23	Na	100.00			3.33	5.43	4.85	4.10	2.72	3.88
24	Mg	78.70			1.48	1.36	1.46	1.03	1.02	1.62
25	Mg	10.13			.74	.90	.49	.51	.68	.39
26	Mg	11.17			1.85	1.81	1.46	2.56	2.04	1.94
27	Al	100.00			7.41	10.86	14.56	24.10	14.63	10.85
28	Si	92.21			1.85	3.62	1.94	2.05	2.04	2.33
29	Si	4.70			2.22	2.71	2.91	2.56	4.08	3.88
30	Si	3.09			.74	.45	1.46	.51	.68	.78
31	P	100.00			.37	.90	.49	.51	.68	.39
32	S	95.00				.45		.51		.39
33	S	.76								
34	S	4.22								
35			Cl	75.53						
36	S	.014	Ar	.337	.74	.90	.97	1.03	.68	.78
37			Cl	24.47	.37	.90	.97	.51	.68	.78
38			Ar	.063	.74	.45	.97	2.05	1.02	1.16
39	K	93.70			28.52	23.08	8.74	7.69	22.79	20.16
40	Ca	96.97	Ar	99.60	(a)	(a)	(a)	(a)	(a)	(a)
41	K	6.88			6.30	4.08	4.85	5.64	5.44	6.20
42	Ca	.64			1.48	.90	.97	1.54	1.36	1.55
43	Ca	.145			1.85	.90	2.43	2.05	3.06	2.33
44	Ca	2.06			.74	.45	.49	.51	.68	.39
45	Sc	100.00			.74	.90	.49	.51	1.36	.78
46	Ti	7.93			.37				.68	.39
47	Ti	7.28			.37					.39

TABLE 1.—*Ion microprobe analysis of Surveyor 3 lower shroud chips*—Concluded

Nominal mass	Element	Natural abundance	Element	Natural abundance	Total Ionization, percent					
					1008-1	1008-2	1010-1	1010-2	1012-1	1012-2
48	Ti	73.94			0.37	0.90			0.34	0.39
49	Ti	5.51				.90		1.03	.34	1.16
50	Ti	5.34	Cr	4.31	.37	.90	0.97	1.03	1.02	.78
51	V	99.76			1.11	.45	1.46	4.10	.68	2.33
52			Cr	83.76	.74	.45	3.88	2.05	3.06	1.55
53			Cr	9.55	1.11	1.36	1.94	1.03	2.04	1.16
54	Fe	5.82	Cr	2.38	.37		.97	1.54	.68	1.94
55			Mn	100.00	1.48	.45	1.94	1.03	1.70	1.16
56	Fe	91.66			.74	1.36	.97	1.03	1.02	1.94
57	Fe	2.19			1.85	1.36	1.94		1.36	.39
58	Fe	.33	Ni	67.88	.74		.49		.34	.78
59	Co	100.00			.37		.97		.34	
60			Ni	26.23	.37		.49			
61			Ni	1.19	.37		.49			
62			Ni	3.66	.37		.49			
63			Cu	69.09	.74		.49			
64	Zn	48.89	Ni	1.08	.74		.49			
65			Cu	30.91	.74		.49			
66	Zn	27.81			.74		.49			
67	Zn	4.11			.37		.49			
68	Zn	18.57			.74		.49			
69	Ga	60.4			.37		.97			
70	Zn	.62	Ge	20.53	.74		.97			
71	Ga	39.6								
72			Ge	27.43						
73			Ge	7.76						
74	Se	.87	Ge	36.54						
75	As	100.00								
76	Se	9.02	Ge	7.76						
77	Se	7.58								
78	Se	23.52								
79			Br	50.54						
80	Se	49.82								
81			Br	49.46						
82	Se	9.19								

ᵃ Not included in the summation.

slight differences between the respective first and second scans of each sample. Scans 1 are more representative of the surfaces; scans 2, taken considerably later, are more representative of the interiors of the samples. There appears to be no obvious general correlation of elemental abundance differences between the chips toward the LM and away from the LM. There is a difference in the aluminum (m/e 27), potassium (m/e 39), silicon (m/e 28), and oxygen (m/e 16) abundances, allowing a qualitative distinction to be made between the chips painted with kaolin ($H_2Al_2Si_2O_8 \cdot H_2O$) and potassium silicate binder and the uncoated chip.

Mirror Scrapings

The high-resolution mass spectral data for the three scraping samples (A, B, and C of fig. 1) from the mirror surface showed no peaks above instrument background. The sample probe was in the ion source at 350°C for approximately 10 min during each run, ample time for sample pyrolysis. It is suspected that the sample (about as fine as lunar dust and highly charged by static electricity) blew out of the well in the probe when it hit the vacuum of the pumps. Further work on another set of mirror scrapings is in progress.

Conclusions

Essentially the entire outside surface of the Surveyor 3 television camera was covered by lunar fines, probably from the following sources: Surveyor landing, lunar transport due to meteoroidal impacts, LM landing, and redistribution during return and subsequent handling. Thus, the evidence of LM and possibly Surveyor 3 descent engine exhaust products on the shroud and mirror was expected. The side toward the LM was heavily sandblasted and more discolored than the side away from the LM. This fact was not too well demonstrated by the organics isolated from the shroud, except that the total ion current for the leeward sample was almost double that of the sample from the side facing the LM.

The same types of organic molecules were found on the mirror and both shroud samples. The sources of the various organic contaminants (ref. 9) are as follows: hydrocarbons from lubricating oils and general terrestrial contamination; dioctyl phthalate probably from polyethylene bagging material (the plasticizer); carboxylic acids from decomposition of grease and general terrestrial contamination; silicones from sources such as lubricating oil; outgassing of electronics and plasticizer; vinyl alcohol and styrene copolymer probably from electronics insulation; and nitrogenous compounds from LM and possibly Surveyor 3 engine exhaust. The organic contamination levels do not seem to contribute to the discoloration of the various surfaces. Analyses for organic contaminants and identification of their sources, even if low in concentration, should be recognized as important criteria in the design of optical or other active instruments for future spacecraft.

References

1. MISSION EVALUATION TEAM: *Apollo 12 Mission Report*, MSC–01855, Houston, Tex., 1970.
2. *Surveyor Program Results*, NASA SP–184, Washington, D.C., 1969.
3. JAFFE, L. D.; AND RENNILSON, J. J.: *Surveyor Program Results*, NASA SP–184, Washington, D.C., 1969, pp. 407–412.
4. BURLINGAME, A. L.: "Data Acquisition, Processing and Interpretation *via* Coupled High Speed Real-time Digital Computer and High Resolution Mass Spectrometer Systems." *Advances in Mass Spectrometry*, vol. 4, The Institute of Petroleum, London, 1968, pp. 15–35.
5. BURLINGAME, A. L.: "Developments and Application of Real-time High Resolution Mass Spectrometry." International Conference on Mass Spectroscopy, Kyoto, Sept. 8–12, 1969, in *Recent Developments in Mass Spectrometry*, University of Tokyo Press, 1970, pp. 105–115.
6. BURLINGAME, A. L.; SMITH, D. M.; AND OLSEN, R. W.: "High Resolution Mass Spectrometry in Molecular Structure Studies. XIV. Real-time Data Acquisition, Processing and Display of High Resolution Mass Spectral Data." *Anal. Chem.*, vol. 40, 1968, pp. 13–19.
7. BURLINGAME, A. L.; SMITH, D. H.; MERREN, T. O.; AND OLSEN, R. W.: "Real-time High Resolution Mass Spectrometry," *Computers in Analytical Chemistry*, vol. 4, *Progress in Analytical Chemistry Series*, Plenum Press, New York, 1970, pp. 17–38.
8. BURLINGAME, A. L.; AND SMITH, D. H.: "High Resolution Mass Spectrometry in Molecular Structure Studies. II. Automated Heteroatomic Plotting as an Aid to the Presentation and Interpretation of High Resolution Mass Spectral Data." *Tetrahedron*, vol. 24, 1968, pp. 5749–5761.
9. SIMONEIT, B. R.; AND FLORY, D. A.: *Apollo 11, 12 and 13 Organic Contamination Monitoring History*, MSC 04350, Houston, Tex., 1971.
10. FLORY, D. A.; SIMONEIT, B. R.; BURLINGAME, A. L.; AND SMITH, I. D.: *Experimental Determination of Potential Lunar Surface Organic Contamination in the Lunar Module Descent Engine Exhaust*, MSC report (in press).
11. SIMONEIT, B. R.; BURLINGAME, A. L.; FLORY, D. A.; AND SMITH, I. D.: "Apollo Lunar Module Engine Exhaust Products." *Science*, vol. 166, 1969, pp. 733–738.

ACKNOWLEDGMENTS

We thank Mrs. E. Scott, Dr. E. Gelpi, Miss P. Wszolek, and F. C. Walls for technical assistance; and Miss M. Petrie for high-resolution mass spectrometry. We also thank N. Nickle (JPL) for valuable discussions; and K. Akiyama, Dr. T. Noda, S. Aoki (Hitachi, Ltd., Naka Works), and Dr. K. Sanbongi (Tohoku University, Sendai) for use of the ion microprobe analyzer.

VI. Micrometeorite Impact Analyses

PART A

PRIMARY AND SECONDARY MICROMETEOROID IMPACT RATE ON THE LUNAR SURFACE: A DIRECT MEASUREMENT

D. Brownlee, W. Bucher, and P. Hodge

The Surveyor 3 television camera was exposed to micrometeoroid bombardment for 31 months on the lunar surface before it was returned to the Earth by the Apollo 12 astronauts. The exposure time was almost an order of magnitude longer than that of any other man-made object ever returned from space for analysis. Determination of the number of micrometeoroid impact craters on the camera provides a unique opportunity to make a very sensitive direct measurement of the flux of interplanetary dust particles impacting the lunar surface.

To make a meaningful flux measurement with this technique, adequate surfaces are required. For many types of surfaces, the crater resulting from the impact of a micrometer-sized hypervelocity particle is highly characteristic and readily distinguishable from pits, particles, and other surface artifacts. Normally, a surface is required that is smooth, that produces distinctive craters, and that is relatively free of surface blemishes which could be confused with craters. Of the television camera surfaces, the optical parts (the mirror and optical filters) are the most appropriate for the detection of small impact craters. This article describes an investigation of the optical filters for micrometer-sized craters.

Filters

The camera contained four filters mounted in a rotatable filter wheel. The filter wheel was located directly below the mirror in a plane perpendicular to the camera axis. The camera axis was tilted 23.5° from the vertical in a direction N 43° W. The elevation of the filter wheel was about 5 m below the lunar terrain around the Surveyor crater. The upward-facing surfaces of the filters were exposed to impacts, but only from a restricted part of the sky because of partial shielding from the mirror and mirror hood. The open area of the hood pointed in the direction N 88° E. (See fig. 1.) The red, green, and clear flint filters were exposed to a segment of sky extending from about the lunar horizon to an elevation of 75°. The blue filter was completely shielded by the camera's internal components except during its brief use.

The filters are made of various types of glass; they measure 4.5 by 4.5 cm and are 0.3 cm thick. The quality of the surfaces is good and there are few scratches, pits, or other crater-like artifacts. When the filters were returned from the Moon, they were covered with a substantial

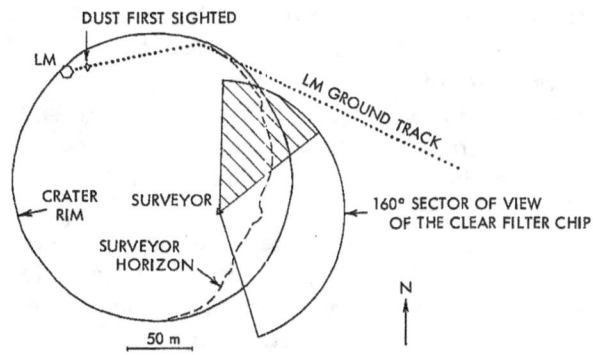

FIGURE 1.—Lunar Module and Surveyor 3 ground tracks. The cross-hatched part of the 160° sector indicates the range of azimuth angle for which the clear filter sees the lunar surface.

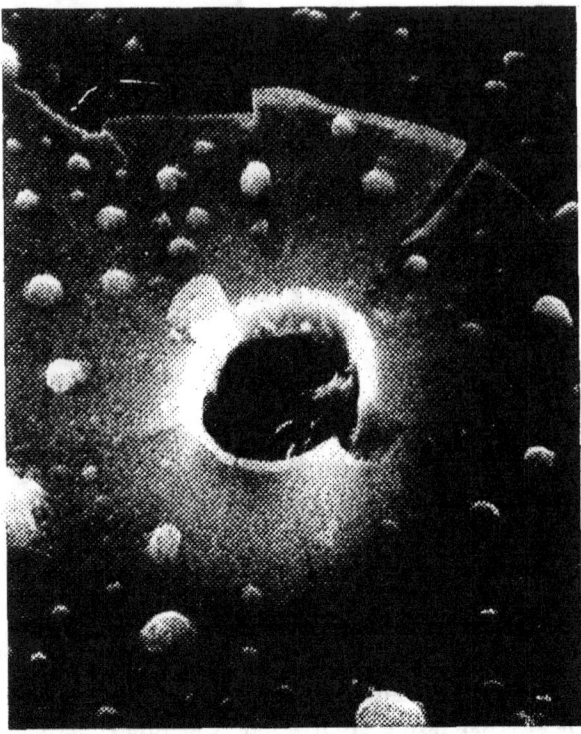

FIGURE 2.—SEM photograph of 3-μm crater produced in glass by a 6-km/sec carbonyl iron sphere.

amount of particulate matter, most probably deposited during Surveyor's landing in 1967. Before the filters were released for analysis, one-half of the top surface of each filter was cleaned of particulate matter at JPL using an acetate strippable film. After stripping, the cleaned portions of the filters were suitable for efficient detection of craters as small as 0.5 μm.

Because they are made of glass, which has cratering properties, the filters are highly suitable for crater searches. When micrometer-sized particles impact glass at velocities in excess of about 2 km/sec, the shock wave produced by the impact produces stresses in the glass exceeding its tensile strength, and extensive fracturing results. Typically, the result of a micrometer-sized medium-density particle impacting glass at 2 to 20 km/sec is a hemispherical cup with fairly smooth walls surrounded by a region of fractured glass. Figures 2 and 3 show scanning electron microscope and optical photographs of typical craters, produced in the laboratory by 6-km/sec iron spheres 1 μm in diameter. The morphology of the crater and surrounding spall zone is determined primarily by the velocity of impact, the angle of impact, and the physical properties of the particle (ref. 1). The fracturing of glass around the crater provides an excellent characteristic facilitating crater detection. Using an optical microscope equipped with upper illumination, light scattered off the fractures surrounding the craters enables efficient detection of craters at low magnification and distinction from particulate matter.

Optical Scan

The first study of the filters was an optical search for craters, conducted with microscopes in a laminar-flow, class 100 clean room. The filters were mounted on 5- by 7.5-cm microscope slides to facilitate handling and to establish a coordinate reference. The scanning for detectable craters was performed at 100 × magnification using a Zeiss GFL microscope operating with upper dark-field illumination. The stripped half of each filter was completely scanned at least once by three different microscopists. The object of the low-power scan was to locate all detectable fractures in the glass. Normally, glass fractures 10 μm and larger can be seen because of their light scattering properties. When a suspected glass fracture was observed, it was examined at higher magnifications. The dark-field illumination control which alters the azimuth angle of the illumination was varied; usually, a glass fracture could be distinguished from other surface features by the manner in which light reflected off the fractures. On difficult features, upper and lower bright-field illuminations also were used. By examining craters produced in glass at the Ames Research Center (ref. 2), it was determined that the scanning technique could reliably detect craters 5 μm or more in diameter.

The optical scan located about 10 glass fractures 10 μm and larger on the cleaned half of each filter. On a statistical basis alone, few of these fractures could be considered the result of hypervelocity impact. The greatest number of fractures was found on the blue filter which, because it was shielded from impact, must be considered a control. An examination also was

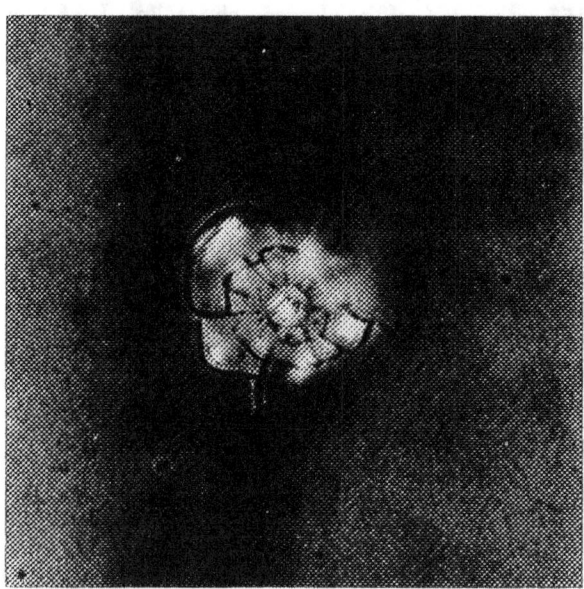

FIGURE 3.—Optical bright-field micrograph of a crater similar to that in figure 2.

made of a set of spare filters identical to the ones sent to the Moon. The same density of fractures was found on these controls as on the flight surfaces.

In the hope that some of the fractures were caused by craters, each one was extensively studied at high power with a Leitz Ortholux microscope and photographed with upper bright-field illumination at 500 × magnification. To be identified as an impact site, the fracture area was required to contain an area that resembled a cup-like crater (possibly greatly elongated) or the remnant of a crater partially spalled away. In at least 90 percent of the craters produced by using microparticle accelerators, the cup-like craters are easily identified. Of all fractures on the Surveyor filters, none contained an identifiable crater. Many of the fractures did, however, possess other features of hypervelocity impact. Many contained radial cracks and conchoidal fractures extending below the surface. These features, however, are not unique to hypervelocity impact and can be produced by simpler processes, for example, by pounding Carborundum grains into a glass surface. It is concluded, therefore, that all filter fractures that were detected in the scanning process are defects in the glass produced by polishing procedures or other processing techniques.

Scanning Electron Microscope Study

After the optical study, the filters were broken into smaller pieces for more destructive analyses. Fortunately, 50 mm^2 of the clear flint filter was saved for scanning electron microscope (SEM) work. Through the generosity of the Planetology Branch of Ames Research Center, we were given sufficient time using their SEM to study a large part of this filter and also to study some artificially produced craters to determine the crater detection limit on the scanning technique. The Surveyor piece of filter was first scanned at 1000 × magnification; possible crater sites were investigated at higher powers. Scanning of glass containing craters produced by a microparticle accelerator established that craters 1 μm in diameter and larger could be spotted reliably using the scanning technique. A second scan was done at 5000 × magnification with a detection limit of 0.2 μm. While scanning the filter, many items were found that were possibly results of low-velocity impact of lunar ejecta. The morphology transition from low-velocity craters to hypervelocity occurs at about the lunar escape velocity, so that impacts of extra-lunar particles can be distinguished from lunar ejecta (ref. 3). To be identified as a hypervelocity crater, an object was required to have at least some of the following properties: cup-like depression, signs of melting or flow within the cup, lip structure, and fracturing around the cup. No features were found that could be identified as hypervelocity impacts.

Low-Velocity Impacts

During the SEM scans, 28 dents were found in the glass; these dents are attributed to low-velocity impacts of particulate matter. The spatial density of these objects is about 200 cm^{-2} in the 0.5- to 10-μm size range. The dents usually are highly irregular and show plastic flow in the glass, suggesting low-velocity impacts of irregular particles. Typically, the dents are depressions in the glass which have a slightly raised rim at one end. The 1500-Å MgF$_2$ anti-reflection coating on the filter is usually chipped away at the edge of the pit near the raised rim. The asymmetrical character of the MgF$_2$ chipping and raised rim suggests low-angle impacts, which is consistent with the restricted angle of view of the filters. For most of the pits, the apparent vertical angle of impact was low enough that an azimuthal impact angle could be estimated from the position of the raised rim and chipped anti-reflection coating. Eighty-five percent of the azimuthal angles estimated for dents larger than 0.5 μm were within the 160° sector of sky not shielded by the camera hood. In view of the uncertainty of determining impact angles, this is considered an excellent correlation and indicates that the dents were produced by particles entering through the open area of the camera hood.

Figures 3 and 4 are SEM photographs of some of the low-velocity impacts. The scale of the pictures is 1.3 μm/cm except for the top two pictures in figure 4 for which the scale is 4.3 μm/cm. The range of possible angles of particle trajectories entering through the open area of the hood is about from the 9 o'clock position through 12 o'clock to the 3 o'clock position. One of the impacts that does not seem to have entered through this exposure window is shown in the lower-right picture in figure 4. This contradiction possibly can be explained as a high zenith angle impact or debris ejected from an impact within the camera hood. The two impacts at the top of figure 5 are the largest located in the SEM analysis and both appear to have been formed by high zenith angle impacts, as no azimuthal angle can obviously be assigned to them. It is believed that these two impacts are low velocity, but the hypervelocity impact of very-low-density particles cannot be completely eliminated. The impact on the right is of particular interest because X-ray analysis showed the existence of iron in its trough-like feature.

Two sources of low-velocity particles are:

(1) Ejecta resulting from meteoroid impact on the lunar surface.

(2) Particles blasted from the lunar surface by the Lunar Module (LM) descent engine.

Jaffe reported extensive sandblasting of the Surveyor by dust generated by the LM landing 155 m away (ref. 4; also see ch. IV, pt. I, of this document). Sharp shadows on the camera produced by this effect indicate that at least most of the dust generation occurred at the point of touchdown. The astronauts first reported seeing dust generation 25 m from the landing site. The camera hood prevented impacts of particles from either of these regions. The LM descent path did, however, pass in front of the exposure window of the filters. The closest approach was at a distance of about 110 m, with the LM at an altitude of about 80 m.

The 160° sector-shaped exposure window of the clear filter chip examined with the SEM and its relation to the LM descent ground track are shown in figure 1. The rim contour of the Surveyor crater, the horizon viewed by the television camera, and the final position of the mirror were taken from Shoemaker et al. (ref. 5), the position of the LM landing site from Jaffe (ref. 4), and the LM ground track from Nickle (ref. 6). The cross-hatched portion of the 160° sector is the range of azimuth angles for which the clear filter sees the lunar surface. This was determined using the relationship of horizon elevation vs. camera azimuth determined by Shoemaker et al. (ref. 5).

Particles generated on the surface along the LM ground track cannot hit the filters at high velocity unless they have a line-of-sight path. Line-of-sight paths exist only inside the azimuth sector where the filter looks at the ground and for points of generation inside the Surveyor horizon. Particles generated outside of this region can hit the filter, but only at velocities on the order of 50 m/sec or less, and cannot produce impact pits. As can be seen in figure 1, the LM ground track is beyond the horizon, except possibly for a small segment just east of north. If high-velocity particles from this region were important, many of the observed low-velocity impacts should aline with a 10° sector pointing north. The estimated azimuth angles of impact are evenly distributed within the 160° azimuth window; it is concluded that, if high-velocity ejecta were produced along the LM ground track, they were effectively blocked by the local terrain and did not hit the filters. The observed low-velocity impacts then must have

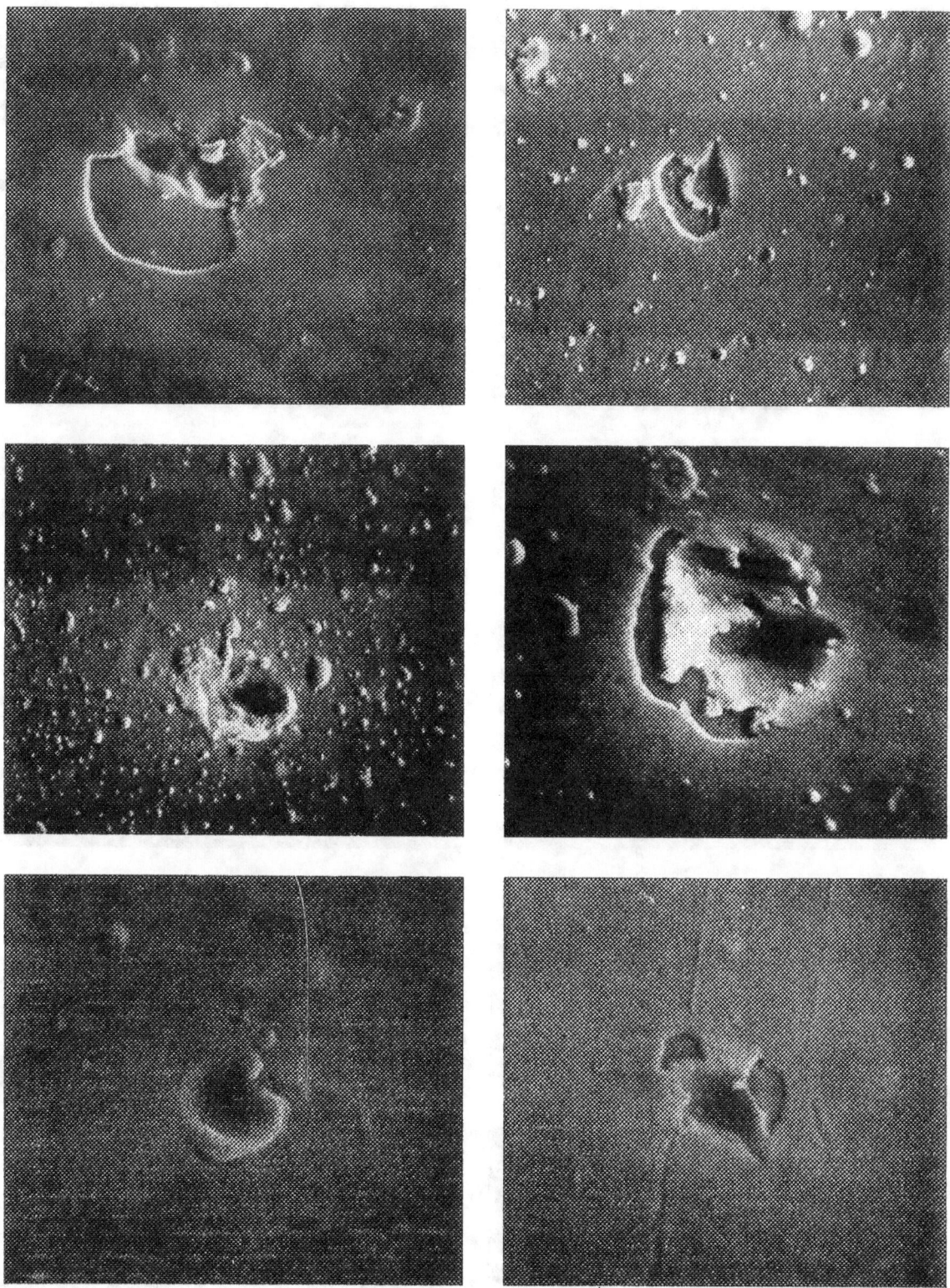

FIGURE 4.—Low-velocity impact craters.

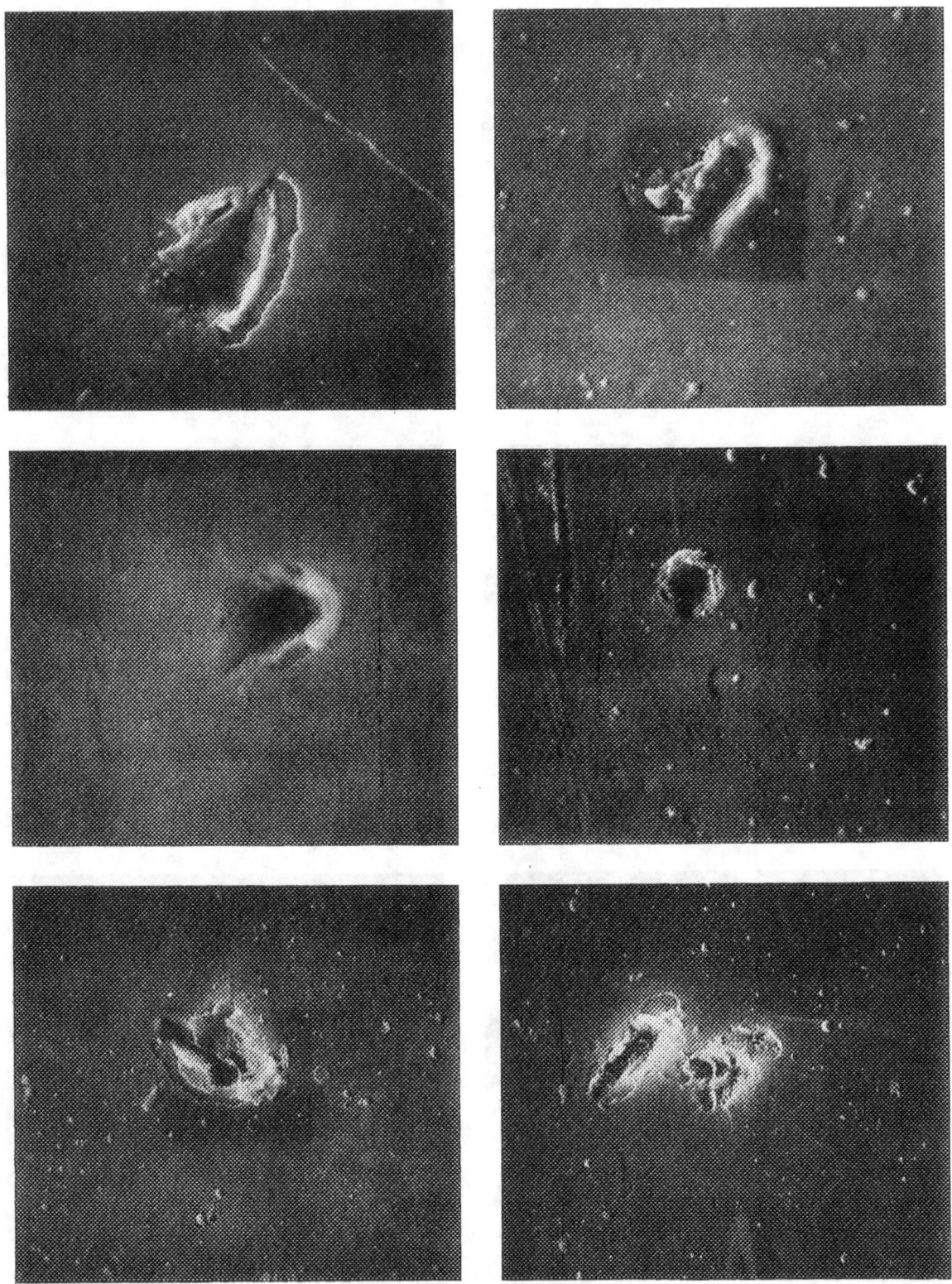

FIGURE 5.—Low-velocity impact craters.

been caused by ejecta produced by the impacts of meteoroids on the lunar surface.

Flux

No craters were found in either the optical or the SEM studies, so only upper limits to the flux can be established. To calculate these limits, the following assumptions are made:

(1) Particle density = 2.5 g cm^{-3} (COSPAR standard).

(2) The particle flux is isotropic. Although usually assumed, there is now strong evidence from Pioneers 8 and 9 (ref. 7) and from zodiacal light doppler shifts (ref. 8) that does not support this assumption, but the error produced by anisotropy does not justify a more sophisticated treatment.

(3) The ratio of crater diameter to projectile diameter is 1.6. This ratio was determined with the help of James Vedder in connection with his analysis of craters produced in soda lime glass with his microparticle accelerator at the Ames Research Center. A study was made on craters in the 1- to 5-μm size that were produced by glass, aluminum, and polystyrene spheres of measured mass and velocity. For particles in the 5- to 10-km/sec velocity range, the crater:particle ratio given was representative. Calibrations using particles of this density probably are more realistic than those using conventional iron spheres.

(4) No hypervelocity impact craters, larger than the detection limits, exist on the scanned surfaces, thus giving only an upper limit. The assumption here is that natural craters are similar to those produced artificially with microparticle accelerators. The predicted crater characteristics used here are based on craters produced by homogeneous spheres. Vedder (ref. 1) has shown that particle shape does affect crater morphology. Little is known about the effects of nonhomogeneity, unusual shape, or low density.

The flux computation was performed by taking the reciprocal of the time-area product (TAP) of the surfaces. This method assumes a 63-percent probability of having one impact (ref. 9). The TAP was computed in the following manner:

$$TAP = TAK_\Omega K_\theta K_D$$

where

T = exposure time
A = area examined, cm^2
$K_\Omega = (2\pi)^{-1}$ (solid angle of sky seen by the filter)[1]
$K_\theta = \cos\theta$, where θ is the average incidence angle of possible impact on the filter[1]
K_D = fraction of the filter not covered by dust

For the three filters exposed to impact, the following factors were used:

Filter	A	K	K$_\theta$	K
Clear flint	9.1	0.35	0.60	0.5
Red	10.0	.078	.42	.7
Green	9.8	.116	.56	.7

Flux for the Optical Scan

Summing the time-area products for the three filters yields an upper limit to the cumulative flux of 7.5×10^{-5} particle m^{-2} sec^{-1} (2π sterad)$^{-1}$. The crater detection limit of 5-μm diameter implies that this limit is for particle masses 2×10^{-11} g and larger.

Flux for the SEM Scan

A total of 10.86 mm^2 of surface was scanned in the 1000 $\times$ magnification scan with a detection limit of 1 μm. The computed flux limit is 1.1×10^{-2} particle m^{-2} sec^{-1} (2π sterad)$^{-1}$ for masses 2×10^{-13} g and larger. The 5000 $\times$ scan covered 0.27 mm^2 with a detection limit of 0.2 μm. The computed flux is 4.5×10^{-1} particle m^{-2} sec^{-1} (2π sterad)$^{-1}$ for masses 2.5×10^{-15} g and larger.

Conclusions

Because of flexibility in analysis, recoverable crater collection experiments are subject to fewer uncertainties in detection of impacts than are remote sensing experiments. Studies like this one and S-10 and S-12 experiments flown on Gemini (ref. 10) provide a permanent record of impact

---

[1] Derived from data provided by Neil Nickle of the Jet Propulsion Laboratory.

events that can be analyzed under laboratory conditions to yield information on particle mass, density, shape, chemical composition, and velocity. Crater collection experiments also record impacts of particles too small or of too low density to register on existing remote sensing experiments. Because of the low-density sensitivity, it is reassuring that the derived optical upper limit to the flux is consistent with the models of both Kerridge (ref. 11) and McDonnell (ref. 12), which are primarily based on remote sensing measurements.

In figure 6, the SEM and optical points represent the upper limits derived in this article. The dashed line is the 1963 average of satellite microphone data (ref. 9) and is included for historical comparison. The "footprint" point is a flux derived from analysis of a Surveyor 3 footprint (ref. 13). The line marked "Kerridge" is an average of experimental data selected by him as reliable measurements of the flux at 1 AU (ref. 11). The line labeled "McDonnell" is a model of the flux at the lunar surface based on controlled experiments (ref. 12).

The SEM points are important because they represent a direct and accurate measurement in a mass range that has not been investigated by means of many other experiments. Pioneers 8 and 9 (ref. 7) indicate a particle cutoff at about 10^{-11} g. A cutoff at this mass is of considerable interest because it corresponds to the dynamical radiation pressure cutoff for particles generated by short-period comets (ref. 14). The discovery by Neukum et al. (ref. 3) of craters on lunar spherules produced by submicrometer-size particles contradicts this cutoff. It is hoped that additional SEM work will yield increased sensitivity and provide additional information on this interesting submicrometer-size particle regime.

Because of the orientation and shielding of the almost horizontal optical filters, they provide a rather unique measurement of the flux of secondary particles impacting the lunar surface. The exposure of the filters to the ground was slight; most of the impacts probably were produced by particles in the 0.3- to 2-km/sec velocity range produced at great distances from the Surveyor crater. The measured rate of secondary impact crater formation on glass is approximately 800 impacts cm^{-2} yr^{-1} (2π sterad)$^{-1}$ for impacts 1 μm

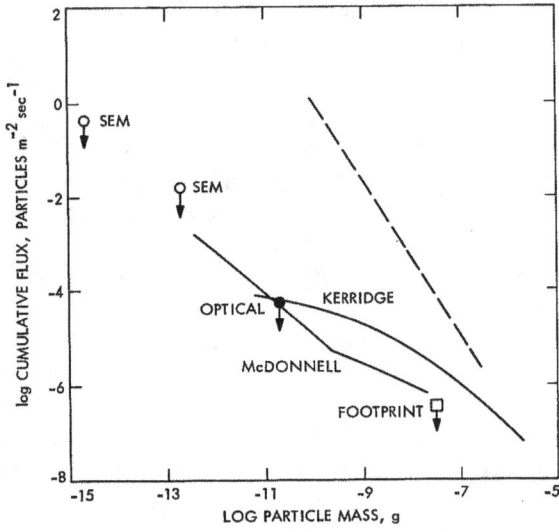

FIGURE 6.—Flux plot.

and larger. This rate is about 10^3 times higher than the crater formation rate expected for primary micrometeoroids estimated using Kerridge's flux curve (fig. 1).

References

1. VEDDER, J. F.: "Microcraters in Glass and Minerals." Paper submitted to *Earth and Planetary Science Letters*, 1971.
2. CUNNINGHAM, B. E.; AND EDDY, R. E.: "Space Environment Simulator for Studies of the Effects of Space Environment on Materials." *Proceedings of the AIAA/IES/ASTM Space Simulation Conference*, 1966, pp. 161–165.
3. NEUKUM, G.; MEHL, A.; FECHTIG, H.; AND ZÄHRINGER, J.: "Impact Phenomena of Micrometeorites on Lunar Surface Material." *Earth and Planetary Science Letters*, vol. 8, 1970, pp. 31–35.
4. JAFFE, L. D.: "Blowing of Lunar Soil by Apollo 12: Surveyor III Evidence." *Proceedings of the Apollo 12 Lunar Science Conference*. To be published.
5. SHOEMAKER, E. M.; BATSON, R. M.; HOLT, H. E.; MORRIS, E. C.; RENNILSON, J. J.; AND WHITAKER, E. A.: "Television Observations From Surveyor III." *J. Geophys. Res.*, vol. 73, 1968, pp. 3989–4043.
6. NICKLE, N.: "Surveyor III Material Analysis Program." *Proceedings of the Apollo 12 Lunar Science Conference*. To be published.
7. BERG, O. E.; AND GERLOFF, U.: *More Than Two Years of Micrometeorite Data From Two Pioneer Satellites*. Paper presented at 13th COSPAR Meeting, Leningrad, May 20–28, 1970.

8. REAY, N. K.; AND RING, J.: "Radial Velocity Measurements on the Zodiacal Light Spectrum." *Nature*, vol. 219, 1968, p. 710.
9. ALEXANDER, W. M.; MCCRACKEN, C. W.; SECRETAN, L.; AND BERG, O. E.: "Review of Direct Measurements of Interplanetary Dust From Satellites and Probes." *Space Research III*, 1963, pp. 891–917.
10. HEMENWAY, C. L.; HALLGREN, D. S.; AND KERRIDGE, J. F.: "Results From Gemini S–10 and S–12 Micrometeorite Experiments." *Space Research VIII*, 1968, pp. 521–535.
11. KERRIDGE, J. F.: "Micrometeorite Environment at the Earth's Orbit." *Nature*, vol. 228, 1970, pp. 616–619.
12. MCDONNELL, J. S. M.: *Review of Insitu Measurements of Cosmic Dust Particles in Space.* Paper presented at 13th COSPAR Meeting, Leningrad, May 20–28, 1970.
13. JAFFE, L. D.: "Lunar Surface: Changes in 31 Months and Micrometeoroid Flux." *Science*, vol. 170, 1970, pp. 1092–1094.
14. HARWIT, M.: "Origins of the Zodiacal Dust Cloud." *J. Geophys. Res.*, vol. 68, 1963, pp. 2171–2180.

PART B

PRELIMINARY EXAMINATION OF SURVEYOR 3 COMPONENTS FOR IMPACT PHENOMENA

D. S. Hallgren, A. T. Laudate, R. P. Schwarz, W. D. Radigan, and C. L. Hemenway

Two pieces of the Surveyor 3 spacecraft recovered during the Apollo 12 mission have been examined for secondary lunar ejecta and micrometeorite impacts. One piece was section E of the aluminum strut; the other was the nickel-coated beryllium television camera mirror. These pieces were exposed to the lunar environment for 31 months. Although both pieces, especially the mirror, had carefully prepared surfaces, examination of the surfaces was compromised to some extent because the vernier descent engines failed to shut down at the proper time in the landing sequence, causing the spacecraft to bounce twice before settling to its final resting place on the lunar surface. The failure of the engines resulted in considerably more contamination of the spacecraft by fine lunar debris than was expected.

Tube Section E

Section E is a piece of aluminum tubing 1.25 cm wide and about 2.5 cm long. Before this sample was received, it had been cleaned to the extent that all loose surface material was removed. A scribed line indicates the general area that, most probably, was facing away from the lunar surface. Light optical examination of the tube at 625 to 1250 × magnification revealed a high concentration of objects which, within the resolution of the light microscope, appeared to be craters. To clearly define these objects, the tube section was mounted in the Stereoscan scanning electron microscope. This examination showed immediately that the objects found in the light microscope were not classical hyperballistic impacts. The craters found can be classified in three categories: (1) round craters with a minimum lip, steep sides, and no debris within the crater; (2) same as (1), but with debris in the crater; and (3) craters that are more shallow and less steep sided. Figures 1 through 3 are representative examples of categories (1) through (3), respectively.

A sample of tubing was provided which was fabricated to the same specification as the flight tube. Examination of this control sample in the Stereoscan showed that the structures shown in figures 1 through 3 were not due to the manufacturing processes. Through the courtesy of Otto Berg of Goddard Space Flight Center, firings were made on the control sample to simulate micrometeorite impacts. Two series of firings were made; these firings included velocities be-

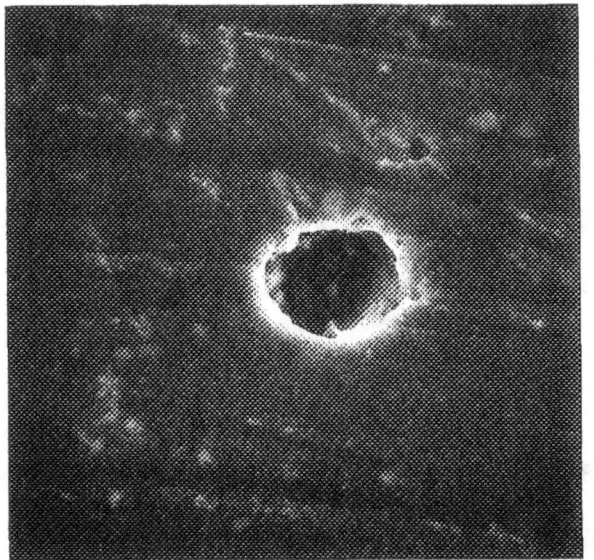

Figure 1.—Crater 4 μm in diameter.

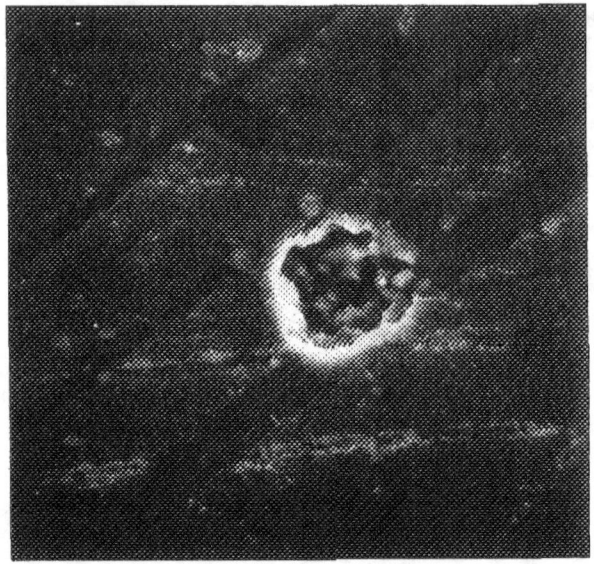

Figure 3.—Crater 2.5 μm in diameter.

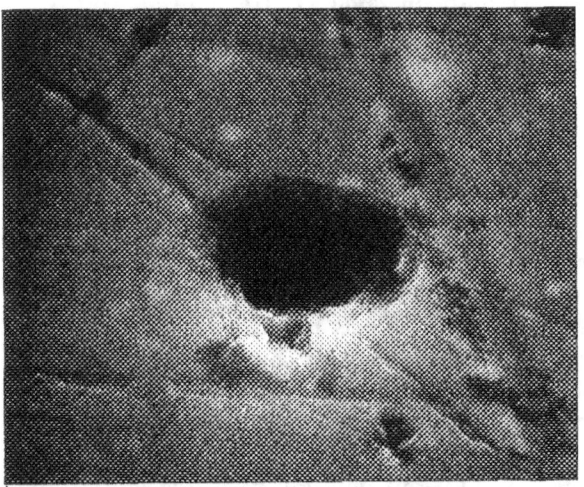

Figure 2.—Crater 4 μm in diameter.

Figure 4.—A 2.5-μm-wide crater in control sample. The crater was formed by iron particles at a velocity of 3 to 6 km/sec.

tween 2 and 3 km/sec and 3 and 6 km/sec. The projectiles used were carbonyl iron spheres. Typical examples of the results of these firings are shown in figures 4 and 5. The appearance of these impacts suggests that the structures observed on the Surveyor 3 sample were produced by particles with less energy than the simulation particles. The impacting particles would have to have been moving with a velocity less than 2 km/sec or have been of lower density and strength than the iron particles used in the simulation. From the structures observed so far, it is difficult to distinguish lunar ejecta craters from micrometeorite impact craters.

On an area of 0.45 mm², 36 craters of the type shown in figure 1 were observed. The impacts range in size from 1 to 4.5 μm, with a peak in the size distribution between 2 and 3 μm.

Television Camera Mirror

The mirror was heavily contaminated with lunar soil, which interfered somewhat with the

FIGURE 5.—A 3-μm-wide crater in control sample. The crater was formed by iron particles at a velocity of 3 to 6 km/sec.

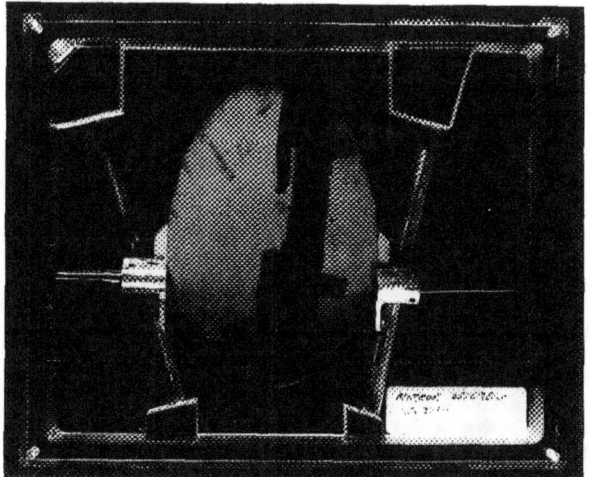

FIGURE 6.—Surveyor 3 mirror assembly.

light microscopy. As shown in figure 6, some areas of the mirror have been cleaned to some extent by replicating the surface. The size of the mirror limited our studies to light microscopy. Even with the high levels of contamination present in most areas, the mirror was a much better surface for study of impacts than the aluminum tube because it had been carefully polished before flight. Examination of the surface, using bright field and 625 to 1250 × magnifications, revealed circular craters with raised lips at the rate of approximately one crater for 2 mm² for the cleaned areas. A 52-mm² area contained 23 impact sites.

It should be noted that the evaluation of these impacts is limited by the resolution of the optical microscope. It is highly recommended that several small sections be cut from this mirror so that they can be examined with higher resolutions in a scanning electron microscope. Probe analyses of the interior of the craters and accurate measurements of the crater morphology may allow determination of the masses and velocities of the impacting particles. This would help to identify the origin of these interesting craters and provide an important and accurate determination of the flux of micrometeorites on the lunar surface.

PART C

EXAMINATION OF SAMPLE OF SURVEYOR 3 STRUT FOR METEOROID IMPACTS

L. Zernow

A 3.8-cm-long sample of the bare 1.3-cm-wide, hollow aluminum alloy strut returned from Surveyor 3 and an equivalent terrestrial sample were examined optically to determine whether any meteoroid impact data could be obtained.

A coupled microscope was assembled with two degrees of translational freedom and one degree of rotational freedom. This equipment permitted large-field, low-power optical scanning (10 to 80 ×) to be transferred to higher magnification scanning (315 ×) at the same centerline site.

Three candidate "craters" were found on the

initial low-power scan. However, more careful examination under high magnification eliminated all three of them as potential hypervelocity impact sites.

The conclusion drawn is that the sample does not reveal any unequivocally identifiable meteoroid impacts, within the magnification and resolution limits used.

PART D

SURFACE STUDIES ON SURVEYOR 3 TUBING SECTIONS

E. A. Buvinger

Two sections (each about 2.5 cm long) of the unpainted, polished aluminum tubing from the strut of the radar altimeter and doppler velocity sensor (RADVS) on the Surveyor 3 spacecraft have been examined in a transmission electron microscope using replication techniques. Section C was received first; section E was received about 3 months later, along with a piece of unused tubing for comparison purposes. The section of comparison tubing, which had been prepared in the same manner as the Surveyor tubing, was useful in determining positively that certain features could be ascribed to polishing and handling procedures.

The purpose of this investigation was to determine the type and degree of microscope surface damage that the tubing incurred during its exposure to the lunar environment. Specifically, the surface was examined for evidence of ion bombardment (sputtering) and micrometeorite damage.

Experimental

Upon receipt, the tubing sections were photographed for record and then washed with acetone to remove possible traces of residue from the soil-peel procedures used by previous investigators. For the replication process, elvanol (polyvinyl alcohol) proved to be the most satisfactory material. The replication procedure used was as follows: a stripe of elvanol (15 percent solution) was dropped along the upper surface of the tubing, was dried, and then was stripped from the tubing. This elvanol stripe provided a "negative" replica of the tube surface. The replica was then shadowed for contrast by coating it with a heavy metal (in this case, platinum) at an oblique angle. A 300- to 500-Å coating of carbon was deposited over the entire surface of the replica to provide support. The thin metal film then was cut into pieces about ⅛ in. square, and the original underlying plastic replica was dissolved. The squares were picked up on 200-mesh grids. The thin-film replica squares were examined in the electron microscope. This procedure was repeated until the entire surface of the tubing had been replicated at least once.

The purpose of metal shadowing at an oblique angle is to cause the relatively higher portions of the plastic replica to shield the areas behind them from the metal. This provides contrast for transmission electron microscopy. The areas in which the metal is thickest will scatter the most electrons, thus causing that area to appear darker on the phosphor viewing screen (or photographic print). Conversely, the areas shielded from the metal will appear bright on photographic prints. It is important to remember that the replica examined is a negative of the original tube surface. Thus, for example, microcraters in the tubing surface will appear in photographs to be rising above the tubing surface. The microcrater will cast a shadow with a length proportional to the depth of the original crater. Most replicas were shadowed at 25°. Some shadowing, however, was performed at 10° to enhance the fine details of the surface structure.

Discussion

Both sections of tubing, when received, were contaminated on one side. One-half to two-thirds of the circumference of the tubing had a dull appearance to the naked eye, while the remainder of the surface appeared bright, as expected with polished aluminum. The replicating plastic lifted much of the contaminant, which appeared globular under low-power optical microscopy. The areas in which the first replicas had been lifted from the dull portion of the tubing were clearly discernible. Large pieces of contaminant that were embedded in the replica could not be supported by the thin metal film when the plastic was dissolved; therefore, they fell away. Smaller pieces adhered to the film and could be observed in the electron microscope. The material was generally opaque to 100-kV electrons. The particles range from a few hundred angstroms to more than 1 μm in size.

The surface of the tubing in the same areas from which the contaminants were removed appears to be eroded extensively. Although this was true for both sections, it was particularly evident in section E. The most likely explanation for this phenomenon would be sandblasting by lunar dust, which could have occurred during the extended landing maneuver of Surveyor 3. The small particles would be lunar debris, which were either embedded in the tubing or which adhered to it with the aid of plume contamination from the retrorockets. The membrane-like material that covers most of the area shown in figure 1 (section C) is assumed to be retrorocket contamination. Although some of this material was also present in the contaminant removed from section E (fig. 2), it was not as prevalent.

Transmission electron diffraction was attempted on the contaminant particles, but meaningful results could not be obtained. The particle thickness exceeds that penetrable by 100-kV electrons. However, the contaminants are most likely a combination of organic material deposited during retro-fire and lunar debris. All of the contaminant material was stable in the electron beam. A comparison of material removed from sections C and E may be seen in figures 1 and 2. Section C contaminants are not so massive as those from section E and contain appreciably more of the membrane-like contamination and less solid particulate material.

The bright parts of the tubing in particular show extensive polishing-and-handling scratches which were apparent even under low-power optical microscopy. Upon examination in the electron microscope, some of these features were so extreme as to cause the thin-film replica to tear, destroying the areas involved. In general, the surface appearance of the bright portions was similar for both sections (fig. 3).

Some microcraters were observed in the bright parts of both sections. More were found from section C than from section E; however, as a

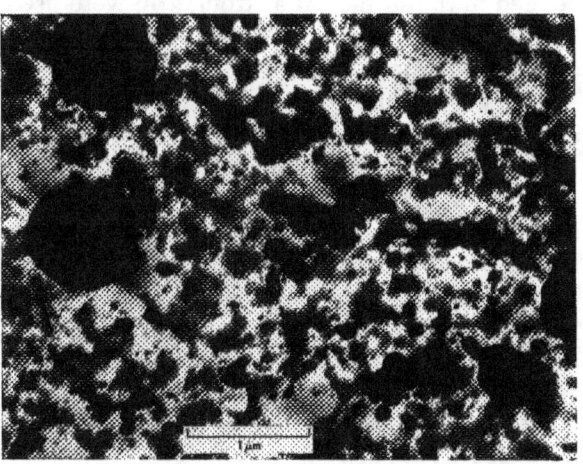

FIGURE 1.—Lower surface of section C.

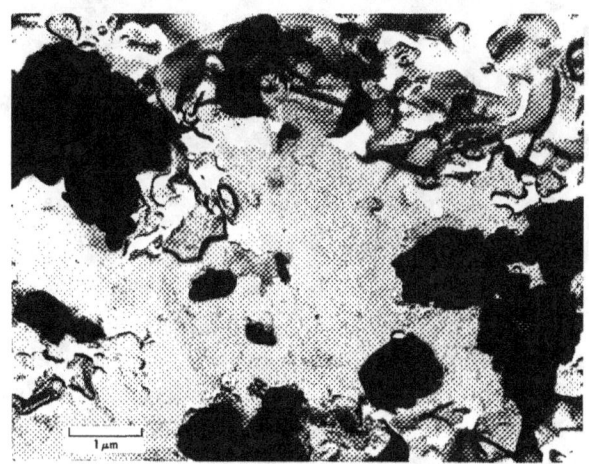

FIGURE 2.—Lower surface of section E.

more thorough examination was conducted on section C, this could account for the apparent difference.

One of the microcraters found was unique (fig. 4). It exhibits a disturbed, raised area around its circumference and a relatively smooth central pit as defined by the shadow. The crater is about 1 μm across its smaller diameter and is about 0.7 μm deep. This microcrater resembles artificial impact craters in aluminum, as described by Weihrauch et al. (ref. 1). Craters with similar features were found in lunar materials by Devaney and Evans (ref. 2).

In view of the many unknown factors involved (temperature at the time of formation, grain structure of the immediate area, size of the impinging particle, etc.), it is difficult to make absolute judgments concerning the cause of the microcrater. However, in the opinion of the author, it is probably the result of a hypervelocity impact.

Figure 5 shows a different type of microcrater. The area around the crater shows moderate disturbance but no definite splash lip. The crater it-

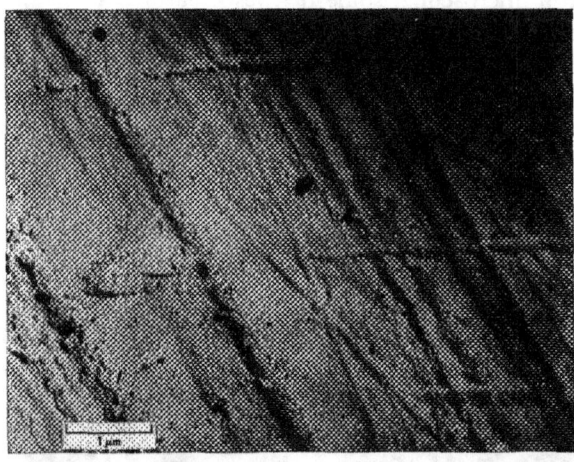

FIGURE 3.—Upper surface of section C. Photograph shows polishing-and-handling scratches common to both sections C and E.

self has a shallow portion around its periphery and a deeper central core with a relatively smooth bottom. The outer edge of the crater is about 1.2 μm in diameter; the maximum depth of the crater is about 1.3 μm.

Another type of crater found is typified by figure 6. The opaque central area is composed of particulate material. The outer diameter of this crater is about 1.3 μm, and its depth is about 0.6 μm. No definite area of disturbed material is apparent around the outside of the crater. The walls of the crater are relatively steep, and the bottom is somewhat rounded.

Apparently, several degrees of violence were involved in the formation of these craters. The size of the impinging particles could be responsible for some of the differences. It is not inconceivable that some of the more shallow craters could have been formed by small, hard (compared with aluminum tubing) particles em-

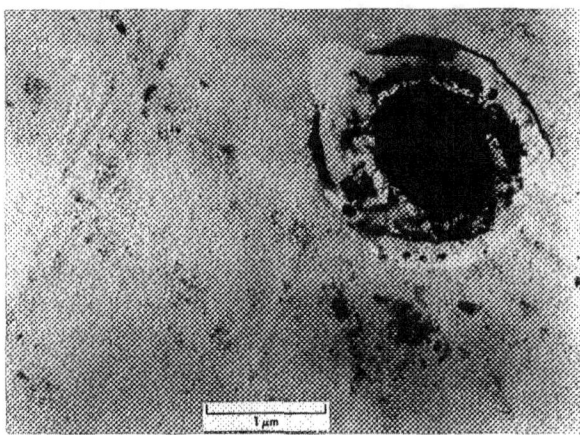

FIGURE 4.—Pit in section C probably caused by hypervelocity impact. Splash lip and smooth central core (as defined by the shadow) are characteristic.

bedded in the surface, if a strong grip were used by the astronaut during the removal of the tubing from the spacecraft.

It has been previously conjectured (ref. 3) that some of the pitting which, on the other Surveyor 3 components examined, could not be attributed to polishing or to high-velocity impact may have been caused by the blowing of dust and debris during the landing of the Apollo 12 Lunar Module (LM). For the case of the polished tubing, however, this is unlikely, since available photographs show that the Surveyor spacecraft was between the RADVS strut and the LM; therefore, the strut would have been protected from such impacts.

All micropits found were from the bright areas

of the tubing. The number found leads to the following approximations: less than 0.2 hypervelocity impact/cm² and, for the other types of craters, less than 2/cm².

It is believed that craters up to about 5 μm

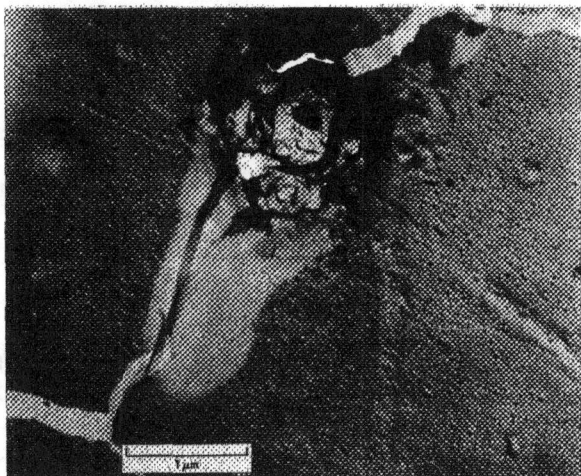

FIGURE 5.—Pit in section E.

in diameter could have been observed; larger features probably would tear out of the replica because of the lack of support on the shadowed side. However, no crater was found that had a diameter greater than 2.5 μm. The maximum depth was 2 μm; most had depths of less than 1 μm.

When considering possible damage due to solar wind sputtering, it must be remembered that, although a relatively smooth surface will develop a higher degree of surface roughness under sputtering conditions because of the slightly different sputtering rates of differently oriented crystallites, this tubing was mechanically buffed with rouge. The resulting smeared surface was probably amorphous, which makes erosion-rate estimates difficult. G. K. Wehner et al. (ref. 4), in an investigation of sputtering effects on the surface of the Moon, calculated an erosion rate of about 0.25 Å/yr due to full solar wind (H and He) striking a smooth, stony surface. This calculation was based upon solar wind data from Mariner 2, Pioneer 6, and Explorer 18.

For the present study, one should consider the possibility of increased erosion due to such factors as increased probability of oblique ion incidence on the 1.37-cm-wide tubing, differences in the sputtering yield for various materials, the presence of heavy ions, etc. However, even if one assumed a factor of 100 increase over the original calculations, the total loss would be less than 65 Å. From the general appearance of the upper surface, it is believed that relatively little material was removed and that the actual loss undoubtedly was no more than 65 Å. The underside of the tubing gives little assistance in these considerations, since the erosion during landing far exceeds that from sputtering by solar wind.

The greatest degree of damage incurred by the tubing during its 31-month stay on the lunar surface was the result of particle impact; this damage is (within the scope of this study) limited to a maximum depth of 2 μm.

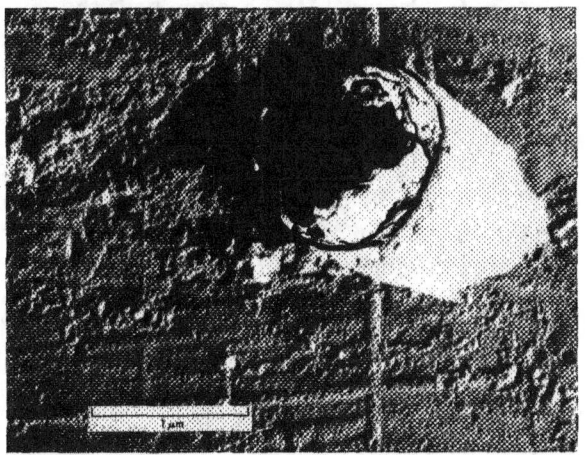

FIGURE 6.—Pit in section C.

References

1. WEIHRAUCH, J. H.; GERLOFF, U.; AND FECHTIG, H.: "Stereoscan Investigations of Metal Plates Exposed on LUSTER 1966, Gemini 9 and 12." *COSPAR Space Research VIII*, North-Holland Publishing, 1968, pp. 566–578.
2. DEVANEY, J. R.; AND EVANS, K.: "Characterization of Hyper-Velocity Particle Impact Craters on Apollo 11 Lunar Soil." *Proceedings of the Electron Microscopy Society of America 28th Annual Meeting 1970*, 1970, pp. 20–21.
3. BENSON, R. E.; COUR-PALAIS, B. G.; GIDDINGS, L. E., JR.; JACOBS, S.; JOHNSON, P. H.; MARTIN, J. R.; MITCHELL, F. J.; AND RICHARDSON, K. A.: "Preliminary Results from Surveyor 3 Analysis."

Apollo 12 Preliminary Science Report, NASA SP-235, Washington, D.C., 1970, pp. 217–223.
4. WEHNER, G. K.; AND KENKNIGHT, C. E.: *Investigation of Sputtering Effects on the Moon's Surface*, NASA CR 88738, 1967.

ACKNOWLEDGMENT

I acknowledge gratefully the helpful telephone conversations with Neil L. Nickle of the Jet Propulsion Laboratory.

PART E

RESULTS OF EXAMINATION OF THE RETURNED SURVEYOR 3 SAMPLES FOR PARTICULATE IMPACTS

B. G. Cour-Palais, R. E. Flaherty, R. W. High, D. J. Kessler, D. S. McKay, and H. A. Zook

The Meteoroid Sciences Branch at the Manned Spacecraft Center (MSC) examined the Surveyor 3 television camera housing and the length of polished aluminum tube retrieved by the Apollo 12 crew. The initial examinations were performed at the Lunar Receiving Laboratory (LRL) during a 6-day period before the return of the camera to Hughes Aircraft Co. (HAC). About 60 percent of the television camera surface area of almost 0.2 m^2 was scanned at $25 \times$ magnification; each suspected impact crater on selected areas of the flat surfaces was recorded. The remainder of the camera surface was scanned at lower magnifications to insure that no significant meteoroid damage had occurred. The polished tube, 19.7 cm long and 1.27 cm wide, was scanned at a general level of $40 \times$ magnification. Local areas of interest were examined at much higher magnifications; typical surface effects and suspected impact craters were photographed for documentary purposes.

Two 2.5-cm sections of the tube from the less uncontaminated ends, sections B and C, were examined in detail by the Meteoroid Sciences Branch after the preliminary examination at the LRL. These sections were optically scanned at $100 \times$ magnification initially; selected areas were later examined with a scanning electron microscope. Typical samples of the polished tubing and the painted surface of the camera housing, supplied by HAC, were also examined optically to determine surface backgrounds. The meteoroid examination of the television camera showed no evidence of meteoroid damage of any consequence by primary or secondary impacts after 950 days of exposure. Five craters were found on the housing, ranging in size between 150 and 300 μm in diameter, that are thought to be characteristic of hypervelocity impact. (However, not all of these may be of meteoroid origin, as three were so closely clustered as to indicate a nonrandom origin.)

Numerous surface chips of probable low-velocity origin were observed on the television camera surface in addition to the possible meteoroid impacts. These were shallow craters generally, and primarily of recent origin, as indicated by their whiteness against the sandy-brown color of the painted surface of the television camera housing. There was a definite concentration (10 to 100 times) of these white craters on the arc of the camera housing facing the Lunar Module (LM) compared with the other side. The distribution of craters peaked at approximately a region directly in line with the LM. Protuberances on the camera such as screw heads, support struts, etc., left dark shadows of unaffected paint on the camera pointing away from the LM. The preliminary examination of the entire polished tube revealed four craters larger than 25 μm in diameter that exhibited some characteristics of hypervelocity impacts at low magnifications. Detailed examination at higher optical magnifications and with the scanning electron

Figure 1.—View of the Lunar Module from the Surveyor 3 spacecraft.

microscope revealed that all of these craters were either low-velocity or polishing artifacts. The lack of meteoroid impacts of these limiting sizes is consistent with current estimates of the micrometeoroid flux on the Moon.

There is a marked concentration of pits on the same side of the tube to which a brown contamination is adhering. The material found in some of the craters is similar in composition to lunar soil.

Location and Geometry of Landing

Apollo 12 landed about 155 m northwest of the Surveyor 3 spacecraft (ref. 1). This closeness is dramatically shown in photographs taken by the astronauts. (See fig. 1.) From such photographs, it is obvious that the LM landed on the rim of the Surveyor crater, and is approximately sitting on the horizon as seen from the Surveyor spacecraft. Note from figure 1 that the front, flat surface of the Surveyor television camera is approximately parallel to a line joining Surveyor 3 and the LM. This is also confirmed by correlating certain craters in figure 1 and those of reference 2. Such a correlation puts the LM at a camera azimuth of about 90°. Also, from reference 2, it is found that the camera is leaning toward the LM, and that the horizon, in the direction of the LM, is at a camera elevation of 25°.

The polished aluminum tube that was sectioned by the astronaut can also be seen on the Surveyor spacecraft in figure 1.

TV Camera Housing

As previously mentioned, the camera housing was examined for evidence of meteoroid impacts during the time the camera was in the Lunar Receiving Laboratory at MSC. The time available permitted only a quick look for obvious impact craters. About 1150 cm² of the surface area was optically examined at 25 × magnification; the other surfaces were scanned at lower powers. Generally speaking, all of the flat surface areas of the housing were covered by the 25 × magnification scan; the cylindrical portions, such as the barrel and the hood, were covered at lower powers. As a result, it is correct to say that there were no damaging impacts on the camera housing. The surface of the mirror also was examined for obvious impacts.

Typical surface effects and suspected impact craters are shown in figure 2. It is interesting to note that the paint surface differs around the periphery of the housing. On the side closest to the Surveyor centerline, the surface appears grainy; on the parts facing outward, the surface is cracked like a dry river bed. Several holes and popped craters appear at the junction of cracks or along the cracks, and these were not included in the total of suspected impacts. There also was evidence of a large number of shallow white craters covering the housing with definite concentration occurring around the periphery. The craters were obviously fresh because the original white painted surfaces had been discolored to a sandy brown and the original color was being displayed. This effect is discussed in greater detail later in this article, as the cause is probably not of meteoroid origin.

The craters identified as of possible meteoroid impact origin because of their hypervelocity also are shown in figure 2. There were five such craters ranging in size from 130 to 300 μm in diameter. However, it is likely that not all of these were caused by meteoroids. This is especially true when it is considered that three of the suspected impacts occurred on the flat mirror gear-box housing, about 25 cm² in area. If the five craters were considered to be of meteoroid origin, then the flux, allowing for lunar shielding (1/2) and spacecraft shielding (1/4), would be $1.49 \times 10^{-6}/m^2/sec$. Allowing for the gravitational attraction of the Earth which, at 20 km/sec, is 1.74, this is a near-Earth flux of 2.62×10^{-6}. The mass associated with the smallest crater found, 150 μm wide, is about $10^{-8.75}$ g using a crater-diameter-to-meteoroid-diameter ratio of 10. The 95 percent of upper and lower limits for five impacts is 11.7 to 1.6 according to reference 3. If this spread in flux is associated with a spread in

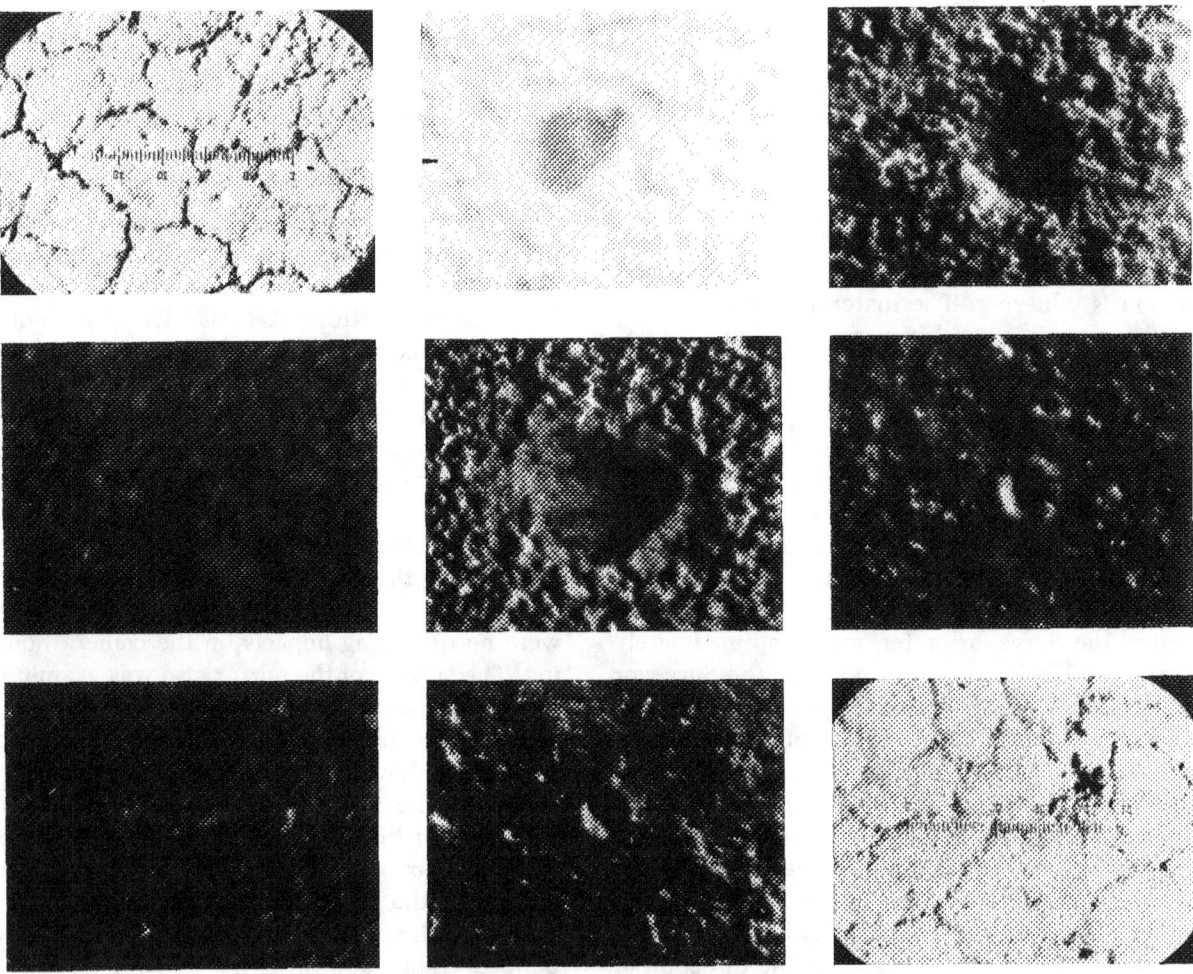

FIGURE 2.—Evidence of impacts on Surveyor 3 camera housing.

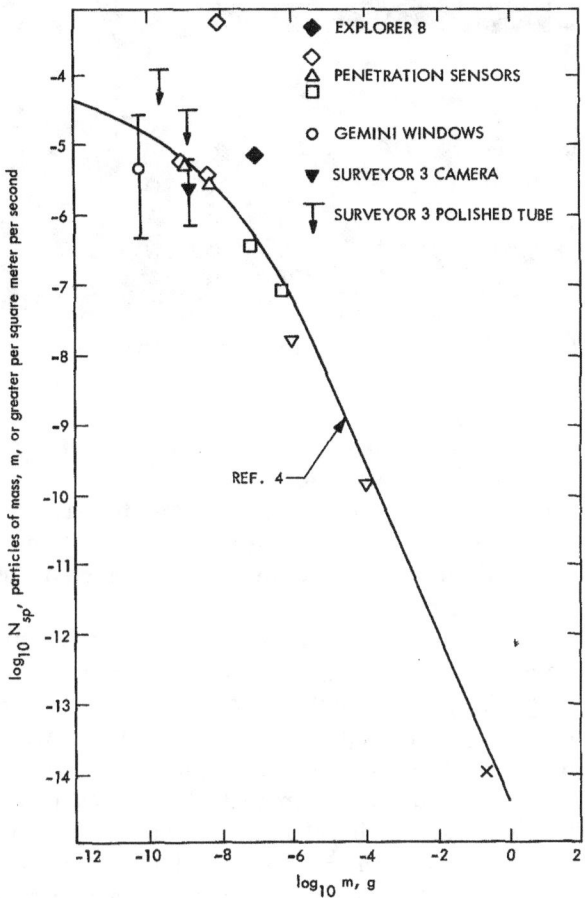

FIGURE 3.—Comparison of Surveyor 3 meteoroid flux with measured flux near Earth.

probable meteoroid mass of $10^{-8.5}$ to $10^{-9.0}$ g, then the corresponding point agrees well with Earth-orbital penetration data as shown in figure 3.

As previously mentioned, in scanning the television camera cratering for meteoroid impacts, certain areas were found to contain numerous "white-bottom" craters, as shown in figure 2. These craters were most numerous on the sides of the camera that were facing the LM. For example, the cylindrical surface just under the mirror head had 255 craters on the surface facing toward the LM, and only two on the side facing away from the LM.

The relationship between the Surveyor camera and the LM was discussed earlier in this article. Figure 4 was taken in the laboratory and is a view of the camera as it would be seen from the LM. When the camera is viewed from this angle, darker "shadows," which are free of a large crater density, are alined with protuberances such as bolts, screw heads, and other parts of the camera. These shadows are noticeable on the mirror hood in figure 4, on the base of the camera where it was partly shielded by a plate, and near the screw heads on the mirror gear box. Figures 5 and 6 are enlargements of two of those screw heads. Note that the shadow in figure 6 is well defined, and that numerous white chipouts are found outside the shadowed region. (The darker shadow extending downward is due to the light source used to take the picture and is not a permanent feature of the surface.) Figure 7 is a top view of the camera and shows the shadow cast by the bolt with the hexagonal head.

Figure 4 shows the apparatus used to hold the camera onto the Surveyor, along with the shadows it casts. The shadows are seen in figure 8, in which the mounting tubes have been removed.

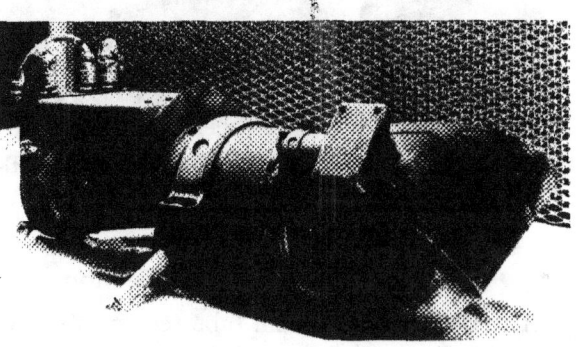

FIGURE 4.—View of the Surveyor 3 camera as seen from the Lunar Module.

The following conclusion is thus inescapable: During the 31 months that Surveyor 3 was on the Moon, the white surface of the camera was discolored; in the final stages of LM landing, lunar dust was accelerated by the LM exhaust. This dust literally sandblasted the Surveyor spacecraft, removing much of the discoloration, except in areas that were shielded. The sharpness of the shadows created by the shielding indicates that the path of the lunar dust was only slightly curved by lunar gravity, indicating the lunar dust was traveling in excess of 100 m/sec. Thus, most craters found on the camera housing are of LM origin.

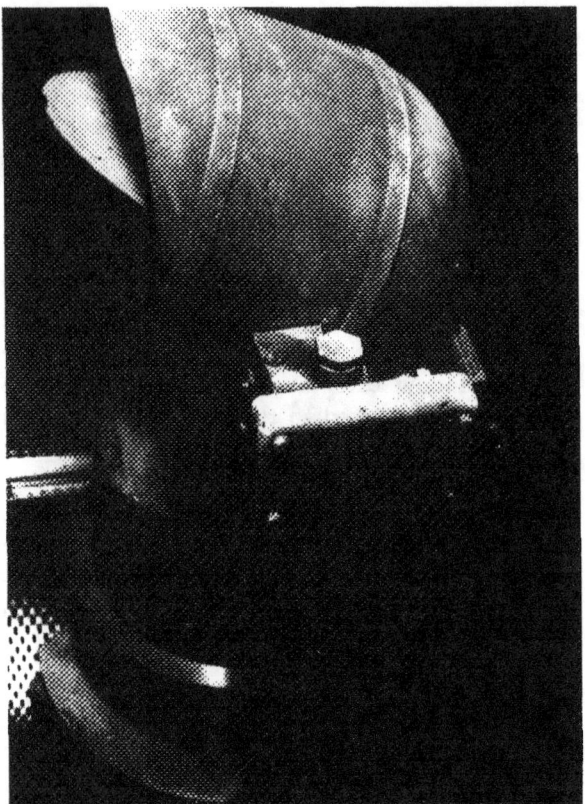

FIGURE 5.—Evidence of particle impact shadows on the camera housing.

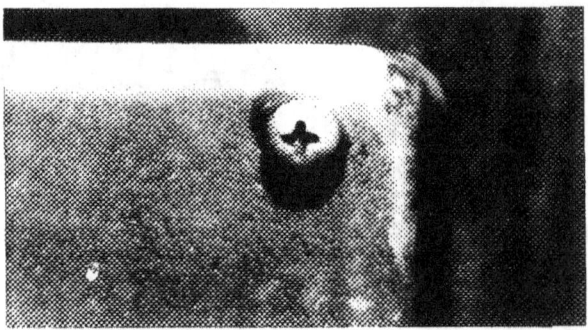

FIGURE 6.—Evidence of particle impact shadows on the camera housing.

Polished Aluminum Tube Section

The polished aluminum tube section obtained from the Surveyor 3 spacecraft was cut from the radar altimeter and doppler velocity sensor (RADVS) support strut adjacent to leg 2 using a pair of long-handled shears (resembling pruning shears) with curved, overlapping blades. The cutting action partially flattened the ends of the tube, as may be seen in figure 9. An increase in contamination also can be seen toward the left end of the tube. This contamination appears brown to the unaided eye. Under a microscope, it also appears brown and seems to be composed, at least partially, of crystals ranging in size up to a few micrometers. As the tube is rotated, there is variation in the amount of the contamination observed.

After the tube was received at the LRL, its entire surface was scanned at a magnification of 40 × for evidence of meteoroid impact. The tube was then cut into six sections and distributed to several investigators for detailed analysis. Sections B and C, two 2.5-cm sections toward the uncontaminated end of the tube, were obtained by MSC and examined in detail for meteoritic impact evidence.

The first part of the examination was a careful optical search for impact craters performed at a magnification of 100 ×. When craters were found, optical magnifications up to 600 × were used to determine whether the craters were caused by meteoroid impact. It was expected that the very high velocities of most impacting meteoroids (averaging 15 to 20 km/sec) would leave characteristic hypervelocity impact craters which would identify them. No hypervelocity impact craters were found; however, many other craters and pits were found.

Figure 10 shows the number of craters with diameters of 20 μm and larger that were observed in the field of view of an optical microscope at 100 × magnification (corresponding to an area of about 1 mm^2). Counts were taken as a function of angle around the tube from the scribe line, which had been ruled along the tube before cutting. This histogram is an average of two trials on section B of the tube. Very high pit densities (up to 40 per field of view) were obtained in two places, but were obviously associated with scratches and so are not included in figure 10. The reduced count rate about 170° from the scribe line is not considered significant.

Also shown in figure 10 is a measure of the relative amounts of brown contamination on section B as a function of angle around the tube. This curve was obtained by photographing the

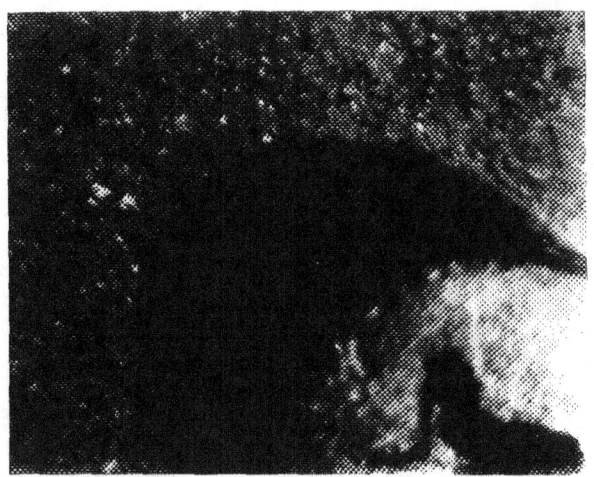

FIGURE 7.—Evidence of particle impact shadows on the camera housing.

FIGURE 8.—Shadows of the attachments on the television camera housing.

tube at each angular position, as the tube was rotated and the lighting held constant. The contamination stood out in the photographs between the angles of 100° and 280° and seemed to peak at about 190°. Outside of these angles, section B was relatively clean. The relative heights of the ordinate of the contamination curve in figure 10 are not quantitatively significant. A high ordinate means that the photograph indicates "high" contamination relative to an angular position with a ordinate. It is immediately evident that there is a close association between the pitting rate and density of the brown contamination.

In addition to the optical work, extensive analyses were performed using a scanning electron microscope (SEM). The SEM was used in three modes of analysis:

(1) To look at higher magnifications of craters found during the optical scan of tube sections B and C in order to determine the origin of these craters.

(2) To perform a spot survey at high magnifications over all of section C.

(3) To determine, by non-dispersive X-ray analysis, the composition of material in the craters and on the surface of the tube.

The results were—

(1) No craters showed evidence of hypervelocity impact origin. (It was not possible, by optical methods alone, to determine whether or not some of the smaller craters had hypervelocity im-

FIGURE 9.—Polished aluminum tube section obtained from the Surveyor.

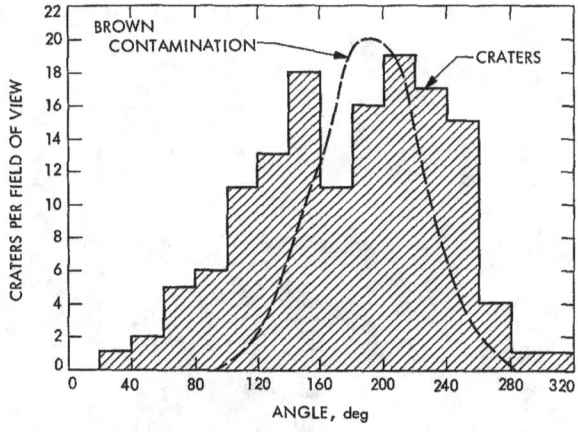

FIGURE 10.—Distribution of brown contamination and of impact craters.

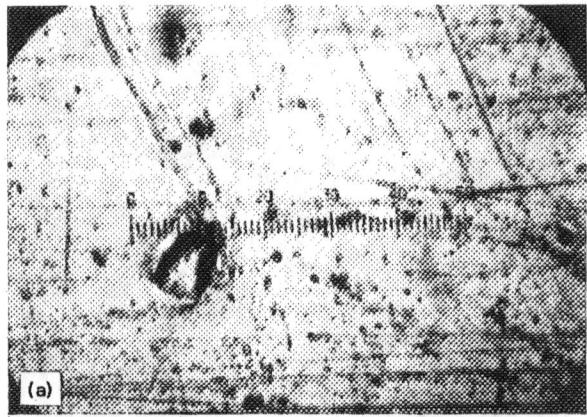
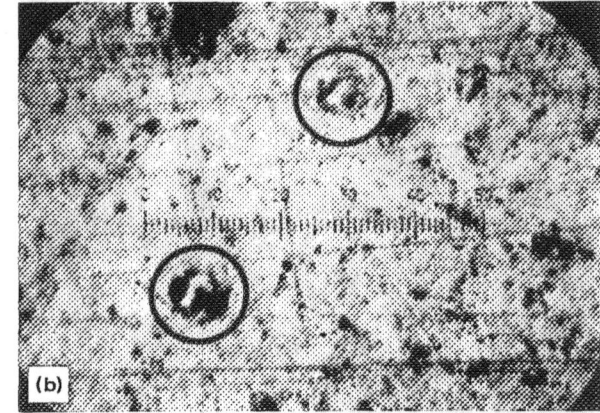
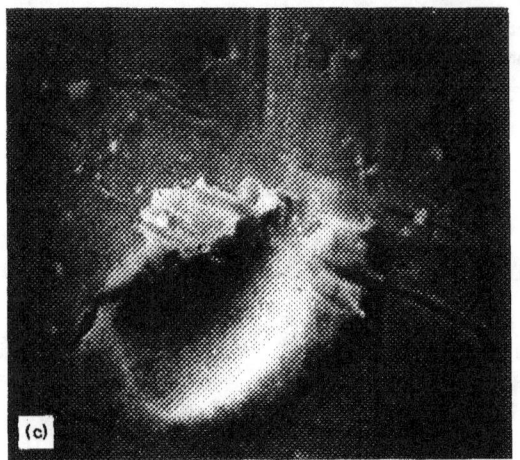

Figure 11.—Optical microscope and SEM views of typical impacts on polished tube section B. (a), (b) Optical microscope views. (c), (d), (e) SEM views.

pact characteristics.) On the contrary, all of the craters examined appeared to have a low-velocity impact origin and many of them had material remaining in them.

(2) The spot survey of section C confirmed the pitting density results of the optical scans, but added little new information.

(3) Analysis of the material in the craters strongly indicated that most of it was of lunar origin.

The brown contamination on the surface did not give any peaks because elements with X-ray energies below about 1 kV are not detectable with the analyzer on this SEM. Thus, elements such as oxygen, nitrogen, carbon, etc., would not have been discovered in this analysis.

Figure 11 shows SEM photographs of three craters on section B, located 280° from the scribe line. The craters obviously are not due to a hypervelocity impact; e.g., there is no smooth, raised lip entirely around the central indentation. However, it is clear that material at relatively low velocity, perhaps a few hundred meters per second, has impacted from the lower left in this photograph. The largest crater is about 30 μm wide, and material is still in the crater. An X-ray pulse height analysis of this material showed it to be composed of silicon, calcium, and iron with significant traces of chromium and titanium.

Figure 12 shows a region of high pitting density at 220° from the scribe line on section B. The

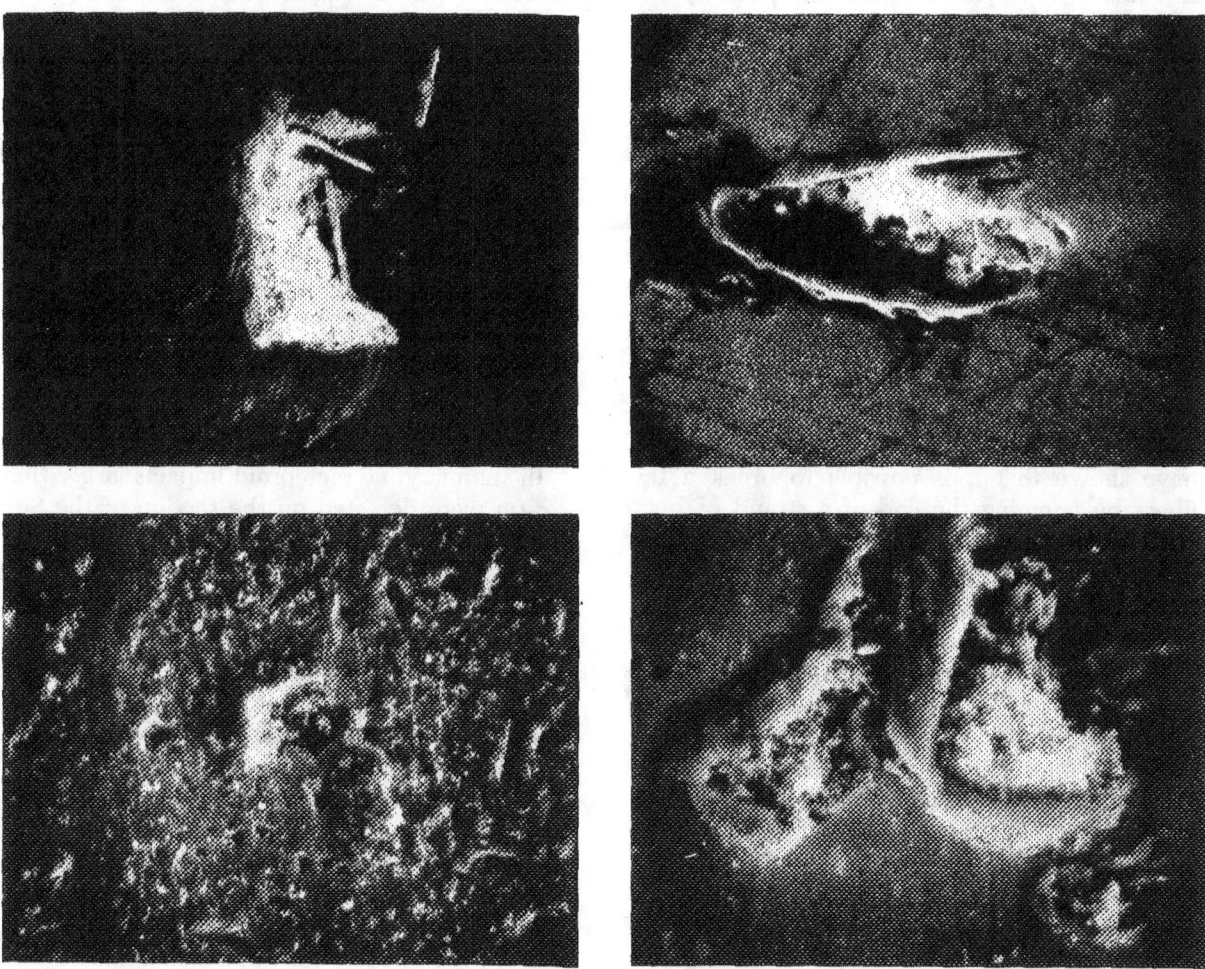

FIGURE 12.—Evidence of typical impact debris found on sections B and C of the Surveyor 3 polished tube.

crater in the center is about 8 μm in diameter; the material in this crater has as major components silicon, iron, calcium, and titanium. Titanium was also found in another crater on this tube. Because only six craters were extensively analyzed by SEM non-dispersive X-ray analysis, the significant amounts of titanium found in three of them are indicative of a lunar origin. From the mineralogical standpoint, at least three phases are present:

(1) A calcium aluminum silicate, which is undoubtedly plagioclase.

(2) A calcium iron magnesium silicate with a trace of titanium, which is consistent with clinopyroxene.

(3) One containing calcium, iron, titanium, and silicon in varying amounts and possibly also containing aluminum and magnesium. This is probably glass and unresolvable mixtures of very fine fragments.

A crater that gave us some difficulty is the one shown in figure 12 at 170° from the scribe line. Its size is about 80 by 110 μm and is one of the largest craters on the tube. The reason for the difficulty was the surprising appearance of "rods" in the crater, which looked very much like glass fibers under an optical microscope. The SEM analysis showed them to be identical in composition to the glass fibers in the astronauts' outer garments and in the back pack in which the Surveyor 3 parts were stowed. Experiments at MSC have shown that it is possible to break a few fibers by jamming the end of a strand of beta-fiber into a crater of this size.

As no meteoroid impacts were found on the tube, it is possible to set upper limits to the meteoroid flux at the Moon. The detection threshold over the entire tube corresponds to craters about 50 μm wide. The highly contaminated region was sufficiently pitted and scarred as to make it impracticable to resolve features of smaller craters. On the non-pitted sides of sections B and C, the detection threshold corresponds to 25 μm and larger craters. The effective non-pitted region is about one-half the area of these sections. If it is assumed that meteoroid impact craters are hemispherical in shape, then the threshold penetration depths are, respectively, 25 μm over the entire tube and 12.5 μm over one-half each of two 2.5-cm sections.

The 50-μm threshold over the entire tube corresponds to a meteoroid 14.5 μm wide and with a mass of $10^{-8.8}$ g. The 25-μm threshold corresponds to a meteoroid 7.5 μm in diameter and $10^{-9.66}$ g in mass. These masses correspond to a 20-km/sec impact velocity and a 1-g/cm density. The area of the entire tube is about 78.5 cm^2; the area of the non-pitted regions of sections B and C is 10.1 cm^2. If it is appropriate to use a shielding factor of one-half due to Moon and another factor of two-thirds due to the fact that the Surveyor spacecraft obliterates about one-third of the remaining solid angle from which meteoroids could approach, the effective area-time exposures are 2.16×10^5 m^2 sec for the entire tube and 2.8×10^4 m^2 sec for the non-pitted regions of sections B and C. Upper confidence limits of 95 percent on the meteoroid flux for no impacts for area-time exposures of 2.16×10^5 m^2 sec and 2.8×10^4 m^2 sec are, respectively, $10^{-4.75}$ impacts/m^2/sec and $10^{-3.88}$ impacts/m^2/sec. To compare these upper limits of the Moon with fluxes of Earth, one must allow for a gravitational flux increase factor of 1.74 at the Earth. Hence, the corresponding upper limits at Earth would be $10^{-4.51}$ impacts/m^2/sec for masses larger than $10^{-8.8}$ g and $10^{-3.64}$ impacts/m^2/sec for masses larger than $10^{-9.66}$ g. These upper limits are in good agreement with penetration measurements but not with older acoustic measurements, as can be seen in figure 3.

In summary, no meteoroid impacts larger than 25 μm were detected on the section of the Surveyor 3 strut returned from the Moon. The close association between the brown contamination and the pits on this section is significant. Also, the fact that there is lunar material in the pits is evidence that this phenomenon occurred while the Surveyor 3 spacecraft was on the Moon. Three possibilities for an origin to the pitting and contamination are—

(1) Lunar secondary and tertiary ejecta disturbed by primary meteoroid impacts bombard the exposed area of the tube, causing the pitting. The contamination is also composed of lunar material. The evidence from the sheared ends of the tube, however, has the contaminated and pitted side of the tube pointing away from a direction from which secondary ejecta is likely to approach. SEM analysis of the contamination

was unable to show any elements with atomic numbers greater than 11 (sodium); thus, it is unlikely that the brown contamination is composed of lunar soil.

(2) The pitting is due to lunar material blasted toward the Surveyor 3 spacecraft by the Apollo 12 LM as it landed. This possibility cannot be discounted, as has been shown previously for the camera housing. Experiments have shown that parts of the tube are visible from the LM. Two problems arise with this hypothesis. One is that the pitting on the tube seems to be more intense than on the camera; the other is that the camera seems to have been brown before the LM landed (and in a somewhat uniform fashion). However, the pitted side of the tube was darkened.

(3) The pitting is due to lunar material blasted toward the tube by the vernier engines; the contamination is due to incompletely burned propellant (unsymmetrical dimethyl hydrazine monohydrate fuel combined with nitrogen tetroxide oxidizer, with some nitrous oxide added as a catalyst). This also is a possible source, as the contaminated side of the tube could point down toward the lunar surface and somewhat in toward the Surveyor spacecraft if the tube is rotated 180° about the astronaut's cutter axis relative to possibility (2).

The Surveyor strut seems to have been pitted by lunar material disturbed by either the LM descent stage or the Surveyor 3 vernier engines. The brown contamination also could have come from either source, as the propellants used are nearly identical. We feel that the Surveyor 3 vernier engines are the more logical source.

Conclusions

The general conclusions arising from the MSC examination of the Surveyor 3 television camera housing and polished tube are—

(1) Meteoroid flux at the lunar surface is as expected from near-Earth measurements.

(2) Lunar ejecta flux related to meteoroid impacts on the lunar surface could not be specifically identified. However, other non-natural sources of low-velocity impacts by lunar surface material were evident.

(3) Lunar surface experiments and hardware must be shielded from the effects of spacecraft jet-exhaust-induced impacts.

Although additional analysis of the data obtained from the samples is continuing, it is not expected that the results given at this time will be altered significantly.

References

1. *Apollo 12 Preliminary Science Report*, NASA SP-235, Washington, D.C., 1970.
2. *Surveyor III—A Preliminary Report*, NASA SP-146, Washington, D.C., 1967.
3. Ricker, W. E.: "The Concept of Confidence or Fiducial Limits Applied to the Poisson Frequency Distribution." *Journal of the American Statistical Association*, vol. 32, 1937, pp. 349–356.
4. *Meteoroid Environment Model–1969 (Near Earth to Lunar Surface)*, NASA SP-8013, Washington, D.C., 1969.

PART F

MICROCRATER INVESTIGATIONS ON SURVEYOR 3 MATERIAL

E. Schneider, G. Neukum, A. Mehl, and H. Fechtig

Two screws from the Surveyor 3 spacecraft recovered during the Apollo 12 mission have been investigated for micrometeorite impact features. A general description of the scientific investigations of Surveyor 3 material is given in reference 1.

The positions of the screws on the Surveyor 3 spacecraft are shown in figure 1. From this pho-

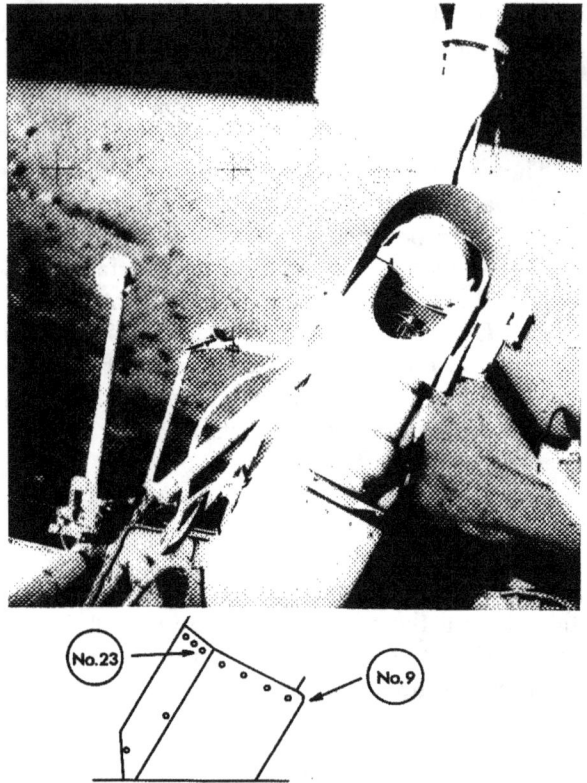

FIGURE 1.—Positions of screws 9 and 23 on the Surveyor 3 spacecraft.

tograph, screw 23 can be seen to point above the Moon's horizon at an angle of 66.6° with respect to the local upward vertical direction. Screw 9 points toward the lunar surface at the same angle with respect to the local downward vertical direction (ref. 2). Therefore, impact craters from extra-lunar particles may be expected primarily on screw 23, possibly together with low-velocity impact craters from secondary lunar debris. Screw 9 should show low-velocity impacts of secondary lunar debris.

Figure 2 shows the two screws including the washers. The investigations were made using a scanning electron microscope (Stereoscan). The scanning magnification was chosen to be 5000 ×, which allowed the identification of craters down to about 0.5 μm in diameter.

The original surfaces of the screws and washers were not specially prepared in any way for scientific investigations. They are rough and probably inadequate to yield reliable results. On screw 2 [1] (see fig. 3), strange features could be observed. Figure 4 shows six interesting objects on screw 1; these objects can be considered as impact phenomena.

The crater objects found on the screws can be compared with artificially produced micrometer-sized impact craters on metal targets. Rudolph (ref. 3) has published photographs of microcraters produced in the laboratory using a 2-MV Van de Graaff dust accelerator. Figure 5 shows some craters produced by impacts of iron projectiles on various metal targets with an impact velocity of 5.2 km/sec. The six objects on screw 1 (shown in fig. 4) appear to be low-velocity impact craters (≤ 5 km/sec). They may have been produced either by interplanetary dust particle impacts or by secondary lunar debris from larger impacts on the lunar surface. The three objects on the surface of screw 2 (fig. 3), however, are considered to be manufacturing artifacts rather than impact craters.

[1] The identification numbers of the screws have been lost. Therefore, we have arbitrarily assigned the numbers 1 and 2 to the screws.

FIGURE 2.—Surveyor 3 screws with washers.

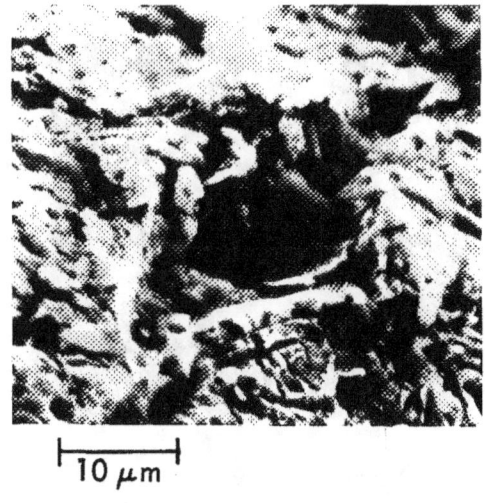
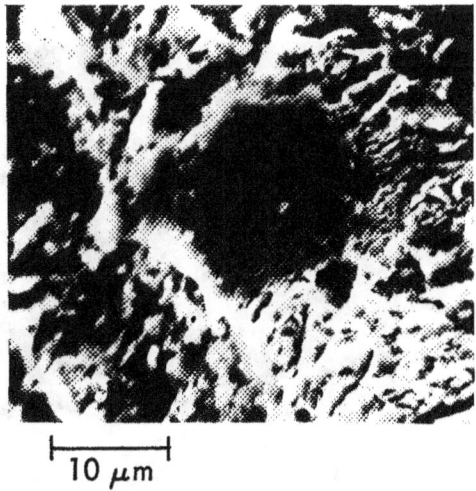
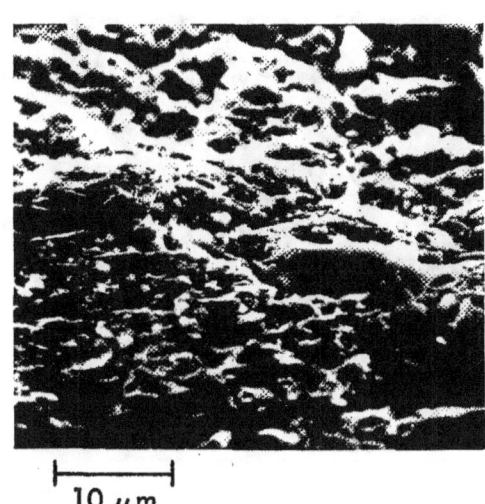

FIGURE 3.—Stereoscan photographs of objects found on the surface of screw 2 (probably not microcraters).

Assuming that the six craters on one of the screws are a result of primary impacts, it is possible to calculate a flux, Φ, for the 31-month exposure time and the surface area of about 0.12 cm^2:

$$\Phi = \frac{N}{Ft}$$

where

    Φ = cumulative flux, m^{-2} sec^{-1}
    N = number of particles/crater
    F = exposed surface area, m^2
    t = exposure time, sec

With the data involved in these investigations, one obtains a flux of $\Phi = 5 \times 10^{-3}$ m^{-2} sec^{-1}.

It seems doubtful to regard this result as interplanetary cosmic dust flux. By comparing this result with the flux obtained from the studies of lunar surface samples (refs. 4 and 5),[2] one should be aware that the particle number density in the interplanetary space at 1 AU shows a deviation in the microcrater distribution in the pit diameter range around 50 μm. This corresponds to a deviation in the microparticle distribution in the particle diameter range of about 25 μm. However, even submicrometer-sized particles exist in the interplanetary space, as indicated by Weinberg (ref. 6) and Hanner[3] from

[2] F. Hörz, J. B. Hartung, and D. E. Gault, Lunar Science Institute Contribution 09, unpublished.
[3] M. Hanner, private communication, 1970.

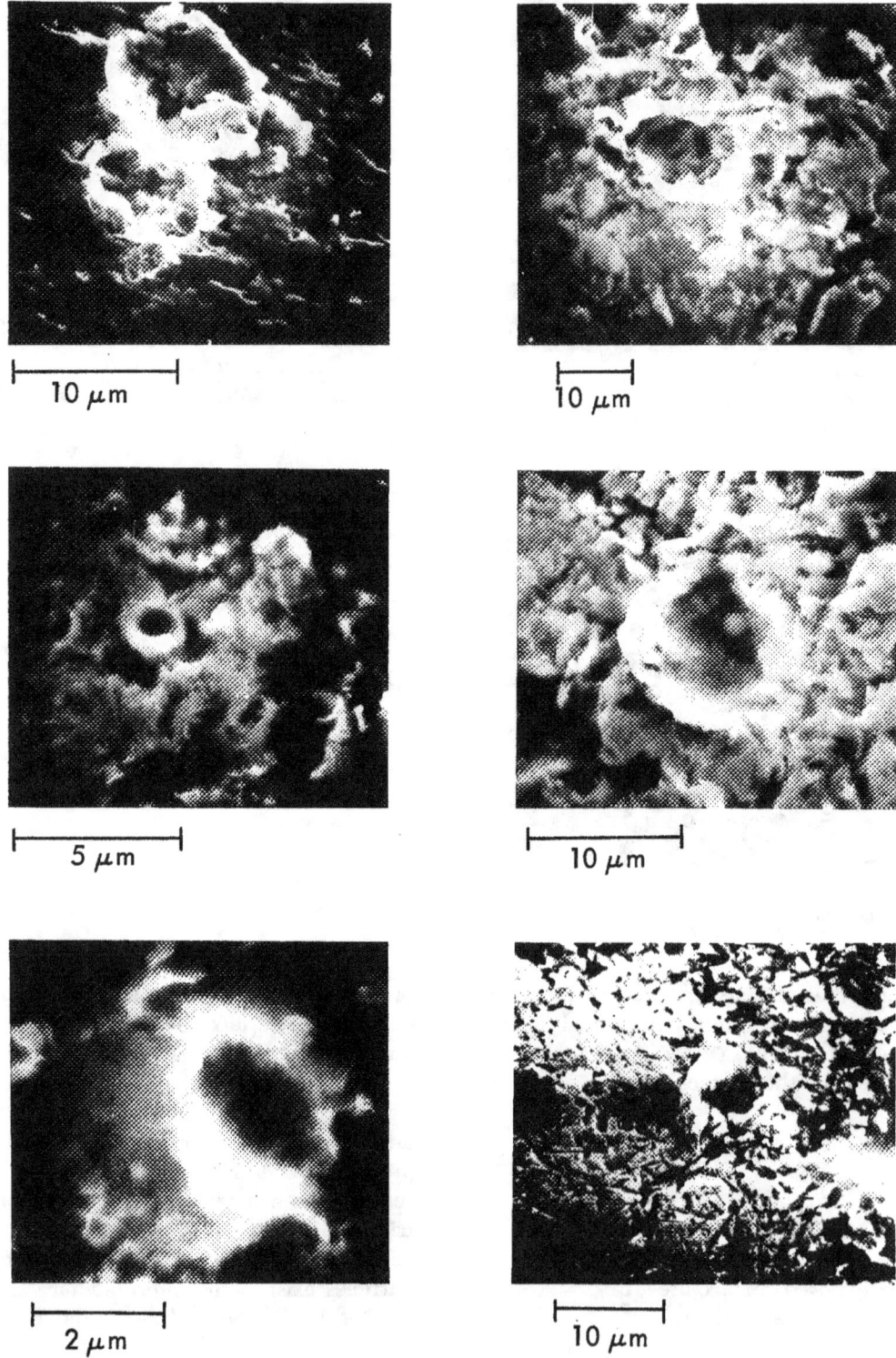

FIGURE 4.—Stereoscan photographs of objects on screw 1; most of the objects are assumed to be microcraters.

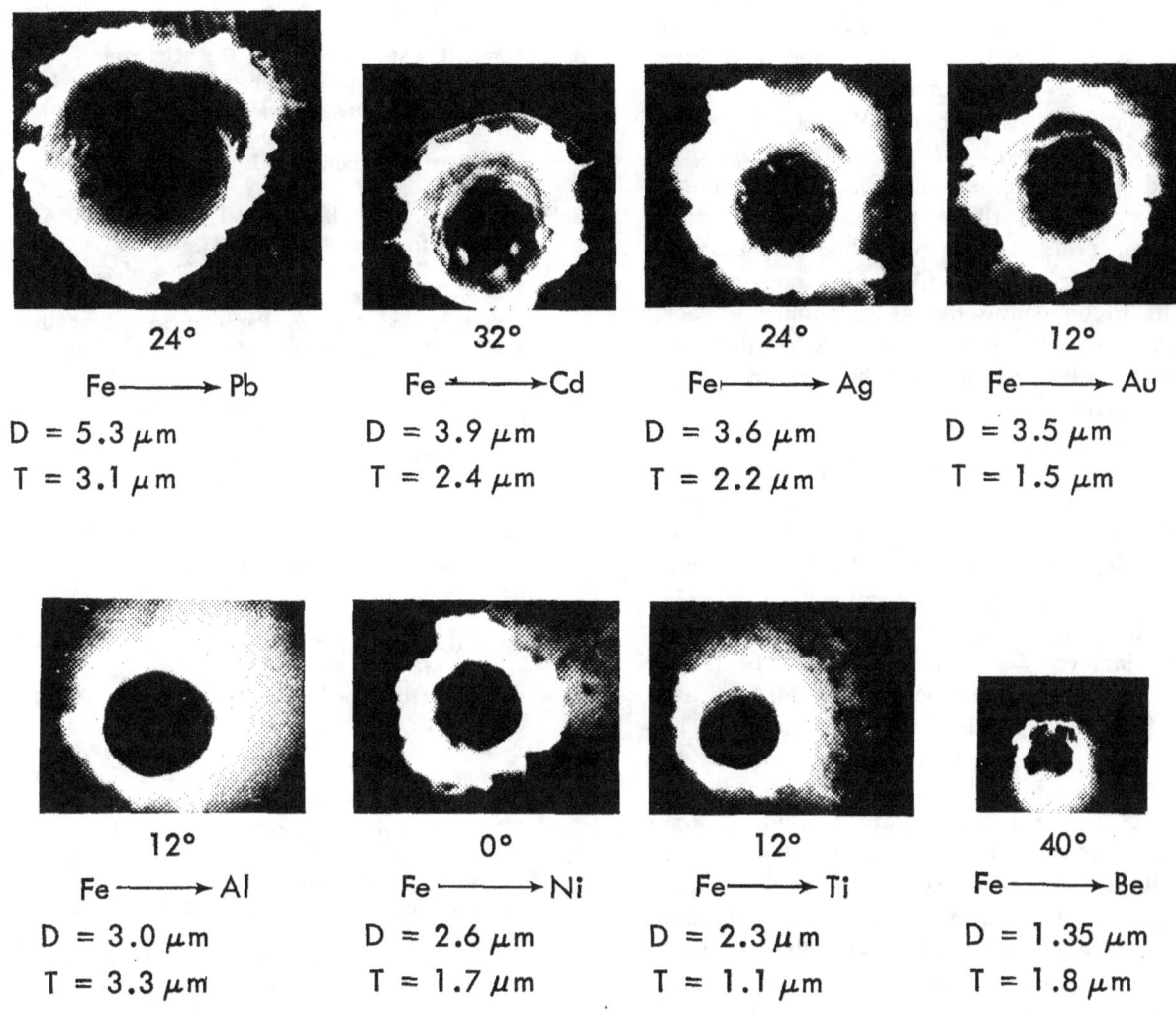

FIGURE 5.—Stereoscan photographs of simulated microcraters caused by iron projectiles on several metal targets at constant impact conditions. Velocity of the projectiles is 5.2 km/sec. d=projectile diameter; D=crater diameter; and T=crater depth. Data from V. Rudolph (see ref. 3).

zodiacal light measurements. Carter (ref. 7) reported the existence of microcraters on lunar glassy spherules down to 300-Å crater diameter. He interprets these craters as produced by secondary particle debris from larger meteoroid impacts on the lunar surface, although these craters found in this investigation can be produced by primary and/or secondary particles. Therefore, we consider the flux of primary particles of

$$\Phi = \frac{N}{Ft} = 5 \times 10^{-3} \text{ m}^{-2} \text{ sec}^{-1}$$

for particle diameter ≥ 1 μm to be an upper

limit. This result is in general agreement with other similar investigations on Surveyor 3 material. Benson et al. (ref. 8) have reported the existence of many dips that have been quoted as produced by the Lunar Module (LM). Only a few craters have been found; none were identified as hypervelocity impact craters. Cour-Palais et al. (ref. 9; also see ch. VI, pt. E, of this report) and Brownlee et al. (ref. 10) have reported a low number of impacts with conclusions similar to those given in this article. Buvinger (ref. 11) has published less than 0.2 hypervelocity impact/cm^2, which suggests our results to be considered as secondary impacts. Zernow (ref. 12) reports negative results for a scanned area with a magnification of 315 $\times$, which seems to be low.

In conclusion, one can summarize that only few impact craters could be detected. As little is known concerning the velocity distribution of interplanetary dust particles, one can interpret the results in two ways. First, the impacts could have been produced by interplanetary particles, then the flux of 5×10^{-3} m^{-2} sec^{-1} for particles with diameter $\geqq 1$ μm would indicate that a deviation from the normal distribution can exist only for particles below 1 μm in diameter. The alternative interpretation is that most of the craters found by different investigators on Surveyor 3 material are due to secondary lunar debris impacts. In this case, the flux of 5×10^{-3} m^{-2} sec^{-1} for particles with a diameter $\geqq 1$ μm must be interpreted as an upper limit for interplanetary particles. This final result is in agreement with recent flux results from lunar samples (refs. 4 and 5) and with the results of the Pioneer dust experiment (ref. 13).

References

1. Nickle, N. L.: *Surveyor 3 Material Analysis Program.* Paper presented at the Apollo 12 Lunar Science Conference, Houston, Tex., 1971.
2. Surveyor Project Science Staff: *Surveyor III Mission Report. Part III: Television Data,* TR 32–1177, Jet Propulsion Laboratory, Pasadena, Calif., 1967.
3. Rudolph, V.: "Untersuchungen an Kratern von Mikroprojektilen im Geschwindigkeitsbereich von 0, 5 bis 10 km/sec." *Z. Naturforsch.,* vol. 24a, 1969, p. 326.
4. Hörz, F.; Hartung, J. B.; and Gault, D. E.: *Lunar Microcraters.* Paper presented at the Apollo 12 Lunar Science Conference, Houston, Tex., 1971.
5. Bloch, R. M.; Fechtig, H.; Gentner, W.; Neukum, G.; and Schneider, E.: *Meteorite Impact Craters, Crater Simulations, and the Meteoroid Flux in the Early Solar System.* Paper presented at the Apollo 12 Lunar Science Conference, Houston, Tex., 1971.
6. Weinberg, J. L.: "The Zodiacal Light at 5300 Å." *Ann. d'Astrophys.,* vol. 27, 1964, p. 718.
7. Carter, J. L.: *Chemistry and Surface Morphology of Fragments from Apollo 12 Soil and Laboratory Produced Craters.* Paper presented at the Apollo 12 Lunar Science Conference, Houston, Tex., 1971.
8. Benson, R. E.; Cour-Palais, B. G.; Giddings, L. E., Jr.; Jacobs, S.; Johnson, P. H.; Martin, J. R.; Mitchell, F. J.; and Richardson, K. A.: "Preliminary Results from Surveyor 3 Analysis." *Apollo 12 Preliminary Science Report,* NASA SP–235, Washington, D.C., 1970, pp. 217–223.
9. Cour-Palais, B. G.; Flaherty, R. E.; High, R. W.; Kessler, D. J.; McKay, D. S.; and Zook, H. A.: *Results of an Examination of the Returned Surveyor 3 Samples for Particle Impacts Conducted at the Manned Spacecraft Center.* Paper presented at the Apollo 12 Lunar Science Conference, Houston, Tex., 1971.
10. Brownlee, D.; Hodge, P.; and Bucher, W.: *Micrometeoroid Flux from Surveyor 3 Glass Surfaces.* Paper presented at the Apollo 12 Lunar Science Conference, Houston, Tex., 1971.
11. Buvinger, E. A.: *Replication Electron Microscopy on Surveyor 3 Unpainted Aluminum Tubing.* Paper presented at the Apollo 12 Lunar Science Conference, Houston, Tex., 1971.
12. Zernow, L.: *Examination of Sample of Surveyor 3 Strut for Meteoroid Impacts.* Paper presented at the Apollo 12 Lunar Science Conference, Houston, Tex., 1971.
13. Berg, O. E.; and Gerloff, U.: *More than Two Years of Micrometeorite Data from Two Pioneer Satellites.* Paper presented at COSPAR Meeting, Leningrad, 1970.

ACKNOWLEDGMENTS

We are grateful to NASA for providing the Surveyor 3 samples, to N. L. Nickle and Dr. O. A. Schaeffer for their help in solving the problems of identification and interpretation, and to Miss O. Kress for conducting much of the scanning work.

PART G

LUNAR SURFACE: CHANGES IN 31 MONTHS AND MICROMETEOROID FLUX

L. D. Jaffe

During the period from April 20 to May 3, 1967, Surveyor 3 sent to Earth thousands of television pictures of the lunar surface near its landing site in Oceanus Procellarum at 23.34° W longitude, 2.99° S latitude (ACIC coordinate system). On November 20, 1969, the site was visited by Apollo 12 astronauts Alan Bean and Charles Conrad, who took many pictures of the lunar surface on 70-mm film using a hand camera. This provided an opportunity to compare pictures of the same small areas of the lunar surface taken 31 months apart.

I have made a preliminary comparison, examining areas that had been disturbed by the Surveyor spacecraft. These disturbances produced markings in the lunar soil which were easily identifiable and simpler in shape than the irregularities, on a scale of centimeters and smaller, characteristic of the undisturbed lunar surface. Accordingly, changes in the disturbed areas should be easier to detect. The surface disturbances studied included groups of imprints produced by two of the footpads of Surveyor during its final (third) landing event (ref. 1), as well as markings made in post-landing operations by the Surveyor soil mechanics surface sampler (ref. 2): four trenches, seven bearing tests, impact tests, and other surface contacts.

About 60 Surveyor pictures and 20 Apollo photographs were examined in detail; the Apollo photographs included several stereo pairs. The material consisted of prints made from copy negatives, in turn prepared from a master positive, on film, of the original 70-mm negative. For Surveyor, prints were made from negatives prepared by digital-computer processing of the television signals recorded on magnetic tapes, and from negatives of photo-print mosaics. Enlargements were up to two-thirds of lunar scale. The view angles and, in general, the Sun angles, in the Apollo photographs were different from those in the Surveyor pictures.

I have found only one definite change in the surface, other than those obviously produced by the astronauts: on the bottom of an imprint made by Surveyor footpad 2, all of the pertinent Apollo photographs show a particle, about 2 mm in diameter, that does not appear in any of the Surveyor pictures (fig. 1, particle 3). Various digital-computer image-processing techniques were tried, without success, to enhance the Surveyor pictures to reveal the object, or its shadow. If the particle had been present when the Surveyor pictures were taken, its shadow, at least, should have been easily detected. (The camera line resolution was 1 mm at the imprint, and the Sun 27° above the horizon at the time fig. 1(a) was televised.) I conclude that the particle was emplaced after the Surveyor pictures were taken. It may have fallen from the rim of the footpad imprint or, perhaps, may have been kicked in by an astronaut as he approached.

The Apollo photographs show that the sides of several steep walls made by Surveyor footpads and surface sampler were still in place. These include the vertical wall of a trench 6 cm deep (fig. 2). The cohesion and internal friction previously reported for lunar soil (refs. 3 and 4) are sufficient, according to standard soil mechanics analysis (ref. 5), to hold such a wall against lunar gravity for an extended time.

Surface areas darkened by ejected fines during the Surveyor landing still appeared dark compared to the undisturbed surface (fig. 3).

On the floor of the footpad imprint shown in figure 1, any crater as large as 1.5 mm in diameter should have been visible in the Apollo photographs. (The line resolution is 0.4 mm or better.) I noted only two pits. One of these, pit 4, is visible in Surveyor as well as Apollo photo-

FIGURE 1.—Imprint in lunar soil made by footpad 2 of Surveyor 3. Waffle pattern in imprint is from the bottom of Surveyor footpad. (a) Part of Surveyor 3 television picture taken April 21, 1967, at 08:24:20 GMT. Sun is in the east, 27° above horizontal. View from north of west. Picture is digitally computer processed. (b) Part of photograph from Apollo 12 hand camera taken November 20, 1969, about 05:22 GMT. Sun is in the east, 23° above horizontal. View from south (from photograph AS12–48–7110). Numbers 1, 2 indicate two particles clearly visible on floor of imprint in each picture; 3 indicates a particle visible only in the Apollo photograph; 4 indicates a pit visible in both pictures; 5 indicates a pit visible in Apollo photograph, tentatively identified in the Surveyor picture; and 6 indicates a small particle next to pit 5.

graphs. The other, pit 5, appears in the Apollo photographs and may also appear in Surveyor pictures. It is immediately adjacent to a small particle, 1 to 2 mm in diameter, and most likely was produced when the adjacent particle fell in during the final landing event of Surveyor 3. Thus, no meteoritic craters as large as 1.5 mm in diameter appeared on the bottom of the imprint, 20 cm in diameter, during the 31 months of exposure. The rate of impact was less than 1.0 particle/m² month or 4×10^{-7} particle/m² sec, for particles producing 1.5-mm-diameter craters. This is for a solid angle of almost 2π.

Braslau (ref. 6) found that a projectile im-

FIGURE 2.—Part of Apollo 12 photograph, taken November 20, 1969, showing trenches made by Surveyor 3 surface sampler. Far corner of nearer trench preserves vertical wall, about 6 cm deep, dug April 22, 1967 (from photograph AS12–48–7108).

FIGURE 3.—Apollo 12 photograph taken November 20, 1969. Note dark ejecta from impact of Surveyor footpad on April 20, 1967 (AS12–48–7110).

pacting dry sand at 6.4 km/sec produced an ejecta mass, plus compression, equivalent to 4700 times the projectile mass. On this basis, a 1.5-mm-diameter crater would be produced by a 3×10^{-7} g micrometeoroid impacting at 6.4 km/sec. At 20 km/sec, a velocity more typical of primary meteorites, the impacting mass for the same energy would be 3×10^{-8} g. A flux of 4×10^{-7} particles/m² sec of this mass is near the lower limit of meteoroid flux derived from spacecraft measurements and many orders of magnitude lower than some previous estimates (fig. 4). It is consistent with zodiacal light and radar meteor data and with some of the more recent spacecraft data obtained with acoustic/capacitance and penetration sensors. The absence of detectable craters in the Surveyor 3 footpad imprint implies, then, a very low micrometeoroid flux on the lunar surface.

References

1. CHRISTENSEN, E. M.; BATTERSON, S. A.; BENSON, H. E.; CHOATE, R.; JAFFE, L. D.; JONES, R. H.; KO, H. Y.; SPENCER, R. L.; SPERLING, F. B.; AND SUTTON, G. H.: "Lunar Surface Mechanical Prop-

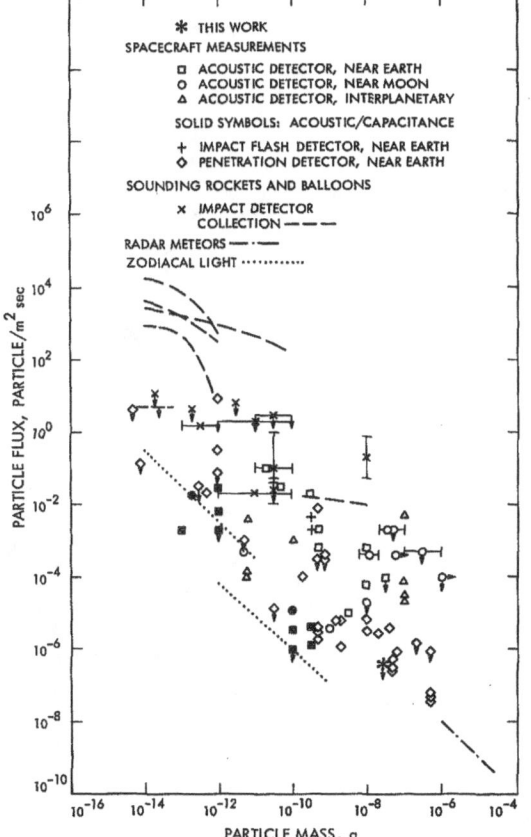

FIGURE 4.—Micrometeoroid flux vs. mass. Based on McDonnell (see ref. 7); the result of this work is added.

erties at the Landing Site of Surveyor 3." *J. Geophys. Res.*, vol. 73, 1968, pp. 4081–4094.
2. SCOTT, R. F.; AND ROBERSON, F. I.: "Soil Mechanics Surface Sampler: Lunar Surface Tests, Results, and Analyses." *J. Geophys. Res.*, vol. 73, 1968, pp. 4045–4080.
3. SCOTT, R. F.; AND ROBERSON, F. I.: "Soil Mechanics Surface Sampler." *J. Geophys. Res.*, vol. 74, 1969, pp. 6175–6214.
4. COSTES, N. C.; CARRIER, W. D.; MITCHELL, J. K.; AND SCOTT, R. F.: "Apollo 11 Soil Mechanics Investigation." *Science*, vol. 167, 1970, pp. 739–741.
5. TERZAGHI, K.: *Theoretical Soil Mechanics*, John Wiley & Sons, New York, 1943, p. 154.
6. BRASLAU, D.: "Partitioning of Energy in Hypervelocity Impact against Loose Sand Targets." *J. Geophys. Res.*, vol. 75, 1970, pp. 3987–3999.
7. MCDONNELL, J. A. M.: "Review of in-situ Measurements of Cosmic Dust Particles in Space." *COSPAR Space Research XI*, Akademie-Verlag, Berlin, 1971, pp. 415–435.

VII. Radioactivity and Radiation Damage Analyses

PART A

EXAMINATION OF RETURNED SURVEYOR 3 CAMERA VISOR FOR ALPHA RADIOACTIVITY

T. E. Economou and A. L. Turkevich

On April 20, 1967, Surveyor 3 landed in the eastern part of Oceanus Procellarum at 23.34° W longitude and 2.99° S latitude (ACIC coordinate system). On November 20, 1970, the same site was revisited by Apollo 12 astronauts Alan Bean and Charles Conrad who brought back the Surveyor 3 television camera; 32 kg of Moon rocks also were returned to Earth. Part of this camera, the visor, was examined by University of Chicago group for the presence of a deposit of alpha radioactivity.

The possibility of such a radioactive deposit on the surface of the Moon was suggested by several authors (ref. 1). Radon isotopes formed by the decay of uranium and thorium diffuse out of lunar material into space where they undergo additional decay, and some of their daughters are deposited on the lunar surface. In the thorium decay series, the daughters have relatively short half-lives, and all had decayed before the visor could be examined. However, the concentration of the alpha-emitting ^{210}Po in the uranium decay series is dependent on the longer half-life (22 yr) of its grandparent ^{210}Pb. A measurement of the amount of ^{210}Po (5.31 MeV) alpha activity on the visor, together with knowledge of the time spent on the Moon, and on the Earth before the measurement, provides a measure of the rate of radon decay product deposition on the lunar surface at Oceanus Procellarum.

The existence of such a deposit would help to provide information on the emanating power of lunar material and on the amount of radon "atmosphere" on the Moon. It also may have an effect on the isotopic composition of the lead in lunar fines.

Experimental Method and Results

Measurements on the Surveyor 3 camera visor (fig. 1) were started at the University of

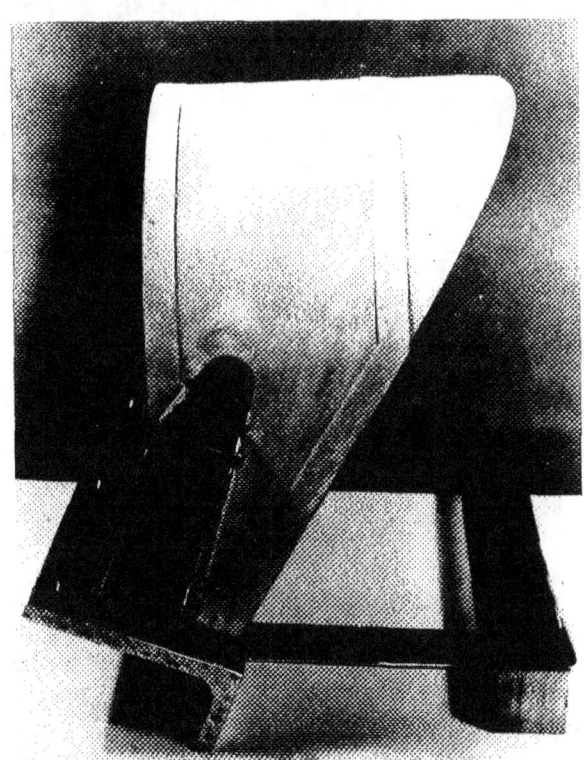

FIGURE 1.—Visor of the Surveyor 3 television camera brought back to Earth by the Apollo 12 astronauts.

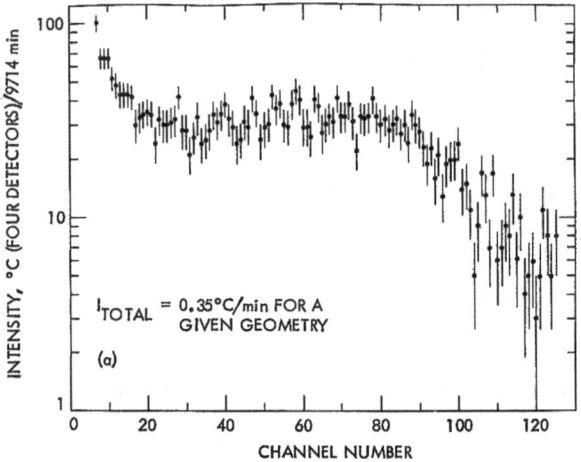

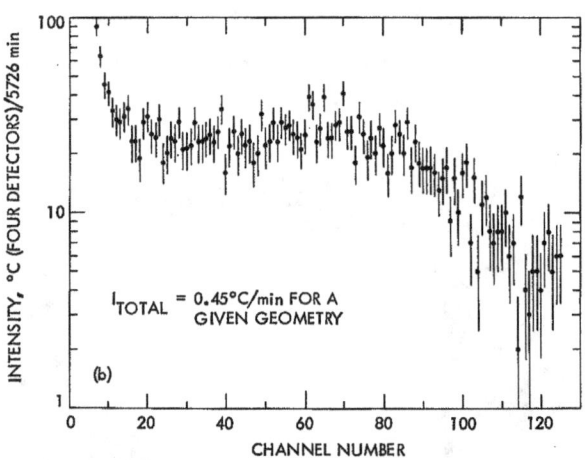

FIGURE 2.—Data taken with alpha-scattering instrument. (a) The alpha spectrum obtained from the Surveyor 3 television camera visor as measured by the alpha-scattering instrument during the period of July 14 to 21, 1970. (b) Background obtained from the paint used on the visor measured during the period of July 31 to August 5, 1970.

Chicago 236 days after it was removed from the Surveyor 3 spacecraft. The visor was placed in a vacuum chamber and examined for alpha radioactivity using the alpha-scattering instrument. (See ref. 2.) To increase the sensitivity, the proton system of the instrument was used because the active area of the proton detectors of this instrument is about 10 times that of the alpha detectors, and the examined visor could be placed closer to the proton than to the alpha detectors. The gold foils in front of the proton detectors, which normally screen the alpha particles in the alpha-scattering instrument, were removed for these measurements and replaced by the VYNS (polyvinylstyrene) films. The visor was measured for a period of 9714 minutes using all four detectors, and an additional period of 4475 minutes with less than the full complement of detectors in order to check on possible asymmetries in the deposit. The backgrounds in the instrument were negligible.

Figure 2(a) shows the experimental data obtained from the visor of the Surveyor 3 television camera. There are several unexpected surprises that characterize these data:

(1) The continuous flat spectrum indicates that the source of alpha activity is not on the surface.

(2) The intensity is too high, several orders of magnitude higher than expected.

(3) The presence of high-energy alpha particles (higher than 6 MeV) indicates that the source probably is due to daughter products of ^{228}Th or ^{234}U.

The surface of the visor, as most of the Surveyor parts, was covered with white paint for thermal-control purposes. Because of the unavailability of a model television camera, the natural background from the visor could not be measured. Figure 2(b) shows the results of measurements made on plates covered with the same paint and made at the same time as the visor itself. In these measurements, the plates were placed very close to each detector in a position where the absolute efficiency of detecting their activity could be calculated.

This spectrum is similar to that obtained from the visor. After comparing the absolute intensities, the conclusion was reached that the gross activity on the visor returned from the Moon was due entirely to the activity of the paint.

Although the presence of this alpha activity of the paint reduced the sensitivity of identifying an alpha radioactive deposit, the data can be used to set upper limits for the ^{210}Po radioactivity on the Moon. Figure 3 shows a comparison of the alpha activity in the region of ^{210}Po (5.3 MeV) for the paint and the visor. Using the gross alpha radioactivity as a measure of the relative efficiencies of detecting radiation from the visor and paint, the paint background could be subtracted from the visor data

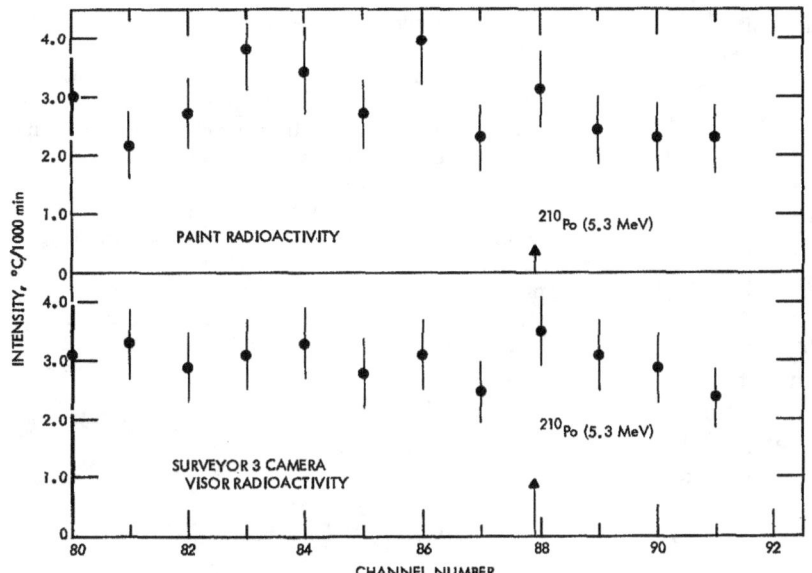

FIGURE 3.—Comparison of alpha-radioactivity from the visor and the paint in the vicinity of ^{210}Po (5.3 MeV).

in the region of interest to obtain a net activity on the visor of $(0.7 \pm 3.5) \times 10^{-3}$ disintegrations (d) min^{-1} cm^{-2}.

To estimate the amount of ^{210}Po activity expected on a square centimeter of the lunar surface after a very long period, this net activity must be corrected for the shadowing of the visor by assorted spacecraft parts (the "view factor" to space was 0.65), for the decay since removal from the Moon (0.64 yr), and for the fact that the visor was on the Moon for only 31 months (keeping in mind the genetic relationship of the ^{222}Rn decay chain). Application of these corrections leads to the measurement implying an activity, after infinite time, of

$$D = (0.88 \pm 4.43) \times 10^{-3} \text{ d sec}^{-1} \text{ cm}^{-2}$$

on the lunar surface at Oceanus Procellarum.

Discussion

The limit on the ^{210}Po alpha radioactivities obtained in this work is compared with the predictions of Kraner et al. (ref. 1) and of subsequent attempts to measure this quantity. (See table 1.) The original prediction was an average for the whole Moon, as was the limit set by Yeh and Van Allen (ref. 3) from Explorer 35 observations. The only reported observation of the presence of the radioactivities is work by Turkevich et al. (ref. 4) from the data obtained by the alpha-scattering instrument on the Surveyor 5 mission to Mare Tranquillitatis.

Lindstrom et al. (see ref. 5) in determining the excess of ^{210}Pb (over that in equilibrium with uranium) on the surface of the rock brought back by Apollo 11 astronauts, also from Mare Tranquillitatis, gave a limit that is 70 times lower than the value reported by Turkevich et al. in reference 4.

The limit set by the present work on such radioactivity in Oceanus Procellarum, a different site, but one at which the uranium content of the soil actually is appreciably higher than in Mare Tranquillitatis, is also lower than the value reported in reference 4.

Although the results of Lindstrom et al. (ref. 5) and the present work appear to contradict the results reported in reference 4, it must be

TABLE 1.—*Equilibrium ^{210}Po alpha radioactivity of lunar surface*

Location	d cm^{-2} sec^{-1}	Reference
Entire Moon	[a] 2.0	1
Entire Moon	< .16	3
Mare Tranquillitatis	.03 ± 0.01	4
Mare Tranquillitatis	< .0004	5
Oceanus Procellarum	< .005	This article

[a] Prediction.

remembered that the radioactive deposit should be confined to the topmost fraction of micrometer on the lunar surface. Any disturbance of the surface, such as the shaking of a dust layer, or abrasion of the surface, also would carry away the deposit. Thus, although these two most recent attempts to detect the alpha radioactivities have failed, and therefore contradict the observation of Turkevich et al. (ref. 4), there is some probability that these recent attempts are not valid checks on the existence of the deposit. In both cases, there is no assurance that the topmost layer was not removed. It may be that the Surveyor 5 mission provided better detection of this fragile deposit than the examinations of samples brought back from the Moon.

References
1. KRANER, H. W.; SCHROEDER, G. L.; DAVIDSON, G.; AND CARPENTER, J. W.: "Radioactivity of the Lunar Surface." *Science*, vol. 152, 1966, pp. 1235–1237.
2. TURKEVICH, A. L.; KNOLLE, K.; EMMERT, R. A.; ANDERSON, W. A.; PATTERSON, J. H.; AND FRANZGROTE, E. J.: "Instrument for Lunar Surface Chemical Analysis." *Rev. Sci. Instru.*, vol. 37, 1966, pp. 1681–1686.
3. YEH, R. S.; AND VAN ALLEN, J. A.: "Alpha Particle Emissivity of the Moon: An Observed Upper Limit." *Science*, vol. 166, 1969, pp. 370–372.
4. TURKEVICH, A. L.; PATTERSON, J. H.; FRANZGROTE, E. J.; SOWINSKI, K. P.; AND ECONOMOU, T. E.: "Alpha Radioactivity of the Lunar Surface at the Landing Sites of Surveyors 5, 6, and 7." *Science*, vol. 167, 1970, pp. 1722–1724.
5. LINDSTROM, R. M.; EVANS, J. C., JR.; FINKEL, R.; AND ARNOLD, J. R.: To be published.

ACKNOWLEDGMENTS

We thank Neil Nickle of the Jet Propulsion Laboratory for making available the Surveyor 3 camera visor, and Edwin Blume for valuable help during this work.

PART B

PRELIMINARY RESULTS ON TRITIUM IN SURVEYOR 3 MATERIAL

E. L. Fireman

Surveyor 3 material exposed on the lunar surface for 31 months and recovered by the Apollo 12 astronauts offered a unique opportunity to measure tritium in the solar wind. The material consisted of thin sheets of aluminum (0.25-g/cm^2 thickness), painted with a coat of white inorganic kaolin paint used to reflect sunlight in order to minimize temperature fluctuations. Bühler et al. (ref. 1) measured the solar-wind-implanted He and Ne in an unpainted aluminum tube from Surveyor 3 and found solar wind ^{4}He contents ranging from 6.30 to 0.40×10^{-6} cm^3 (STP)/cm^2, depending on the site of the measurement. Because it is possible with present techniques to measure tritium contents as low as 0.005 dpm/cm^2 in the painted aluminum and probably smaller amounts of tritium in the unpainted aluminum, tritium in the solar wind can be determined in concentrations as low as ^{3}H:^{4}He of about 3×10^{-10}.

To determine the presence of solar wind tritium, it is necessary to establish a correlation between tritium excesses and the solar wind exposure. The solar wind He and Ne in painted aluminum cannot be determined easily because of the adhesion of small amounts of lunar soil to the paint, as the lunar soil contains enormous amounts of solar wind He and Ne. The tritium, however, can be determined in the painted aluminum because the paint contains little hydrogen and the lunar soil does not contain excessive amounts of tritium. As the amount of exposure of the Surveyor samples to sunlight is proportional to the solar wind exposure, the presence of solar wind tritium could be established for painted aluminum material by correlating tritium excesses with the exposure to sunlight.

Measurements

The apparatus used for hydrogen extraction and tritium counting was identical to that used for lunar samples (see ref. 2), except that alumina rather then molybdenum crucibles were used as samples and that smaller hydrogen counters with lower backgrounds were used for counting. Alumina crucibles, when tested by melting 0.21 g of zone-refined aluminum, released only 0.05 cm^3 (STP) of hydrogen, which was less than the molybdenum crucibles released.

Various hydrogen counters with different volumes were built. The counter most appropriate for painted sample 1011,2 which had an area of 1.3 cm^2 and a weight of 0.335 g, was a proportional counter of 7-cm^3 volume with a background of 0.0261 ± 0.0014 count/min and an efficiency of 50 percent. Except for its smaller volume, this counter had the same design as that shown in reference 2, figure 3. Its resolution with an ^{55}Fe source was 24 percent.

The sample was placed in the alumina crucible in a quartz furnace and the system was evacuated. Carrier hydrogen, approximately 0.30 cm^3 (STP), was added to the furnace. The furnace was heated to 270°C for 2 to 3 hr; the gas was removed from the furnace, with the furnace and sample at 270°C. The volume of the gas was measured; it was then transferred to a section of the system with finely divided vanadium metal powder at 800°C. The hot vanadium removed the chemically active constituents. The vanadium was slowly cooled to room temperature to absorb hydrogen as vanadium hydride. No measurable amount of gas remained. The vanadium was reheated; the evolved hydrogen was collected and its volume measured. The hydrogen was passed through a charcoal trap at dry-ice temperature to remove any possible radon contamination and added to the proportional counter that contained 400-torr pressure of P–10 gas (counting gas with 90 percent argon and 10 percent methane). The counter was then removed from the system and counted in a low-level unit where the tritium activity and its energy spectrum were measured. The sample was melted by induction heating in the presence of a similar amount of hydrogen carrier; the gases were removed and the hydrogen was purified in the same manner and counted. The sample was remelted; deposits on the furnace walls were severely heated until no tritium remained.

Table 1 gives the results for three painted aluminum blanks of the same material as that used in the Surveyor 3 camera shroud and for sample 1011,2. The amounts of hydrogen released from the blanks were between 0.36 and 0.47 cm^3 (STP)/cm^2. The hydrogen was counted in three counters of different sizes. No tritium was observed in the hydrogen from the blanks. Surveyor sample 1011,2 had a larger amount of hydrogen than the blanks; its total hydrogen content was 1.65 cm^3 (STP)/cm^2; 0.48 cm^3 (STP)/cm^2 was released in the 270°C heating. Measurable amounts of tritium were released in the 270°C heating, in the melting, and in the heating of the wall deposits and remelting, but not in the reheating of the wall deposits and second remelting. In the 270°C heating, there was 0.018 ± 0.004 dpm of tritium; in the melt and the heating of the wall deposits and first remelt combined, there was 0.045 ± 0.006 dpm. On a weight basis, there was a total tritium activity of 188 ± 21 dpm/kg; on an area basis, there was 0.048 ± 0.005 dpm/cm^2.

The ^{3}H activity in sample 1011,2 is compared in table 2 with the amounts observed by R. W. Stoenner and R. Davis [1] in samples 931, 932, and 937. The hydrogen was reduced from water collected on charcoal and counted with carrier hydrogen by Stoenner and Davis, who obtained the charcoal from a melt by O. A. Schaeffer in an extraction system of a mass spectrometer. These samples had been subjected to a 150°C bakeout with an additional Surveyor sample. The water collected from the 150°C bakeout of these four samples had 0.33 dpm of tritium.

Results

On an area basis, sample 1011,2 had approximately the same amount of tritium as sample 931, a factor of 2 less than sample 932, and a factor of 50 less than sample 937. There appears

---
[1] Private communication, 1971.

TABLE 1.—*Hydrogen and tritium contents measured*

Sample	Mass, g	Area, cm²	Extraction	Hydrogen released, cm³ (STP)	Effective counter volume, cm³ (percent)	Counter background, counts/min	Activity, counts/min	^3H, dpm
Painted aluminum blank	0.320	1.3	Melt	0.45	42 (62)	0.140 ±0.005	0.139 ±0.005	<0.012
Painted aluminum blank	.255	1.0	Melt	.47	14 (55)	0.048 ±0.003	0.042 ±0.003	<0.008
Painted aluminum blank	.335	1.3	20° to 270°C	.09 [a]	7 (50)	0.0261±0.0014	0.0269±0.0017	<0.005
			270°C to melt	.30 [a]				
1011,2	.335	1.3	20° to 270°C	.62	7 (50)	0.0261±0.0014	0.0353±0.0016	0.018±0.004
			270°C to melt	.82	7 (50)	0.0261±0.0014	0.0375±0.0016	0.022±0.004
			(Wall deposit heating and remelt.)	.70	7 (50)	0.0261±0.0014	0.0383±0.0016	0.023±0.004
			(Wall deposit reheating and second remelt.)	.09	7 (50)	0.0261±0.0014	0.0252±0.0016	<0.004
Total								0.063±0.007

[a] Combined hydrogen for counting.

TABLE 2.—*Tritium activity per unit area in painted Surveyor 3 samples*

Sample [a]	Area, cm^2	Sample location	^3H, dpm/cm^2	Estimated sunlight dosage
931	1.3	Side of lower shroud toward LM	0.048 ± 0.005 [b]	Only late evening
1011,2	1.3	Front side of lower shroud perpendicular to LM.	0.48 ± 0.005 0.035 ± 0.004 [c]	Only morning
932	1.17	Side of lower shroud away from LM	0.086 ± 0.009 [b]	Morning and noon
937	.08	Top visor	2.5 ± 0.3 [b]	Almost all day

[a] Order of samples is according to increasing amounts of sunlight received.
[b] Activity after 150°C bakeout obtained by R. W. Stoenner and R. Davis.
[c] Activity after 270°C extraction.

to be a correlation of tritium content with exposure to sunlight. Sample 937 was taken from the top visor, which protected the camera mirror from the glare of direct sunlight, and therefore received the maximum amount of sunlight. The other samples were taken from different sides of the lower shroud of the camera housing. These sides were approximately perpendicular to the lunar surface. The side away from the Lunar Module (LM) was tilted slightly toward the sky; the side toward the LM was tilted slightly toward the lunar surface. Sample 932 was taken from the side away from the LM, and sample 931 from the side toward the LM. Sample 1011,2 was taken from the front side, which was perpendicular to these two sides and more closely perpendicular to the lunar surface. From photographs of Surveyor 3 and the sample locations, it is estimated that sample 931 received sunlight only during the late evening; sample 1011,2 received sunlight only during the morning; sample 932 received sunlight during the morning and noon; and sample 937 received sunlight for almost the entire day.

The amount of tritium in Surveyor 3 sample 1011,2 (188 ± 21 dpm/kg) exceeds the amount expected from the tritium content of lunar rock 12002 by at least a factor of 3. At the top of rock 12002 (0- to 0.8-cm depth) there was 392 ± 11 dpm/kg (ref. 2). If this sample were bombarded for only 31 months by the same intensity of cosmic rays and solar flares, it would have had only 57 ± 2 dpm/kg. Although the Surveyor material was a surface sample only 0.26 g/cm^2 thick, it was almost perpendicular to the lunar surface so that about half the solid angle was shielded by the camera. Although there were several large solar flares during the 31 months of Surveyor 3 exposure, it is unlikely that its bombardment was more than three times as intense as the average of the past 30 years, which included the active 1958 to 1961 period. A flux of $10^5/cm^2$ yr of solar wind tritium would account for the excess tritium in sample 1011,2. If retained in lunar material, a solar wind tritium flux of this magnitude would have contributed 25 percent of the tritium observed in the top sample of rock 12002. However, the location of sample 1011,2 was such that it was exposed to the solar wind only during the lunar morning; thus, the top of a lunar rock such as 12002 would have received a much greater solar wind exposure than this Surveyor sample. This leads to the conclusion that if there is solar-wind-implanted tritium on lunar rocks, it is not retained by them.

Some of the tritium from the Surveyor 3 samples is loosely bound; 28 percent of the tritium from sample 1011,2 came off at 270°C, which may indicate that some of its tritium was artificial contamination. However, some of the solar wind tritium may be loosely bound because it is implanted within 1 μm of the surface and the paint contains fine clay particles of kaolin. To establish the presence of solar wind tritium, the tritium must be measured in more samples, and the correlation with solar wind exposure must be more firmly established.

References

1. Bühler, F.; Eberhardt, P.; Geiss, J.; and Schwarzmüller, J.: "Trapped Solar Wind Helium and Neon in Surveyor 3 Material." *Proceedings of the Apollo 12 Lunar Science Conference.* To be published.
2. D'Amico, J.; DeFelice, J.; and Fireman, E. L.: "Solar-Flare Bombardment of the Moon." *Proceedings of the Apollo 11 Lunar Science Conference*, vol. 2, 1970, pp. 1029–1036.

ACKNOWLEDGMENTS

We are grateful to J. DeFelice and J. D'Amico for their work on this experiment.

PART C

HIGH-VOLTAGE TRANSMISSION MICROSCOPY OF SURVEYOR 3 CAMERA SHROUDS

R. M. Fisher, W. R. Duff, L. E. Thomas, and S. V. Radcliffe

Successful retrieval of the television camera and other components from the Surveyor 3 spacecraft by the Apollo 12 astronauts (ref. 1) has provided a unique opportunity to study the effects of a known and relatively extensive exposure to the lunar environment. Microstructural effects, including those produced by micrometeorite impact, radiation damage (by both the solar wind and cosmic rays), and solar heating, could be expected in the materials used to fabricate the spacecraft. High-voltage transmission electron microscopy (HVEM) is a good method of examining the internal microstructure at high resolution because the higher penetrating power permits an examination of thicker specimens, thus providing a more representative sampling and minimizing the uncertainty of obtaining a suitable specimen from a limited supply of material. Even if no effects due to exposure on the lunar surface can be found, high-voltage examination will at least serve to categorize the internal structure to aid in the interpretation of the results obtained from other studies.

Samples

Samples received were in the form of 1 cm² of painted and unpainted aluminum alloy sheet from the top of the camera visor and the sides and bottom of the lower camera shroud. (See fig. 1.) The exact location of each sample is shown by the white areas in the photographs in figure 2. The JPL code identification and brief description of the samples are given in table 1. Complete information about the samples and their handling, etc., may be found in references 2 through 4.

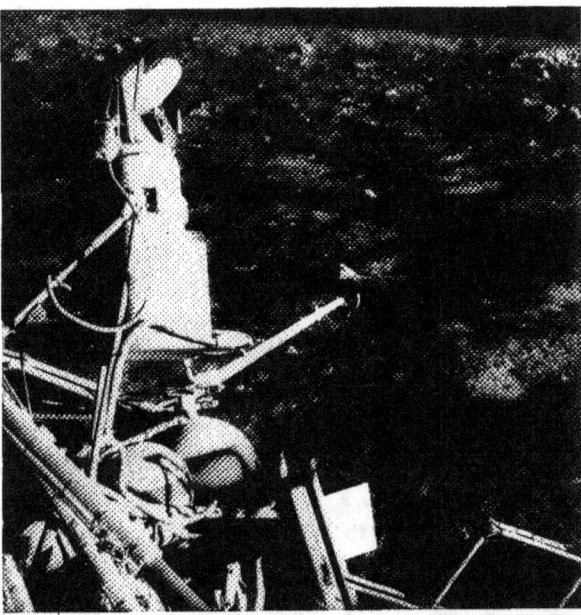

Figure 1.—Samples from the Surveyor 3 camera visor (at top and bottom shroud) examined by high-voltage electron microscopy (NASA photo AS12–48–7105).

RADIOACTIVITY AND RADIATION DAMAGE ANALYSES 185

UPPER SHROUD BONNET

TOWARD LM

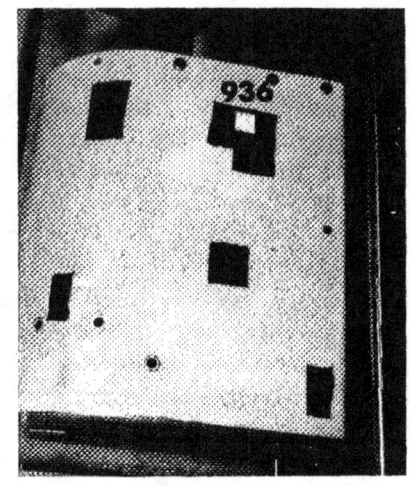

LEEWARD

LOWER SHROUD

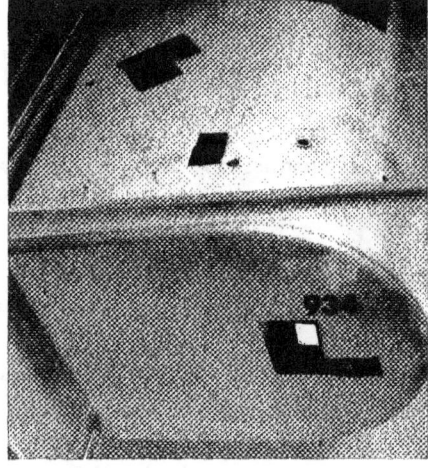

BOTTOM

FIGURE 2.—White areas indicate position of samples of camera shrouds examined by high-voltage electron microscopy.

TABLE 1.—*Identification of Surveyor 3 camera samples*

Log number [a]	Description	Radiation time,[b] hr	Reflectivity,[c] percent
933 (1/2 of 906)	Upper shroud visor:	2121	
	White paint exterior		33 to 52
	Optical black interior		3.5 to 4
934 (894)	Lower shroud bottom: White paint exterior	689	47.5 to 62
935 (898)	Lower shroud side toward LM: White paint exterior	814 to 1440	40 to 58
936 (900)	Lower shroud side away from LM: White paint exterior	5322	61 to 75

[a] Parentheses indicate original number before redistribution; number indicates position designation in ref. 3.
[b] Detailed description of radiation exposure is in ref. 2.
[c] For λ of 0.5 and 1 μm (from fig. J–15 and table J–34 of ref. 3).

The aluminum alloy is given as type 6061–T4, which has the approximate composition 1.0 percent Mg, 0.6 percent Si, 0.25 percent Cu, 0.25 percent Cr. This alloy is commonly considered as a binary of aluminum with Mg_2Si. The T4 designation means that the alloy was quenched from a temperature of 515° to 540°C so that the Mg_2Si will be dissolved in solid solution. However, the alloy will age "naturally" at room temperature to form extremely small spherical zones, which somewhat increases the hardness. Unidentified chromium-rich particles are not dissolved at the T–4 temperature, so that the quenched alloy also contains a relatively coarse dispersion of spherical chromium particles.

The varying exposure to solar radiation each lunar day of the different samples as determined by Nickle is shown in figure 3. The time of exposure for each lunar day is approximately double the number of Sun angle degrees shown as the Sun traverses the lunar sky at 0.51°/hr. The total exposure during the 31.9 lunar days varied from 689 to 5322 hr. (See table 1). Also shown are the approximate values for the reflectivity given in references 3 and 4. The reflectivities of the painted surfaces are considerably less than values of more than 90 percent obtained at the time of launch because of the presence of a coating of fine lunar dust particles and degradation by solar radiation. There is some recovery in reflectivity during exposure in the Earth's atmosphere, so that the reflectivity at the time of the Apollo 12 mission was probably even lower. These matters are discussed in some detail in references 3 and 4.

Before preparing the samples for transmission microscopy, they were examined optically to determine the nature of the surfaces. A few impact sites were noted on the polished bottom surface of the lower shroud. One-half of several stereo pairs of one of these sites is shown in figure 4 at different magnifications and after ultrasonic cleaning. Elemental analysis in the scanning electron microscope showed that the embedded particle has the composition of lunar pyroxene. Whether this particle was embedded in the surface of the aluminum while the camera was on the Moon or during subsequent handling and shipment is not readily determined. However, appearance of the impact site and its surroundings suggests that the particle struck the surface at high velocity and was not "ground in."

Transmission Electron Microscope Examination

Specimens were prepared for transmission electron microscopy by first hand-grinding with abrasive paper to a thickness of 150 μm. The edges were lacquered and the sample electropolished in 10 percent perchloric methanol using the "window" method, to a thickness of about 25 μm. Final thinning was accomplished by polishing 3-mm punched disks in an acetic-phosphoric-nitric acid solution. In all cases, there was no difficulty in obtaining foils suitable for HVEM.

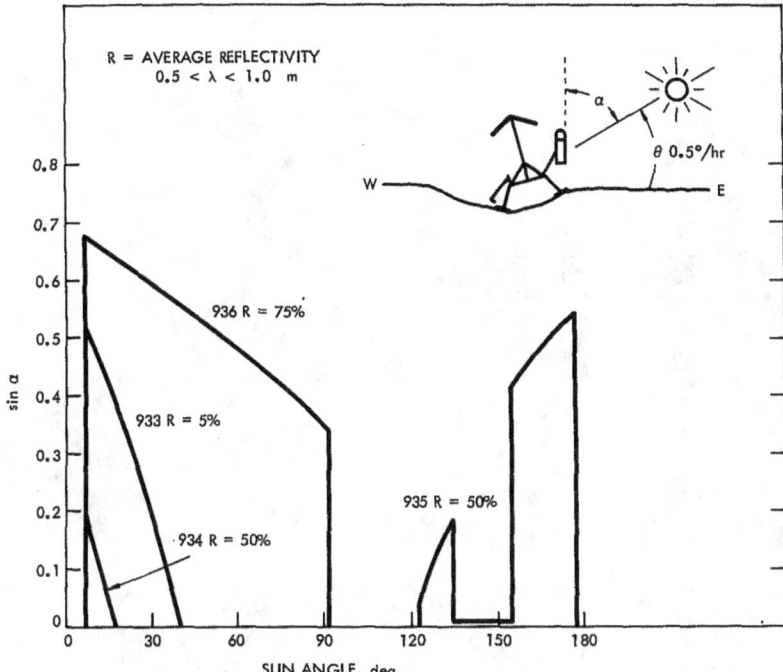

FIGURE 3.—Solar radiation exposure of Surveyor 3 camera shrouds during each of 31.9 lunar days obtained from analysis by N. L. Nickle (ref. 2). Reflectivity measurements by Hughes Aircraft Co. (ref. 3).

Sample 933: Camera Visor. The transmission electron micrograph in figure 5(a) reveals the presence of laths and plate-like Mg_2Si precipitate. This microstructure corresponds to a fairly advanced state of aging, as discussed in a subsequent part of this article. The sample contains a fairly high dislocation density, probably introduced when the aluminum alloy sheet was formed into the visor hood.

Sample 934: Polished Camera Bottom. As seen in figure 5(b), this sample contains a very high density of extremely fine needles of the intermediate form of Mg_2Si. Because of the high density of precipitates, the dislocation content is extremely difficult to determine in the micrograph, but tilting into other contrast conditions showed that relatively few dislocations were present.

Sample 935: Painted Lower Shroud Toward LM. The electron micrograph in figure 6(a) shows that the sample contains a large number of lath-like particles of Mg_2Si and some evidence for the plate-like form. This represents a stage of aging intermediate between samples 933 and 934.

Sample 936: Painted Aluminum Side Away From LM. The microstructure in this sample is similar to 935 (see fig. 6(b)), except that the needles are a little smaller, indicative of slightly less aging.

Three of the four samples supplied were painted before launch with an inorganic paint about 125 μm in thickness, so that there was little point in examining the metal surface below the paint. However, the bottom of the lower shroud was not painted, and it was possible to determine the nature of the structure site at the surface of this sample by polishing from only the inside. The procedure does not produce a good foil; however, it did succeed in revealing the presence of an extremely high dislocation density in the first 2 or 3 μm from the surface, as shown in figure 7(a). Subsequent slight polishing showed (as in fig. 7(b)) that the dislocation structure is somewhat less dense at 5 or 6 μm below the surface. At a depth of 25 μm, the dislocation structure is definitely less dense (fig. 7(c)); the fine precipitate needles found in the interior of the sample are not evident. It is possible that the magnesium was lost from this region by preferential oxidation during the T-4

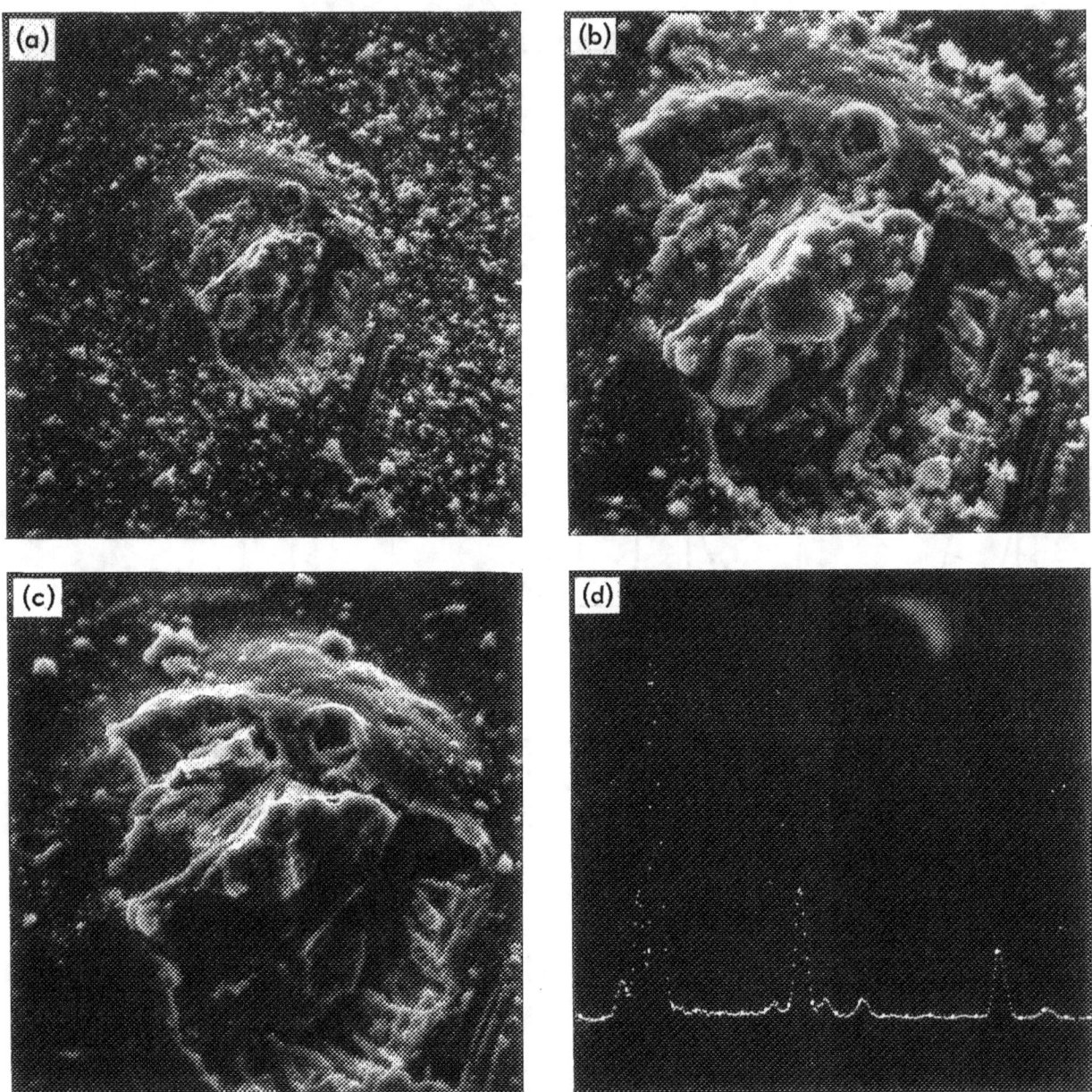

FIGURE 4.—Scanning electron micrograph of lunar dust embedded in bottom of lower shroud. (a) As received; 500 ×. (b) as received; 1000 ×. (c) after ultrasonic cleaning; 1000 ×. (d) spectrochemical analysis identified particle as pyroxene.

heat treatment so that the Mg_2Si precipitate did not form upon aging.

Simulated Microstructures

To aid in the interpretation of the microstructure observed in the Surveyor 3 samples, specimens of commercial alloy 6061 were given the T–4 temper (quenched from 530°C) and isothermally aged over a range of times and temperatures. Representative microstructures developed during some of the heat treatments used are shown in figure 8. All results of the

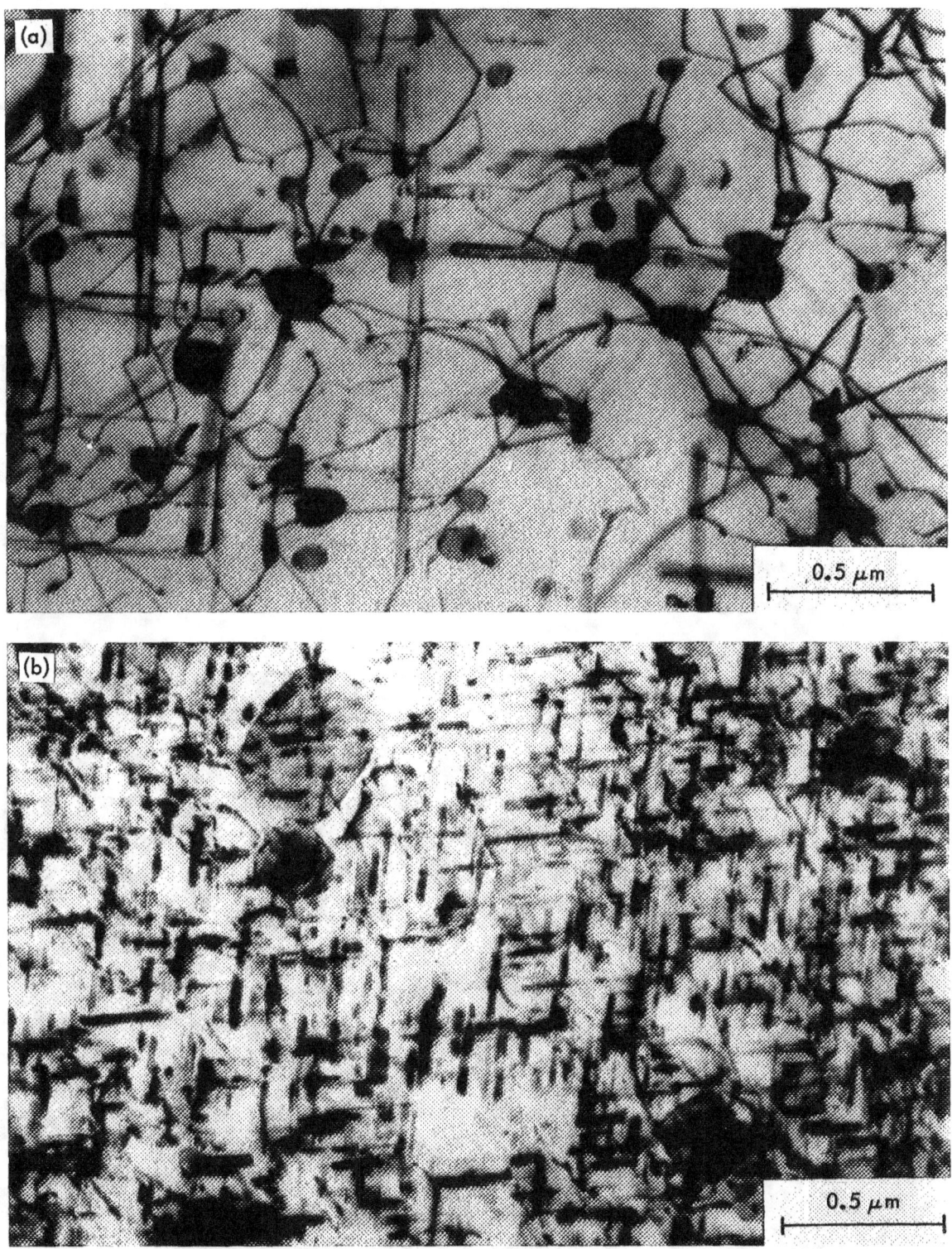

FIGURE 5.—High-voltage (1000 kV) transmission electron micrographs of Surveyor 3 camera shrouds. (a) Upper shroud visor: sample 933. (b) Lower shroud bottom: sample 934.

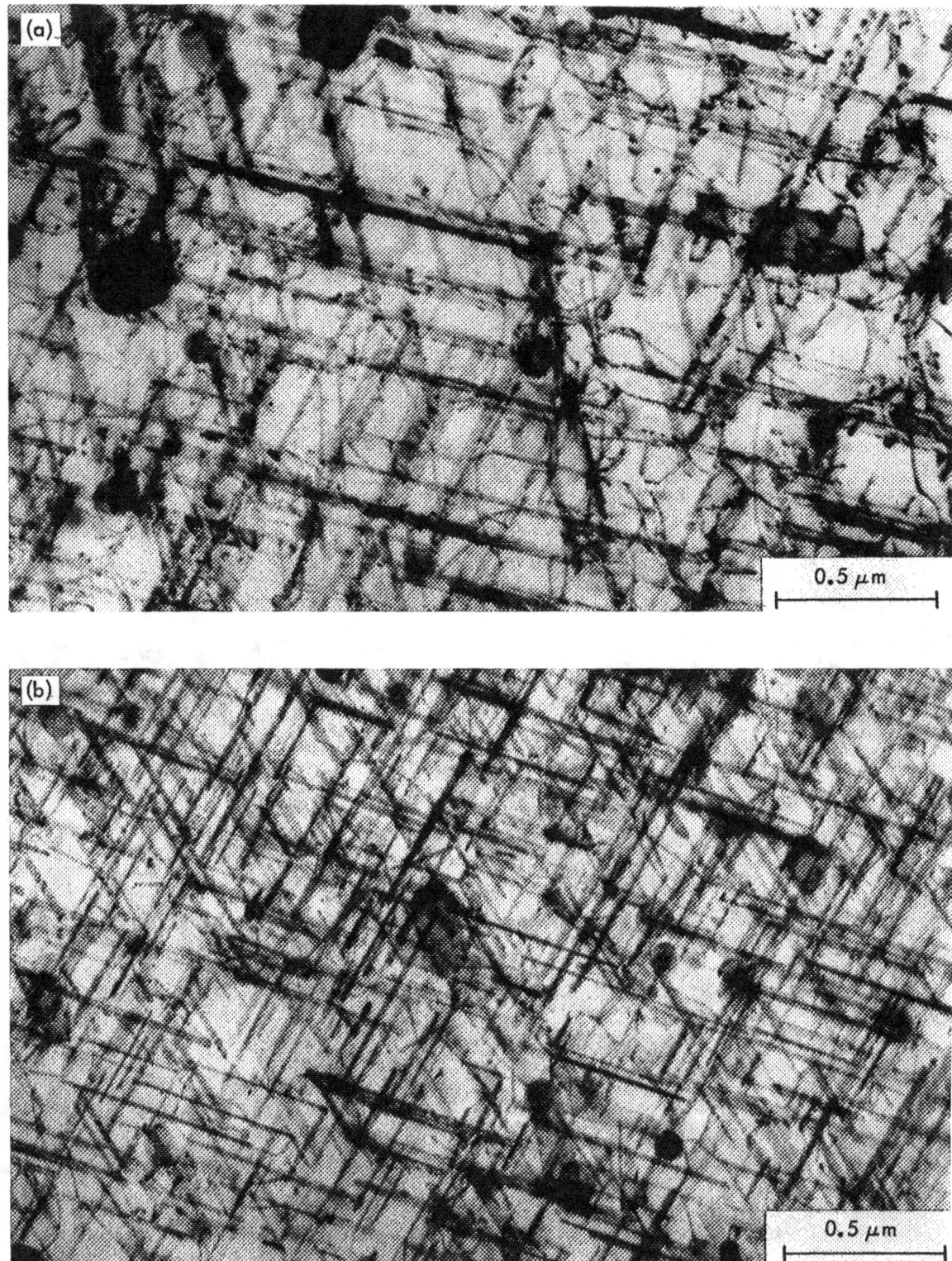

FIGURE 6.—High-voltage electron microscopy (1000 kV) of Surveyor 3 lower shroud showing Mg_2Si precipitates. (a) Side toward LM: sample 935. (b) Side away from LM: sample 936.

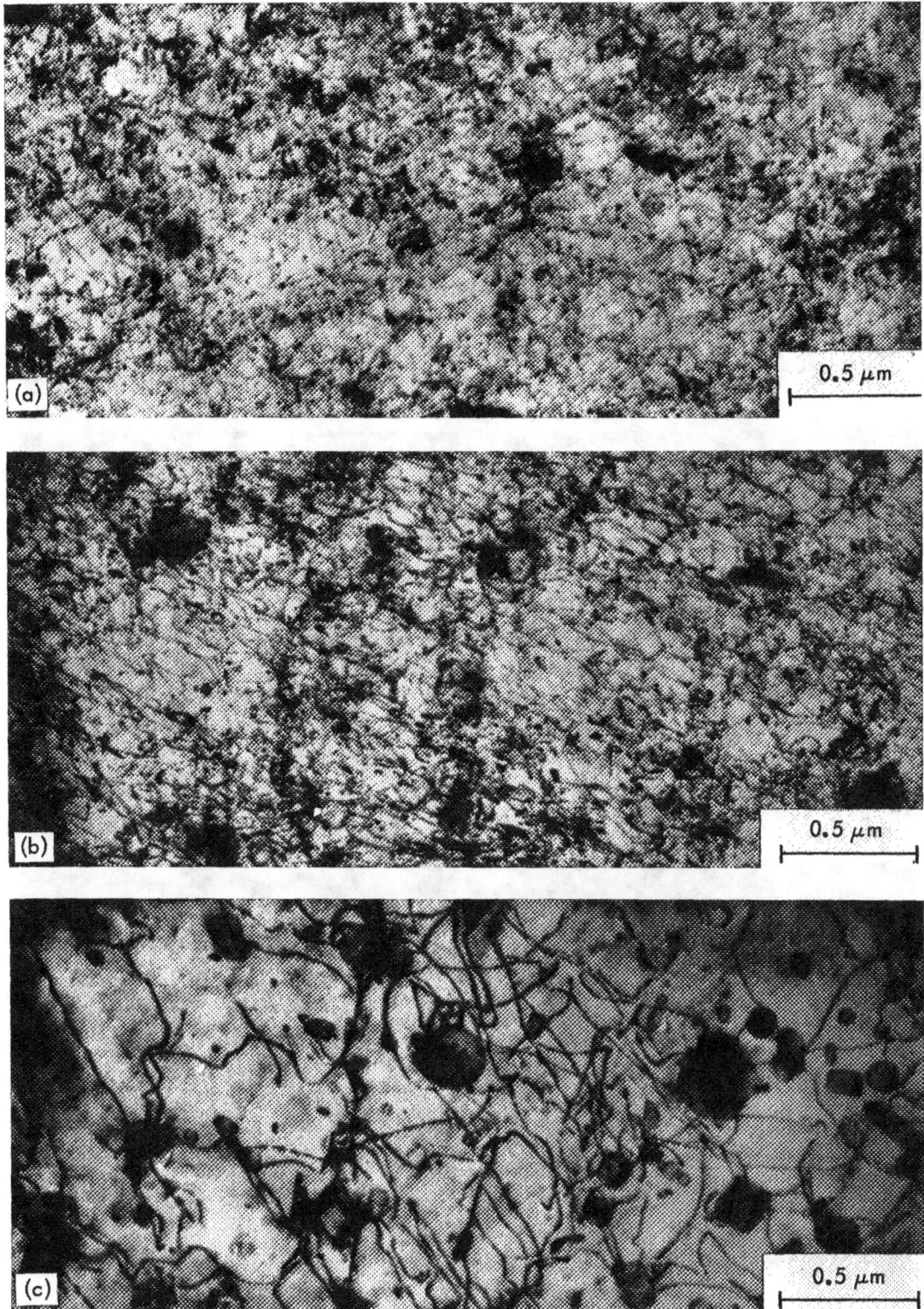

FIGURE 7.—Dislocation structure near exterior bottom surface of polished aluminum alloy camera shroud: sample 934. At 1000 kV. (a) 0 to 2 µm. (b) 5 to 8 µm. (c) 25 µm.

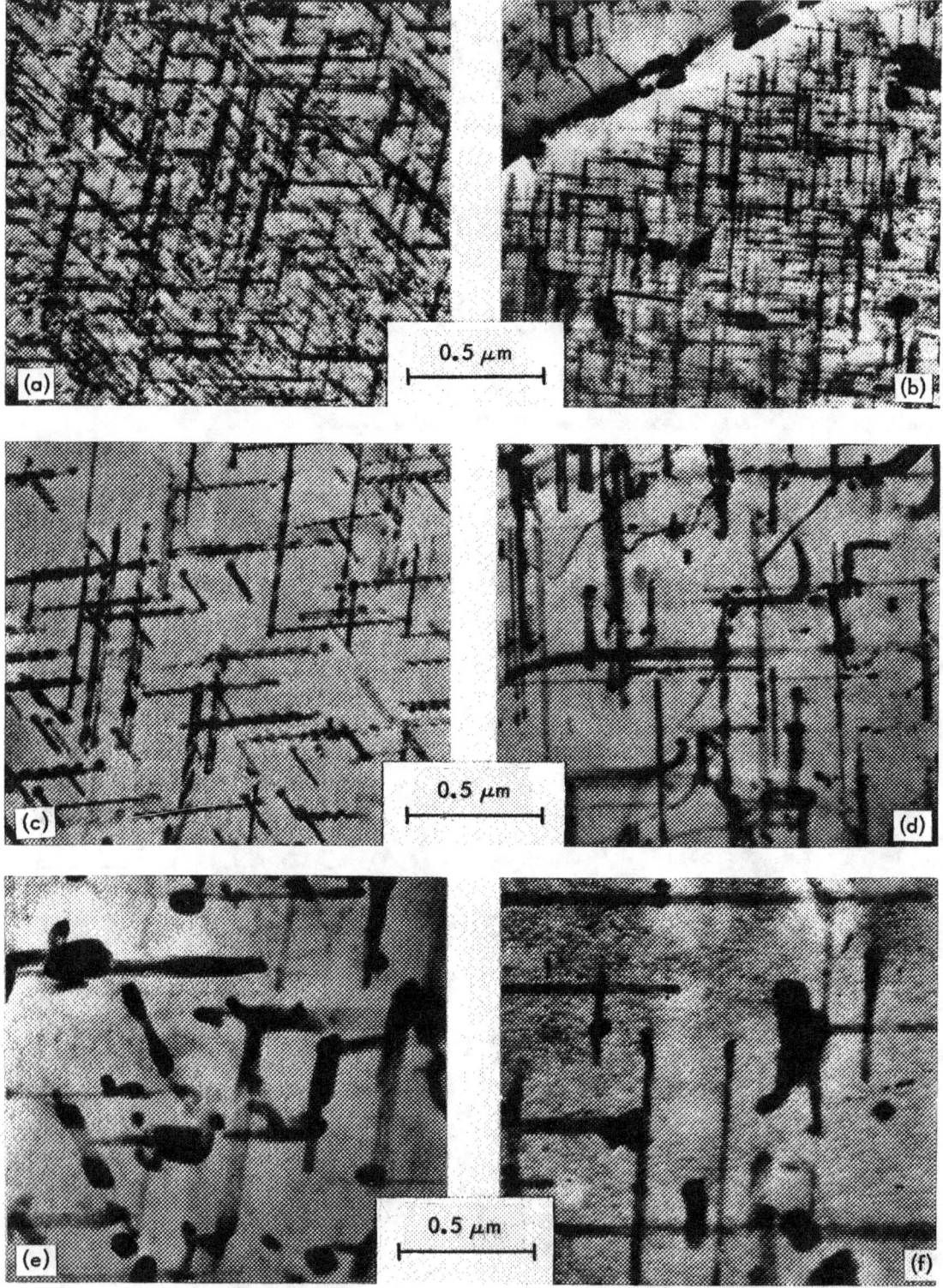

FIGURE 8.—Transmission electron micrographs of quench-aged samples of Al-Mg$_2$Si (Alloy 6061–T4). At 100 kV. (a) 175°C; 216 hr. (b) 250°C; 3 hr. (c) 300°C; 64 hr. (d) 325°C; 1 hr. (e) 350°C; 19 hr. (f) 375°C; 2¼ hr.

isothermal aging are summarized in the customary log time vs. $1/T$ plot in figure 9. Although few of the simulated microstructures were exactly the same as the various Surveyor samples, it was possible to draw lines corresponding to each case with reasonable accuracy. Samples 935 and 936 are shown together as a dotted line, as the uncertainty is greatest in this case.

The temperature coefficient, i.e., "activation energy," for structures corresponding to the early stage indicated by sample 934 is about 29 kilocalories. This corresponds to that obtained for the production of peak hardness during quench aging (ref. 5). The temperature exponent from sample 933 is approximately double the above value and is equal to 58 kilocalories. This apparent "activation energy" corresponds to coarsening of the Mg_2Si particles and is higher than the diffusion value for Mg in aluminum of about 35 kilocalories. The heat of solution of the Mg_2Si precipitate of about 7 kilocalories is to be added to the diffusion figure. However, it is probable that the coarsening process is controlled by the rate of the movement of the coherent interface. At any rate, regardless of the precise meaning of the temperature exponent, it can be used to describe the kinetics of the precipitate process as long as the comparison is made on the basis of equivalent microstructures.

Results

Radiation Effects

Unlike the case of nonmetallic materials, ionization tracks are not produced in metals by high-energy electrons, protons, or heavier particles. Thus, structural evidence for radiation damage is in the form of aggregates of point defects originally produced that occur either as interstitial or vacancy loops (both types may be present) or voids. Nucleation of defect clusters is a complex process determined by the instantaneous concentration of defects and the presence of impurity

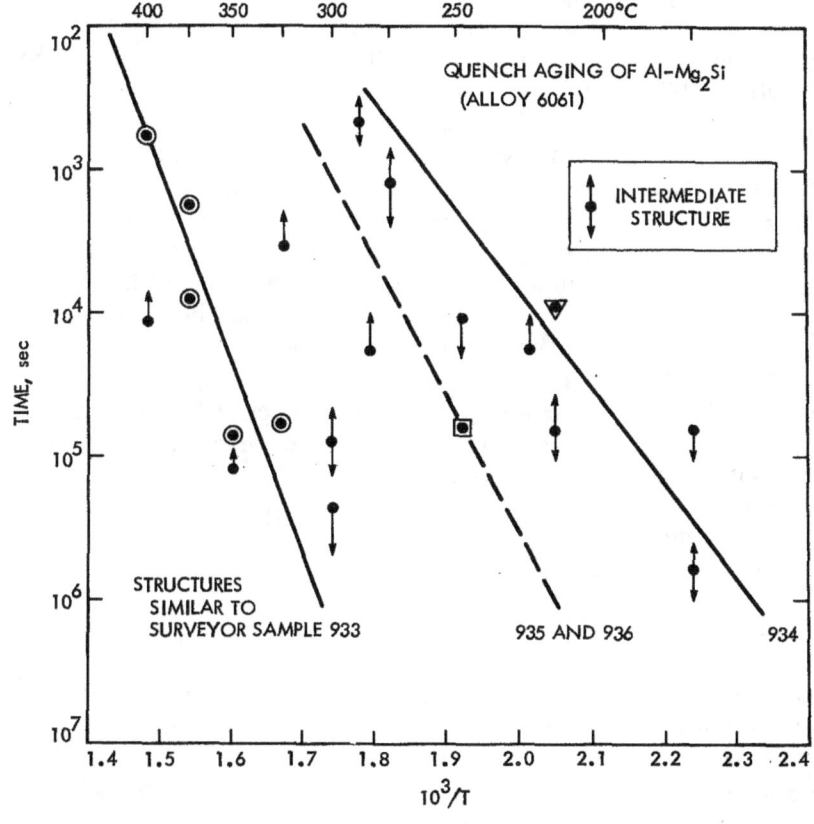

FIGURE 9.—Quench aging of Al-Mg_2Si to produce microstructures similar to those in samples of Surveyor 3 camera shrouds.

atoms, dislocations, grain boundaries, and precipitate surfaces that may act as sinks. Low dose rates may never produce sufficient concentration to result in visible clusters.

No clear-cut evidence of radiation damage effects was observed in the samples examined. However, the total dose of high-energy protons (1 to 80 MeV) was no more than 10^{12} (refs. 6 and 7), which is substantially below levels that have been used to investigate proton damage in metals. Any gas bubbles that could have formed would have been stable to temperatures of 450°C. To check on the possibility that the very low dose rate could modify the precipitation kinetics, simulation samples of alloy 6061 were irradiated with 17-MeV protons in both the fully aged and unaged condition. No detectable differences in microstructure resulted, nor was there any effect on the subsequent aging process.

The particle density of 1- to 8-kV protons and helium ions in the solar wind is higher, although the penetration is limited to less than 1 μm. Thus, only the unpainted sample is suitable for examination. This piece received a total of 600 hr in exposure to the solar wind at close to grazing incidence so that penetration would be even less. The high dislocation density of about 10^{12}, which is produced by polishing, rules out any possibility of observing gas bubbles or dislocation loops. An interaction between helium and other gases with these dislocations should be considered in interpreting rare gas evolution effects.

Solar Heating

The noticeable differences in microstructures between the upper visor and the sides and bottom of the lower shroud suggest a different thermal history in each case. In terms of the extent of the aging, they rank in order as 933, 935, 936, and 934. This observation can be interpreted quantitatively to a limited extent by comparison with the simulation samples illustrated in figure 8 and summarized in figure 9. However, it is necessary to make several basic assumptions in order to estimate the thermal history. The samples did not receive any heat treatment subsequent to the T-4 temper (quench) other than ambient temperature natural aging which occurred before launch. The differences found between the sides and bottom of the lower shroud suggest that this is a plausible assumption. The kinetics of the initial stages of precipitation depend on quenching rate, and the simulation samples were not given a constant period at room temperature after quench prior to aging. Both of these factors introduce some uncertainty into the comparison with the Surveyor material; the resulting errors are not likely to be large. By comparing the microstructures found in the Surveyor shroud and visor with similar types in the simulation specimens, it is possible to estimate the thermal history without detailed knowledge of the time dependence, specific atomic mechanisms, etc.

If the Surveyor samples remained at constant temperature during the whole period of solar heating, the respective temperatures could be read directly from figure 9, as the total time of exposure is known. However, as shown in figure 3, the relative solar energy incident per unit area changed drastically as the Sun rose and set on the lunar surface, and the temperatures may be expected to follow. Telemetry data of the temperatures of various Surveyor 3 components (ref. 8) showed very little lag in temperature change during an eclipse or when the sensor was shadowed by some other part of the spacecraft. Presumably, this is a result of the relatively low heat capacity of the components and approximately blackbody conditions. Thus, it is a reasonable assumption that the temperature changes with Sun angle in accord with the angle of incidence, as shown in figure 3, which can be taken as approximately linear with time.

Following the customary simple analysis of reaction kinetics, the number of "events," N, required to produce a given microstructure in a given time, t, is

$$N = \dot{R} t$$

where the reaction rate is

$$\dot{R} = \nu_0 \, e^{-Q/RT}$$

and ν_0 and Q are the usual pre-exponential "frequency" factor and apparent "activation" energy, respectively.

In the case of continuous heating (or cooling), the structure is produced by an accumulation of events occurring in each time and temperature interval during the total lunar exposure; i.e.,

$$N = n\,v_0\,e^{-Q/RT_1} t_1 + e^{-Q/RT_2} t_2 +, \text{etc.}$$

where $n =$ the number of lunar cycles; i.e., about 32.

Adopting the linear relationship between time and temperature discussed above,

$$T = \frac{T_M - T_0}{t_e} t + T_0$$

where

T_M = maximum temperature
T_0 = temperature at end (or beginning of exposure)
t_e = time of solar exposure per lunar day

It does not seem realistic to assume that $T = 0°K$ when $\sin \alpha = 0$, as the various components of the spacecraft were in contact and the lunar surface remains at about 100°C during the lunar day. For this reason, it was assumed that $T_{\sin \alpha_{=0}} \approx 100°C$; i.e., 373°K. The aging that occurs at this low temperature is not great, but this assumption has a significant effect on $\Delta T/\Delta t$, and thus the actual time during heating (or cooling) between 95 and 100 percent of T_M, where more than 90 percent of the aging occurs.

The assumed situation is essentially reversed for sample 935; in this case, the initial brief period of solar heating was ignored.

The value of T_M for each case was found by numerical methods on a computer in the following manner. Equating "structures" produced during continuous and isothermal heating

$$n t_e\,v_0\,e^{-Q/RT_c} = \frac{n\,t_e\,v_0}{T_M - 373} \int_0^{T_M} -\frac{Q}{RT_M}\,dT$$

where T_c is the "constant" equivalent temperature corresponding to the total exposure time nt_e and was derived from figure 9 (by extrapolation) for each case.

The apparent maximum temperatures obtained in this way are—

Sample 933: about 319°C
Sample 934: about 164°C
Sample 935: about 217°C
Sample 936: about 179°C

As expected, the sample with the lowest reflectivity (camera visor 933 with an optical black interior) apparently reached the highest temperature. The polished aluminum bottom of the shroud shows the lowest maximum temperature, although it is somewhat higher than the values of 100° to 125°C indicated by telemetry for some of the Surveyor compartments and electronic packages (ref. 8). The precipitation microstructure observed would have formed if the sample material remained at about this temperature for the full 10 000 hr of solar exposure on the Moon. Thus, the temperature in this case can be bracketed only between 100° and 164°C. The effect of low-temperature aging is not great for the other samples and can be neglected.

Samples 935 and 936 have similar microstructures between the two extremes, although 935 could be slightly more advanced. This sample was exposed for a much shorter period than sample 936, so that analysis indicates a maximum temperature almost 40°C higher. This could be due to the much lower reflectivity of sample 935 compared with 936. As discussed in detail in references 3 and 4, the reflectivity of the Surveyor surface decreased markedly as a result of lunar exposure due to a coating of lunar dust and degradation by solar radiation. Some recovery of the latter change was noted during the Hughes study so that the reflectivity on the surfaces on the Moon probably was even less than the values measured 6 months after the camera was returned to Earth.

Because of deterioration by solar radiation, the reflectivity would be less and the solar heating greater during each succeeding lunar day. This could explain the discrepancy between values transmitted from the spacecraft during the first few months compared with the temperatures apparently reached later.

If possible, the maximum temperature values obtained in this investigation should be checked against other estimates. As stated, the results depend entirely on the validity of the assumptions. The apparent temperature rise due to solar heating will not affect the structural integrity of spacecraft components unless very long periods of exposure result in appreciable deterioration in reflectivity. However, interpretation of solar wind rare gas studies may be affected if the indicated possibility for substantial thermal diffusion is not considered.

References

1. *Apollo 12 Preliminary Science Report*, NASA SP–235, Washington, D.C., 1970.
2. NICKLE, N. L.: "Surveyor III Material Analysis Program." *Proceedings of the Apollo 12 Lunar Science Conference*. To be published.
3. *Test and Evaluation of the Surveyor III Television Camera Returned From the Moon by Apollo XII*, vols. I and II, SSD 00545R, Hughes Aircraft Co., Culver City, Calif., 1971.
4. *Surveyor III Parts and Materials; Evaluation of Lunar Effects*, P–70–54, Hughes Aircraft Co., Culver City, Calif., 1971.
5. SHCHEGOLEVA, T. V.: "Aging Mechanisms of the Alloy Al-Mg-Si." *Fiz. Metal. Metalloved.*, vol. 25, 1968, pp. 246–254.
6. CROZAZ, G.; AND WALKER, R. M.: "Solar Particle Cracks in Glass From the Surveyor III Spacecraft." *Science*, vol. 171, 1971, p. 1237.
7. FLEISCHER, R. L.; HART, H. R.; AND COMSTOCK, G. M.: "Very Heavy Solar Cosmic Rays." *Science*, vol. 171, 1971, p. 1241.
8. *Surveyor III Mission Report. Part II: Scientific Results*, TR 32–1177, Jet Propulsion Laboratory, Pasadena, Calif., 1967.

ACKNOWLEDGMENTS

We acknowledge gratefully the cooperation and helpful assistance of Neil Nickle of JPL. Thanks are due to the staff members of the USS Fundamental Research Laboratory, C. E. Brickner for the SEM micrographs and analysis, J. L. Bomback and N. Louat for numerical integration of the reaction kinetics, and to A. Szirmae for assistance with the HVEM. We also thank Professor Cohen of the University of Pittsburgh for irradiating some specimens with 17-MeV protons. Helpful information about the microstructures of alloy 6061 was supplied by D. L. Robinson of the Alcoa Research Laboratories.

PART D

SOLAR AND GALACTIC COSMIC-RAY EXPOSURE OF SURVEYOR 3 AS DETERMINED FROM COSMOGENIC RADIONUCLIDE MEASUREMENTS

L. A. Rancitelli, R. W. Perkins, N. A. Wogman, and W. D. Felix

The Surveyor 3 spacecraft had spent 31 months on the lunar surface before the landing of Apollo 12. During this period, it served as a target for the production of spallogenic radionuclides from both the solar and galactic cosmic-ray flux. The uniqueness of the return of pure material that had been exposed to the cosmic-ray flux was unparalleled by any event other than that of the first returned lunar samples. Components from the television camera, the surface sampler scoop, and various other parts of Surveyor 3 have undergone an intensive investigation in order to characterize the solar and galactic cosmic-ray flux from measurements of induced radionuclides in the spacecraft materials.

Procedure

Nondestructive gamma-ray spectrometric techniques developed for the radionuclide analysis (ref. 1) of lunar material have been used to determine the radionuclide content of selected Surveyor 3 parts. Most of the Surveyor 3 components were thin samples and of specific geometric configuration in marked contrast with the lunar rocks. The determination of radionuclide disintegration rates from gamma-ray spectrometric measurements of the samples was made in the following manner. Mockups of each component that contained known amounts of ^{22}Na on the outer and inner surfaces were prepared and counted in the same geometry as the actual Surveyor 3 sample. A comparison of mockup and Surveyor 3 sample photopeak count rates provided a direct method for determining the disintegration rate of the sample. The Surveyor 3 components on which these radionuclide measurements have been made are summarized in table 1 with their observed radionuclide concentrations. The error listed by each measurement reflects the uncertainty in the aluminum content of the sample and in the detector calibration, as well as uncertainties in the mockups and the counting statistics associated with the counting measurements. The value reported in table 1 for the scoop is based on the assumption that it was composed of 50 percent aluminum by weight. The principal target elements for ^{22}Na production in lunar surface materials are Na, Mg, Si, and Al; the target element in Surveyor 3 components was aluminum. Therefore, the discussion of radionuclide concentrations is concerned primarily with ^{22}Na production from Al by high-energy proton reactions.

Radionuclide Production Rate Calculations

To interpret the ^{22}Na concentration in terms of the cosmic-ray exposure, it is essential to know the expected production rate as a function of depth in the Surveyor 3 material. Radionuclide production on the lunar surface is due primarily to solar flare protons in the first few millimeters, but the contribution decreases rapidly with depth.

A calculation of the production rate depth gradient for formation of ^{22}Na from aluminum on the lunar surface requires a knowledge of both the excitation function for the reaction and the energy spectrum of the incident solar and galactic protons. The well-characterized excitation function for ^{22}Na production in aluminum was used.[1] The shape factor for the solar cosmic-ray energy spectrum was determined from satellite data collected during the November 1968 and April 1969 flares.[2] Expressed in the kinetic power law form, the solar proton energy distribution can be stated as

$$\frac{dJ}{dE} = ke^{-\alpha}$$

where J is the proton flux (P/cm² sec sterad MeV); E is the particle energy (MeV), and k is a constant determined from the flare intensity. The shape function, α, was 3.1 for two of the flares and 3.5 for the third flare. For our production rate calculation, an α value of 3.1 was used.

[1] R. L. Brodzinski, BNWL, personal communication, 1971.
[2] T. Hsieh and T. Simpson, University of Chicago, personal communication, 1970.

The calculations were performed assuming that each Surveyor part consisted of an infinite plane. The lamina thickness within the plane was set at 0.05 mm and given a cross section compatible with thin target calculations. Activation within the laminae due to secondary particles was assumed negligible. For a unit incident flux within a specific angular distribution, energy attenuation was calculated as a function of the depth of the laminae within the plane for 2-MeV increments from 10 to 400 MeV. Activation within each lamina was computed as a function of the energy-dependent target element cross section and the proton flux in that lamina as calculated from the primary flux by considering attenuation of the overlying laminae.

Production rates were calculated based on Sun angle irradiation, which assumes that all of the bombarding particles arrive at the lunar surface along the lines of sunlight, and 2π isotropic irradiation. The actual exposure on the lunar surface is probably a combination of both Sun angle and isotropic irradiation. However, an isotropic irradiation flux, which produces a slightly steeper production rate gradient than a Sun angle irradiation, was assumed in the analysis of observed concentrations. The solar production gradient for ^{22}Na from Al is shown in figure 1. The production rate decreases by about a factor of 2 through the first 2 mm, then decreases at a somewhat lower rate with depth. One would, therefore, expect to find a significantly lower average ^{22}Na concentration in the thicker samples than in thin materials. As indicated in table 1, the ^{22}Na concentration in the Al components lies within a fairly narrow range except for the relatively high concentration in the thin-walled aluminum tubing. This is inconsistent with the calculated concentration gradient in aluminum.

Discussion

Calculations were made of the contribution of the nine most prominent solar flares, which include the major flares of April and November 1969, to ^{22}Na production in aluminum components of Surveyor 3. During its lunar residence time, more than 99 percent of the ^{22}Na in Surveyor 3 was produced by the nine flares. About 75 percent of the ^{22}Na present at the time of the

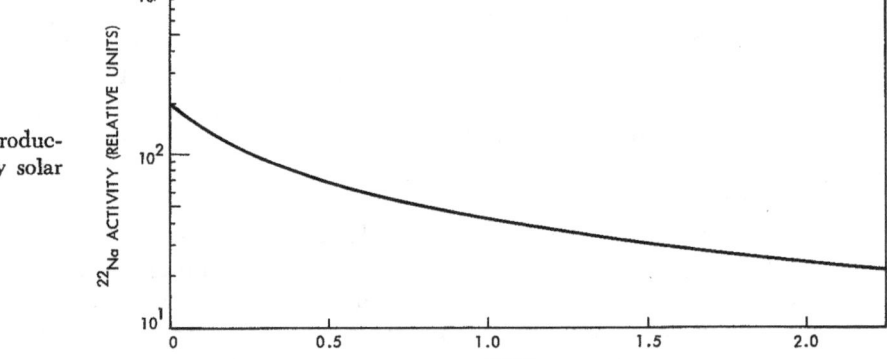

FIGURE 1.—Calculated ^{22}Na production gradient in aluminum by solar protons.

Apollo 12 landing originated during the large flares of April and November 1969; about 20 percent was produced by the November 1968 flare. The remaining prominent flares during Surveyor's residence time produced only about 5 percent of the residual solar-proton-induced ^{22}Na.

Because the energy spectrum of solar protons is much less energetic than galactic protons, their nuclear interactions are limited to near-surface areas, and the effects of the solar proton bombardment would be most readily observable in the thinner Surveyor 3 parts. The ^{22}Na content of the thin (35 mil thick), unpainted aluminum tubing was calculated as described and compared with the observed ^{22}Na content. The solar cosmic-ray-produced ^{22}Na in the thin aluminum tubing was estimated to be 18 dpm/kg, while the observed ^{22}Na activity was 62 ± 10 dpm/kg, indicating a galactic cosmic-ray contribution of about 44 dpm/kg.

Thus, even in the thinnest samples where solar cosmic-ray effects should be most readily observable, galactic cosmic-ray production accounts for more than two-thirds of the total ^{22}Na. In thicker Al samples such as the support struts and support collar, about 95 percent of the ^{22}Na was produced by galactic cosmic-ray bombardment.

The galactic cosmic-ray contribution to the ^{22}Na production in Surveyor 3 can be estimated from known production rates on meteorites. The Lost City meteorite, which had an orbit extending to 2.35 AU (ref. 2), had a ^{22}Na content of 88 dpm/kg in its most heavily shielded fragment.[3] From observed secondary buildup with depth in the St. Severn meteorite, it is apparent that surface concentrations are about one-half those at depth (ref. 3). Also, a meteorite has had a 4π bombardment to saturation of ^{22}Na, while the Surveyor 3 situation had a 2π bombardment for one half-life of ^{22}Na (2.58 yr). Thus, the production of ^{22}Na from aluminum by galactic cosmic rays in Surveyor 3 would be about one-eighth of that from aluminum in meteorites. The ^{22}Na is produced in meteorites from spallation of Al, Mg, and Si; in Surveyor 3, it arises solely from the spallation of Al. Thus, the meteorite value must be adjusted further for the abundances (ref. 4) of these target elements and their relative cross sections for ^{22}Na production.

By this analysis, the galactic cosmic-ray pro-

[3] L. A. Rancitelli, unpublished data, 1971.

TABLE 1. — *Radionuclide content of Surveyor materials*

Material	^{22}Na acitvity,[a] dpm/kg
Struts:	
156	40 ± 5
422	43 ± 5
423	32 ± 4
Support collar:	
44	42 ± 2
45	36 ± 2
Aluminum tubing	62 ± 10
Bipod	26 ± 6
Scoop	36 ± 6
Mirror	1.4 ± 0.2 [b]
	5.3 ± 0.2 [c]

[a] Corrected for radioactive decay to Nov. 20, 1969.
[b] ^{22}Na activity, dpm.
[c] ^{60}Co activity, dpm.

duction rate of ^{22}Na in Al of 42 dpm/kg Al is obtained. This value agrees well with the value obtained (44 dpm/kg) for galactic cosmic-ray production in the thin aluminum tubing and is also in accord with the total ^{22}Na content of strut numbers 156, 422, and 423 and support collars 44 and 45. (See table 1.) This excellent agreement between the galactic cosmic-ray production of ^{22}Na in Surveyor 3 and the Lost City meteorite offers strong evidence that the galactic cosmic-ray flux is almost the same at 1 AU, the location of Surveyor, and in the Lost City orbit, which extends to 2.35 AU.

As the solar cosmic-ray contribution to the ^{22}Na content of the thick Al members is small, this ^{22}Na content provides an excellent means of estimating the incident galactic particle flux for the 31-month lunar residence period of Surveyor 3. Using an average ^{22}Na content of 40 dpm/kg Al and a cross section of 15 millibars for ^{22}Na production by galactic protons (p), we estimate a galactic cosmic-ray flux of about 4 ± 1 p/cm^2 sec on Surveyor 3.

The scoop from the Surveyor 3 spacecraft is constructed of a variety of materials including aluminum, iron, and plastic. The ^{22}Na concentration in the scoop is consistent with the composition of its construction materials. In addition to the ^{22}Na that was observed in the scoop, there were measurable concentrations of Th and U and detectable quantities of ^{26}Al and ^{60}Co. The ^{60}Co could be an impurity in the iron or other construction materials as is also possible for the U and Th. The ^{26}Al may be the result of some residual lunar material remaining in the scoop at the time of our measurements. The television mirror contained a relatively low concentration of ^{22}Na, which is consistent with its composition. The mirror was composed mainly of Be with an aluminum frame and a nickel surface on which a reflective coating was deposited. Although it is possible that the observed ^{60}Co could have been produced in the nickel or in Co impurities in the nickel, the detection of ^{60}Co in a mirror from a Surveyor 3 terrestrial counterpart indicates that it could be an impurity in the construction materials. About 2 g of lunar material was obtained from this scoop and was made available by the Jet Propulsion Laboratory for our analysis. The ^{26}Al and ^{22}Na concentrations in this soil are shown in table 2, where they are compared with ^{26}Al and ^{22}Na concentrations as a function of depth in the double core tube 12025 (ref. 5). From these measurements, it is evident that the lunar material remaining in the scoop was from an average burial depth of about 3.5 cm.

References

1. PERKINS, R. W.; RANCITELLI, L. A.; COOPER, J. A.; KAYE, J. H.; AND WOGMAN, N. A.: "Cosmogenic and Primordial Radionuclide Measurements in Apollo 11 Lunar Samples by Nondestructive Analysis." *Proceedings of the Apollo 11 Lunar Science Conference*, vol. 2, 1970, pp. 1455–1469.
2. MCCROSKY, R. E.: "The Lost City Meteorite Fall." *Sky and Telescope*, vol. 39, 1970, pp. 2–6.
3. MARTI, K.; SHEDLOVSKY, J. P.; LINDSTROM, R. M.; ARNOLD, J. R.; AND BHANDARI, N. G.: "Cosmic-Ray Produced Radionuclides and Rare Gases Near

TABLE 2.—*Comparison of ^{22}Na and ^{26}Al concentrations in core tube 12025 with Surveyor scoop soil*

Sample	Average depth, g/cm^2	^{22}Na,[a] dpm/kg	^{26}Al,[a] dpm/kg	^{26}Al:^{22}Na
12025,14	0.4	197 ± 80	246 ± 69	1.25 ± 0.62
13	1.4	109 ± 36	241 ± 31	2.21 ± 0.78
13,12	1.8	70 ± 35	202 ± 29	2.89 ± 1.50
12	2.7	62 ± 62	176 ± 51	
11	3.8	53 ± 29	169 ± 33	3.18 ± 1.85
10	5.2	56 ± 25	86 ± 21	1.54 ± 0.78
9,8,7	8.5	35 ± 8	54 ± 7	1.54 ± 0.41
6,5,4	13.5	34 ± 7	62 ± 8	1.82 ± 0.44
Surveyor soil		44 ± 15	85 ± 14	1.93 ± 0.73

[a] See ref. 5.

the Surface of Saint Severin Meteorite." *Meteorite Research*, Springer-Verlag, Berlin, 1969, p. 22.
4. CLARKE, R. S., JR.; JAROSEWICH, E.; AND NELEN, J.: "The Lost City, Oklahoma, Meteorite: An Introduction to Its Laboratory Investigation." *J. Geophys. Res.*, to be published.
5. RANCITELLI, L. A.; PERKINS, R. W.; FELIX, W. D.; AND WOGMAN, N. A.: "Erosion and Mixing of the Lunar Surface from Cosmogenic and Primordial Radionuclide Measurements in Apollo 12 Lunar Samples." *Proceedings of the Apollo 12 Lunar Science Conference*. To be published.

ACKNOWLEDGMENTS

We thank R. M. Campbell, D. R. Edwards, J. G. Pratt, and J. H. Reeves of the Battelle Memorial Institute, Pacific Northwest Laboratories, for their aid in standards preparation and in data acquisition.

VIII. Solar Wind Rare Gas Analysis

TRAPPED SOLAR WIND HELIUM AND NEON IN SURVEYOR 3 MATERIAL

F. Bühler, P. Eberhardt, J. Geiss, and J. Schwarzmüller

On April 20, 1967, Surveyor 3 landed on the lunar surface in Oceanus Procellarum. Thirty-one months later, on November 20, 1969, the Apollo 12 astronauts Charles Conrad and Alan Bean recovered several pieces of this Surveyor spacecraft and returned them to Earth. Among these returned parts was a section of a support strut, a 12.7-mm-diameter tube of polished, unpainted aluminum (alloy 2024) with 1.2-mm wall thickness. This unpainted tube was salvaged to investigate implanted solar wind particles. The location of the returned section of the strut on the Surveyor spacecraft is shown in figure 1. A small ring from this tube was received from the Jet Propulsion Laboratory, Pasadena, Calif. The ring, designated B-1, was about 2 mm wide and was located about 41 mm from the A-end of the aluminum tube.

Measurements on section B-1 represent a preliminary investigation with the aim of establishing the presence of trapped noble gases from the solar wind in the aluminum surface and of measuring the abundances of the light noble gases. Contamination by lunar dust and the distribution of the trapped solar wind around the tube were studied. Our measurements have provided the necessary information for subsequent, more detailed studies of the implanted solar wind gases in Surveyor 3 materials.

Procedure

Our analytical sensitivity is sufficiently high to allow ^{4}He and ^{20}Ne determinations in a 1-mm^2 sample. Consequently, we cut the ring B-1 into two rings of about equal width (fig. 2) using a wire saw with a 0.2-mm diamond-impregnated

FIGURE 1.—Original location of returned part of Surveyor 3 strut (from NASA photograph AS12–48–7114).

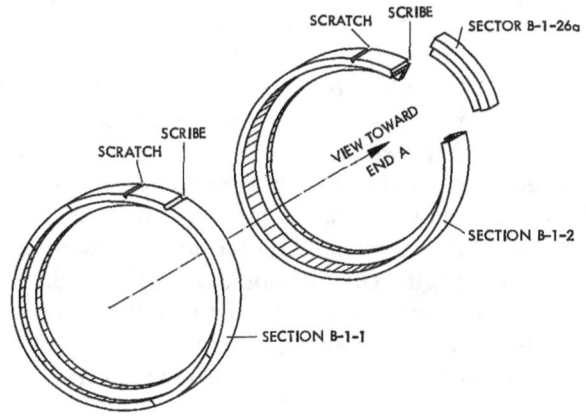

FIGURE 2.—Orientation of sectors cut from ring section B-1.

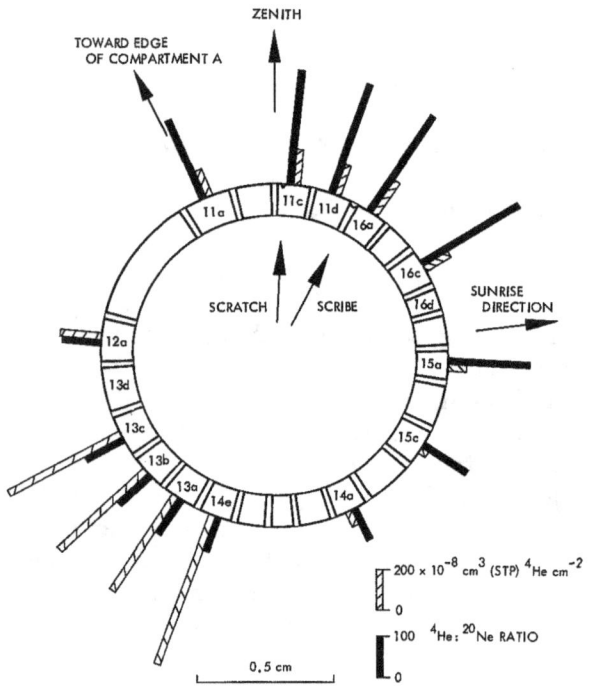

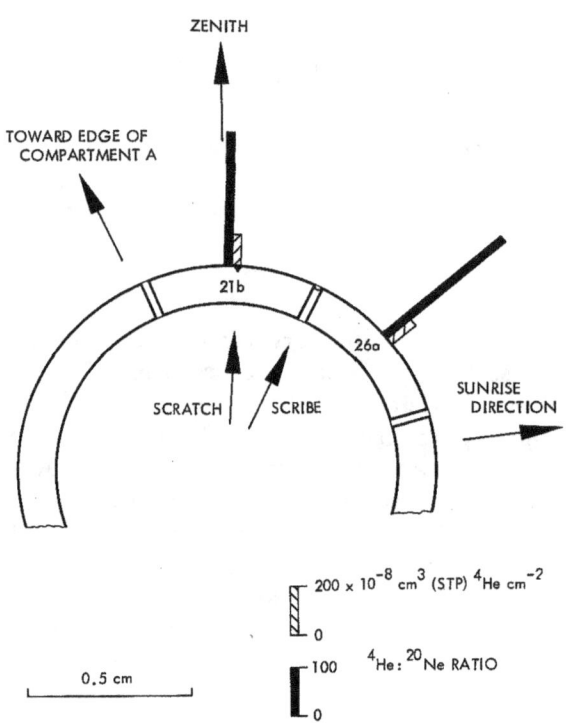

FIGURE 3.—^{4}He concentrations and ^{4}He:^{20}Ne ratios measured in small sectors cut from ring section B-1-1 (same view as in fig. 2). No corrections for Ne blanks of the aluminum tube have been made. The orientation of the tube on the lunar surface is according to Carroll. (See text footnote 1.)

FIGURE 4.—^{4}He concentrations and ^{4}He:^{20}Ne ratios measured in larger sectors cut from ring section B-1-2 (same view as in fig. 2). No corrections for Ne blanks of the aluminum tube have been made. Orientation of the tube is according to Carroll. (See text footnote 1.)

stainless-steel wire. To lower the noble gas blank from the aluminum, we removed the aluminum from the inside of these rings, reducing the weight by about 50 percent. The rings were then cut into sectors (figs. 2 through 4) and thoroughly cleaned by repeated ultrasonic treatment in acetone. Inspection under the optical microscope and investigation with a scanning electron microscope revealed that about one-half of the B-1 ring shows surface alterations; it is contaminated with fine crystalline particles, presumably of lunar origin. The ultrasonic treatment reduced this contamination, but did not completely eliminate it. No dust particles were found after the ultrasonic treatment on sector 16d taken from the uncontaminated side of B-1.

So far, the noble gases He and Ne have been determined in a number of sectors. The measurement procedure and the analytical blanks were the same as for the foil analyses of the solar wind composition (SWC) experiment (ref. 1). The results are given in tables 1 and 2, and the distribution around the ring is shown in figures 3 and 4. No corrections were applied for He and Ne from lunar dust contamination or for blanks in the aluminum.

The probable orientation of the ring B-1 at the lunar surface given by Carroll[1] was adopted for figures 3 and 4. Accordingly, the contaminated side of the aluminum tube essentially faced the lunar surface, whereas the clean side was sunlit, and thus exposed to the solar wind. We assume that the contamination and surface alteration occurred during landing as the vernier engines were cut off only 34 sec after the initial touchdown.

Results

Our He and Ne data do essentially support the probable orientation given by Carroll. In the

[1] W. Carroll, JPL, private communication by N. Nickle, JPL, October 1970.

TABLE 1.—*Results of ^{4}He and ^{20}Ne measurements in sectors cut from unpainted aluminum tube of Surveyor 3*

[See figs. 1 through 3][a]

Sample	Area, mm^2	Mass, mg	^{4}He	^{20}Ne	^{4}He:^{20}Ne
			10^{-8} cm^3 (STP)/cm^2		
B-1-11a	1.06	1.94	140	0.66	210
	±0.03	±0.02	±6	±0.17	±50
B-1-11c	0.98	1.49	180	0.62	290
	±0.04	±0.02	±10	±0.04	±10
B-1-11d	0.88	1.42	165	0.58	285
	±0.06	±0.02	±13	±0.04	±15
B-1-12a	1.29	2.00	204	2.15	95
	±0.04	±0.02	±9	±0.30	±13
B-1-13a	1.38	2.00	570	5.60	102
	±0.05	±0.02	±30	±0.30	±5
B-1-13b	1.48	2.00	590	5.50	107
	±0.08	±0.02	±40	±0.40	±4
B-1-13c	1.16	1.65	630	5.90	107
	±0.05	±0.02	±30	±0.40	±4
B-1-14a	1.43	1.97	80	0.98	83
	±0.06	±0.02	±5	±0.06	±5
B-1-14b [b]	1.33	1.82	128	1.54	83
	±0.04	±0.02	±5	±0.06	±5
B-1-14e	1.26	1.71	780	8.50	93
	±0.05	±0.02	±40	±0.40	±4
B-1-15a	1.34	1.93	104	0.50	210
	±0.04	±0.02	±5	±0.04	±15
B-1-15c	1.55	2.06	40	0.31	130
	±0.06	±0.02	±2	±0.02	±7
B-1-16a	1.28	1.95	194	0.71	275
	±0.05	±0.02	±7	±0.04	±15
B-1-16c	1.68	2.31	148	0.52	285
	±0.06	±0.02	±13	±0.03	±30

[a] Values not corrected for Ne blank of aluminum tube.
[b] No ultrasonic cleaning used.

TABLE 2.—*Results of noble gas measurements on two larger sectors cut from unpainted Surveyor 3 aluminum tube*

[Figs. 1, 2, and 4][a]

Sample	Area, mm^2	Mass, mg	^{4}He	^{20}Ne	^{4}He:^{3}He	^{20}Ne:^{22}Ne	^{22}Ne:^{21}Ne	^{4}He:^{20}Ne
			10^{-8} cm^3 (STP)/cm^2					
B-1-21b	8.00	6.21	143	0.51	2780	13.2	30	280
	±0.25	±0.02	±6	±0.03	±40	±0.4	±5	±10
B-1-26a	8.45	7.23	133	0.42	2770	13.3	32	315
	±0.40	±0.02	±7	±0.03	±120	±0.4	±6	±15

[a] Values not corrected for Ne blank of aluminum tube.

Apollo 12 lunar fine material (ref. 2) and also in lunar dust adhering to the Apollo 12 SWC foil, the He:Ne ratios are below 100. Thus, He and Ne found on the lower part of the ring (fig. 3) are readily explained as resulting from residual dust contamination. On the sunlit side, we find ^{4}He:^{20}Ne $\approx$ 300, which is a much higher ratio than in the lunar fine material and closer to the ^{4}He:^{20}Ne ratio in the solar wind (ref. 1). Considering the observed high He:Ne ratio, the smoothness of the He distribution obtained on the sunlit side, and the fact that a thorough investigation with the scanning electron microscope on sector 16d did not reveal any lunar dust particles after ultrasonic treatment, we conclude that the He data obtained on the sunlit side of the ring represent the solar wind particles implanted in the aluminum and that, in this area, the contamination from the dust is minimal or absent.

In table 3, averages are given for the trapped solar wind He and Ne in the Surveyor 3 material. The maximum ^{4}He surface concentration is the average of sectors 11c, 11d, and 16a (fig. 3). The ^{4}He:^{20}Ne ratio is the average of the six sectors with the highest ^{4}He:^{20}Ne ratios (tables 1 and 2). All other ratios are averages of the two large sectors 21b and 26a (table 2). The values in table 3 were corrected for the Ne blank of the aluminum tube. This blank correction is based on the noble gas concentration found in aluminum turnings removed from the inside of ring B-1 [measured concentration: $(19 \pm 5) \times 10^{-8}$ cm^3 (STP) ^{4}He/g; $(0.4 \pm 0.1) \times 10^{-8}$ cm^3 (STP) ^{20}Ne/g]. As the ^{4}He:^{20}Ne ratio of 50 indicates contamination with lunar dust, only 30 percent of the measured ^{20}Ne concentration was assumed to be blank. The blank correction was always smaller than 7 percent.

The ^{4}He:^{20}Ne ratio found in the trapped solar wind gas in the Surveyor 3 material is almost a factor of 2 lower than the ratio observed in the aluminum of the Apollo 12 SWC foil (ref. 1). Differences in the trapping efficiencies of the SWC aluminum foil and the Surveyor 3 aluminum tube are expected to be small and cannot account for the low ^{4}He:^{20}Ne ratio found. A surface contamination by terrestrial ^{20}Ne on the polished Surveyor 3 aluminum tube could explain the difference. However, the required Ne

TABLE 3.—*Maximum surface concentration and elemental and isotopic abundance ratios of trapped solar wind He and Ne in section B–1 of returned Surveyor 3 aluminum tube*[a]

^{4}He$_{max}$	$(180 \pm 20) \times 10^{-8}$ cm^3 (STP)/cm^2
^{4}He:^{3}He	2770 ± 120
^{4}He:^{20}Ne	295 ± 15
^{20}Ne:^{22}Ne	13.3 ± 0.4
^{22}Ne:^{21}Ne	31 ± 5

[a] The figures given are corrected for the Ne blank in the aluminum. No correction for a possible residual lunar dust contamination or for diffusion loss was applied.

concentrations of more than 10^{-9} cm^3 (STP)/cm^2 are orders of magnitude larger than the surface blanks observed on different kinds of aluminum foils investigated in connection with the SWC experiment. Furthermore, the necessary correction for such a high terrestrial Ne blank would lead to a ^{20}Ne:^{22}Ne ratio for the trapped solar wind particles considerably higher than the values observed in the Apollo 11 and 12 SWC experiments (ref. 1). Diffusion loss of He or a residual contamination with lunar dust could explain the low ratio ^{4}He:^{20}Ne $=$ 295 found on the clean, sunlit side of the Surveyor 3 strut. We estimate that, around lunar noon, the temperature of the strut has reached 120° to 140°C. Trapped solar wind He begins to diffuse out of aluminum at these temperatures (ref. 3), and we cannot exclude that a sizable fraction of the trapped ^{4}He was lost. The subsequent discussion of our Surveyor 3 results must thus consider the possibility of preferential ^{4}He diffusion loss or a residual dust contamination.

For comparison, the expected ^{4}He distribution around the tube has been calculated; it is plotted in figure 5, taking into account the oblique and variable angle of incidence of the solar wind ions and the passage of the Moon through the Earth's tail. The following assumptions were made:

(1) The trapping probability of He is proportional to the cosine of the angle of incidence, α (ref. 3).

(2) The aberration and co-rotation of the solar wind is 3° (refs. 4 and 5).

(3) The Earth's tail is assumed to be 50 Earth radii wide at the lunar orbit; i.e., the Surveyor 3 landing site is in the Earth's tail for solar wind

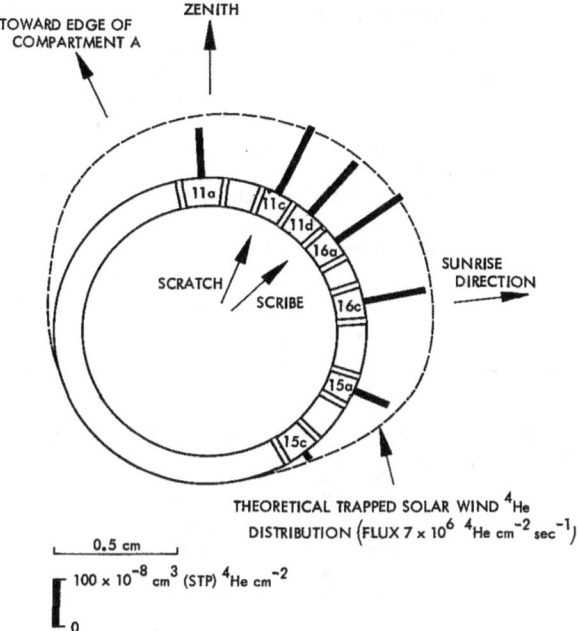

FIGURE 5.—Comparison of theoretical trapped ^{4}He distribution with measured concentrations. The shape of the left side of the theoretical curve depends strongly on the exact location of compartment A. Orientation of the tube has been changed from the orientation given by Carroll to obtain best fit.

zenith angles from 54° to 5° (ref. 6). A negligible ^{4}He flux was assumed for the Earth's tail (ref. 7; also see the subsequent discussion).

(4) The ^{4}He solar wind flux was adjusted to agree with the observed maximum concentration of ^{4}He.

The measured ^{4}He concentration was corrected for a possible lunar dust contamination by assuming ^{4}He:^{20}Ne = 460 (average of Apollo 11 and 12 SWC experiments) in the trapped solar wind, and ^{4}He:^{20}Ne = 90 for the lunar dust. The resulting correction for possible lunar dust ^{4}He is small (<30 percent) for all sectors except 15c. Virtually the same relative angular distribution is obtained if a different ^{4}He:^{20}Ne ratio is assumed for the trapped solar wind. ^{4}He diffusion loss also would not change the shape of the ^{4}He angular distribution curve, as the temperature of the aluminum ring is estimated to be uniform within 1°C. To obtain the best possible fit of the experimental data with the theoretical curve, the probable orientation of the returned aluminum tube on the lunar surface had to be changed somewhat. The best agreement is obtained by rotating the tube clockwise (as seen from end G) by 20° relative to the probable orientation given by Carroll. (See footnote 1.) The scribe line corresponds then to a solar zenith angle of 48° toward lunar east. The agreement obtained between the theoretical and the measured angular distribution of the implanted solar wind ^{4}He is satisfactory, especially if one takes into account the uncertainty in the exact location of the Earth's tail during the 31 months of exposure of the aluminum tube.

For the most strongly irradiated sectors (11c, 11d, and 16a) the integrated exposure is equivalent to 75 days of exposure at orthogonal incidence. The average implanted ^{4}He concentration of sectors 11c, 11d, and 16a, corrected for a possible lunar dust contamination, as outlined above, corresponds to an average solar wind ^{4}He flux of 7×10^6 cm^{-2} sec^{-1} (trapping efficiency 0.9 cos α). Because of the specific assumption made for the correction of a possible residual lunar dust contamination and because diffusion loss may have occurred, this flux value has to be considered as a lower limit of the true average ^{4}He flux during the exposure time. An appropriate upper limit of 13×10^6 cm^{-2} sec^{-1} is obtained if we assume that the low ^{4}He:^{20}Ne ratio is due to preferential diffusion loss of ^{4}He from the aluminum tube, with virtually no Ne loss, and that the true ratio for the trapped solar wind particles is ^{4}He:^{20}Ne = 460 (average of Apollo 11 and 12 SWC experiments). The solar wind ^{4}He fluxes measured by other experiments are well within the possible flux range deduced from the Surveyor 3 material. (See table 4.)

In table 5, the isotopic compositions of the solar wind during the exposure of the Surveyor 3 material are given, as derived from our measurements compiled in table 3. It was assumed that the Surveyor 3 aluminum had the same trapping properties as the SWC aluminum foil (ref. 1). Isotopic fractionation due to diffusion loss and the effects of a possible residual lunar dust contamination were neglected and will be discussed in detail. For comparison, the solar wind compositions, as measured by the Apollo 11 and 12 SWC experiments, are given.

The ^{4}He:^{3}He ratio obtained from the Surveyor 3 material is higher than the ratios measured

TABLE 4.—*Comparison of average ^{4}He solar wind fluxes*

Experiment	Time period	Average ^{4}He flux	Reference
Surveyor 3	Apr. 20, 1967, through Nov. 20, 1969.	Between 7×10^6 and 13×10^6 cm^{-2} sec^{-1}	This article
Apollo 11 SWC	July 21, 1969; 03:35 to 04:52 GMT.	$(6.2 \pm 1.2) \times 10^6$ cm^{-2} sec^{-1}	Ref. 1
Apollo 12 SWC	Nov. 19, 1969, at 12:35 GMT, to Nov. 20, 1969, at 07:17 GMT.	$(8.1 \pm 1.0) \times 10^6$ cm^{-2} sec^{-1}	Ref. 1
Vela 3A and 3B	July 1965 to July 1967	9×10^6 cm^{-2} sec^{-1}	Ref. 16

TABLE 5.—*Isotopic composition of solar wind derived from trapped gases in the unpainted aluminum tube recovered from Surveyor 3* [a]

Ratio	Surveyor 3 [b]	SWC experiments [c]	
		Apollo 11	Apollo 12
^{4}He:^{3}He	2700 ± 130	1860 ± 140	2450 ± 100
^{20}Ne:^{22}Ne	13.3 ± 0.4	13.5 ± 1.0	13.1 ± 0.6
^{22}Ne:^{21}Ne	31 ± 5		26 ± 12

[a] All values represent averages over respective exposure periods.
[b] Surveyor 3 data were not corrected for effect of diffusion loss or for possible residual lunar dust contamination (see text).
[c] SWC data from Geiss et al. (ref. 1).

with the Apollo 11 and 12 SWC experiments. It could be that time variations of this ratio (refs. 1 and 8) are responsible for this difference; i.e., the long-time average of the ^{4}He:^{3}He ratio is higher than the two values found during the Apollo 11 and 12 missions. However, the ^{4}He:^{3}He ratio in the Surveyor 3 strut could have been altered and, thus, does not necessarily represent a true solar wind average. The effects that must be considered are discussed in the subsequent paragraphs.

(1) *Spallation by Cosmic Rays or Solar Protons.* The spallation rate induced by cosmic rays at the lunar surface is $\sim 10^{-14}$ cm^3 ^{3}He/g year. Even if we assume that solar protons produce on the average 10 times more ^{3}He, the relative contribution to the observed ^{3}He is still only on the order of 10^{-4}. Also, for ^{21}Ne, the contribution from spallation is negligible.

(2) *Stripping by Cosmic Ray or Energetic Solar Alpha Particles.* It is readily estimated that He produced from ^{4}He in stripping reactions in the Surveyor 3 material can be neglected.

(3) *Recycling of Solar Wind He and Radiogenic Ne.* Released lunar radiogenic ^{4}He and trapped solar wind He could be recycled and retrapped in solid material at the lunar surface in the same way as ^{40}Ar is retrapped (refs. 9 through 11). Estimates show that the influence of this process on the ^{4}He:^{3}He ratio in the Surveyor 3 strut should be negligible. The efficiency of the process for He is even smaller than for Ar because most of the He is lost from the Moon by gravitational escape before it is ionized. Moreover, the orientation of the strut is such that it is a poor collector for accelerated lunar ions that should arrive nearly horizontally from the southern or northern direction.

(4) *He From the Terrestrial Atmosphere.* It is possible that the terrestrial atmosphere loses most of its helium by way of the polar wind (ref. 12). If these helium ions escape from the Earth through the magnetospheric tail, then the Moon would encounter a flux of ^{4}He$^+$ of terrestrial origin for a few days each month. With Axford's (ref. 12) estimate of the helium flux in the polar

wind, we obtain an upper limit for the terrestrial ^{4}He$^+$ flux in the tail of 10^4 cm^{-2} sec^{-1}. The effect of this flux in the Surveyor 3 material is negligible.

(5) *Mass Discrimination Near the Moon.* Mass discrimination could result from disturbances of the electromagnetic field in the solar wind near the Moon and also from the static 35γ field found near the Apollo 12 site by Dyal et al. (ref. 13). However, the equality of the ^{4}He:^{3}He ratios found in sectors 21b and 26a renders any significant mass discrimination unlikely.

(6) *Mass Dependence of Trapping Probability.* For small angles of incidence, the trapping probability of ^{4}He in aluminum is 90 percent (ref. 3); i.e., 10 percent of the incoming ^{4}He ions are backscattered. In evaluating the ^{4}He:^{3}He ratio, we have assumed a somewhat larger backscattering coefficient (12 percent) for ^{3}He. An error in the estimate of such a small backscattering coefficient does not affect significantly the trapping probability. However, at large angles of incidence, the trapping probability for helium falls below 50 percent, and here the difference between ^{4}He and ^{3}He could be appreciable.

(7) *Diffusion.* We have concluded that some of the helium on the sunlit side of the strut may have been lost by thermal diffusion. This could have led to a depletion of ^{3}He relative to ^{4}He. The reasons are twofold: The average depth of implantation d is smaller for ^{3}He than for ^{4}He, and the diffusion constant of ^{3}He is larger than that of ^{4}He. From the range formula given by Nielsen (ref. 14), we estimate $d_3 \approx 0.91 d_4$. Assuming diffusion constants inversely proportional to the square root of the mass, the characteristic parameter $d/D^{1/2}$ for ^{3}He is 15 percent smaller than for ^{4}He. The resulting isotopic enrichment is a function of the loss fraction and of $d/D^{1/2}$. For a loss of 50 percent of ^{4}He, we estimate an isotopic fractionation of 5 to 10 percent.

(8) *Contamination by Lunar Dust.* The lunar fine material at the Apollo 12 landing site has a ratio ^{4}He:^{3}He ~ 2300 (ref. 15 and unpublished Bern data). It is expected that this ratio will depend on the grain size, similar to the observations made for the Apollo 11 fine material (ref. 10). The very fine material, which has to be considered as possible source of a remaining lunar dust contamination, should have a higher ^{4}He:^{3}He ratio. The maximum possible lunar dust contamination, as deduced from the ^{4}He:^{20}Ne ratios, would necessitate a correction of approximately 3 percent of the measured ^{4}He:^{3}He ratios (sectors 21b and 26a, lunar dust ^{4}He:^{3}He = 2300 assumed). The true solar wind ^{4}He:^{3}He ratio, averaged over the Surveyor 3 exposure time, could thus be as high as 2800 ± 130.

Conclusions

Effects (4) through (7) all enrich ^{4}He relative to ^{3}He. The combined effect can hardly be more than 10 percent and the true solar wind ^{4}He:^{3}He ratio, averaged over the Surveyor 3 exposure period, must be higher than 2400. As an upper limit, for the case of the maximum possible lunar dust contamination, we obtain a value of 2800. The Surveyor ^{4}He:^{3}He ratio is thus distinctly higher than the value measured during the Apollo 11 EVA and probably also higher than the value determined during the Apollo 12 mission. The Surveyor ^{4}He:^{3}He ratio agrees quite well with the value derived from the ilmenite of the Apollo 11 lunar fine material (ref. 10). However, the Surveyor 3 exposure time is only a small fraction of a solar cycle. We may expect that the ^{4}He:^{3}He ratio varies with the solar cycle and the Surveyor results are not necessarily a good long-time average of the present-day solar wind ^{4}He:^{3}He ratio.

The neon isotopic composition obtained in the Surveyor 3 material agrees within the limits of error with the results of the Apollo 11 and 12 SWC experiments. If the relatively low ^{4}He:^{20}Ne ratio found should be due to a residual dust contamination, then the resulting correction would raise the ^{20}Ne:^{22}Ne ratio by a few percent. A comparison of the ^{20}Ne:^{21}Ne:^{22}Ne ratios shows that the large difference in the isotopic abundances of neon in the terrestrial atmosphere and in the solar wind is mainly due to mass fractionation and not to nuclear reactions. This confirms a conclusion which was drawn from data obtained in lunar fine material (ref. 10).

References

1. GEISS, J.; EBERHARDT, P.; BÜHLER, F.; MEISTER, J.; AND SIGNER, P.: "Apollo 11 and 12 Solar Wind

Composition Experiments: Fluxes of He and Ne Isotopes." *J. Geophys. Res.*, vol. 75, 1970, pp. 5972–5979.
2. THE LUNAR SAMPLE PRELIMINARY EXAMINATION TEAM: "Preliminary Examination of Lunar Samples." *Apollo 12 Preliminary Science Report*, NASA SP-235, Washington, D.C., 1970, pp. 189–216.
3. MEISTER, J.; BÜHLER, F.; EBERHARDT, P.; AND GEISS, J.: *Apollo Solar Wind Composition Experiment: Trapping of Low Energy Ions in Aluminum*. To be published.
4. HUNDHAUSEN, A. J.: "Direct Observations of Solar Wind Particles." *Space Sci. Rev.*, vol. 8, 1968, pp. 690–749.
5. AXFORD, W. I.: "Observations of the Interplanetary Plasma." *Space Sci. Rev.*, vol. 8, 1968, pp. 331–365.
6. MIHALOV, J. D.; COLBURN, D. S.; AND SONETT, C. P.: "Observations of Magnetopause Geometry and Waves at the Lunar Distance." *Planet. Space Sci.*, vol. 18, 1970, pp. 239–258.
7. SNYDER, C. W.; CLAY, D. R.; AND NEUGEBAUER, M.: "The Solar-Wind Spectrometer Experiment." *Apollo 12 Preliminary Science Report*, NASA SP-235, Washington, D.C., 1970, pp. 75–81.
8. BAME, S. J.; HUNDHAUSEN, A. J.; ASBRIDGE, I. R.; AND STRONG, I. B.: "Solar Wind Ion Composition." *Phys. Rev. Letters*, vol. 20, 1968, pp. 393–395.
9. HEYMANN, D.; AND YANIV, A.: "^{40}Ar Anomaly in Lunar Samples from Apollo 11." *Geochim. et Cosmochim. Acta*, Suppl. I, vol. 2, 1970, pp. 1261–1267.
10. EBERHARDT, P.; GEISS, J.; GRAF, H.; GRÖGLER, N.; KRÄHENBÜHL, U.; SCHWALLER, H.; SCHWARZMÜLLER, J.; AND STETTLER, A.: "Trapped Solar Wind Noble Gases, Exposure Age and K/Ar-age in Apollo 11 Lunar Fine Material." *Geochim. et Cosmochim. Acta*, Suppl. I, vol. 2, 1970, pp. 1037–1070.
11. MANKA, R. H.; AND MICHEL, F. C.: "Lunar Atmosphere as a Source of Argon-40 and Other Lunar Surface Elements." *Science*, vol. 169, 1970, pp. 278–280.
12. AXFORD, W. I.: "The Polar Wind and the Terrestrial Helium Budget." *J. Geophys. Res.*, vol. 73, 1968, pp. 6855–6859.
13. DYAL, P.; PARKIN, C. W.; AND SONETT, C. P.: "Apollo 12 Magnetometer: Measurement of a Steady Magnetic Field on the Surface of the Moon." *Science*, vol. 169, 1970, pp. 762–764.
14. NIELSEN, K. O.: "The Range of Atomic Particles with Energies about 50 keV." *Electromagnetically Enriched Isotopes and Mass Spectrometry*, Butterworths Scientific Publications, London, 1956, pp. 68–81.
15. LUNAR SAMPLE PRELIMINARY EXAMINATION TEAM: "Preliminary Examination of Lunar Samples from Apollo 12." *Science*, vol. 167, 1970, pp. 1325–1339.
16. ROBBINS, D. E.; HUNDHAUSEN, A. J.; AND BAME, S. J.: "Helium in the Solar Wind." *J. Geophys. Res.*, vol. 75, 1970, pp. 1178–1187.

ACKNOWLEDGMENTS

We thank the National Aeronautics and Space Administration for providing Surveyor 3 material for analyses, and especially the Apollo 12 astronauts Charles Conrad, Jr., Alan L. Bean, and Richard F. Gordon. We are grateful to N. Nickle and W. Carroll of the Jet Propulsion Laboratory for their support.

We thank Dr. H. U. Nissen and R. Wessicken from the Laboratorium für Elektronenmikroskopie, ETH, Zürich, and Dr. N. Grögler from our Institute for the supporting scanning electron microscope investigations. The help of H. Wyniger in sample preparation is acknowledged.

IX. Particle Track Analyses

PART A

SOLAR PARTICLE TRACKS IN GLASS FROM SURVEYOR 3

G. Crozaz and R. M. Walker

The extraordinary pinpoint landing of Apollo 12 has provided samples of Surveyor 3 parts that were exposed for 31 months on the Moon. Nuclear track studies have been made of a piece of clear filter glass used to cover the lens of the television camera. The results of the studies have provided information about low-energy nuclear particles from the Sun and have provided a basic calibration for nuclear track studies in the surfaces of lunar rocks. (See refs. 1 through 4.)

The Surveyor 3 television camera was mounted vertically inside a shroud open on one side. Images were obtained from a mirror above the lens, which was covered by a horizontal filter wheel. After 2 weeks of operation, the clear filter was left directly in front of the shroud opening until recovery.

A piece of this filter glass, about 0.35 cm² in area and 0.3 cm thick, was found by a microprobe scan to contain a large amount of Pb and a smaller amount of K. From the density (3.60) and the index of refraction ($n = 1.61$), the Pb content is estimated as close to 43 wt % (ref. 5). The microprobe also showed a small amount of Mg, presumably from a $\lambda/4$ coating of MgF_2.

The geometry of the glass with respect to the opening was specified as follows:

(1) A reference line $\overline{AB}$ was drawn on the surface of the glass. This line ran from left to right for an observer standing in front of the camera.

(2) A wire was placed perpendicular to the line $\overline{AB}$ at the point where the sample was taken. The wire then was rotated in a vertical plane until it bisected the opening from top to bottom.

(3) A piece of clear plastic was placed perpendicular to the wire, and the outline of the opening was traced.

Specification of the coordinates in the plane of the plastic (x, z), the elevation of the wire, and the distance from the sample to the XZ reference plane fix the geometry.

The sample was cut into three parts, with the first cut made parallel to $\overline{AB}$ at 60° to the exposed surface in order to place the tracks incident at a steep angle. One piece (section I) was put in epoxy, polished, and etched (1 percent hydrofluoric acid solution) to study the depth dependence. The other pieces were used to study tracks on the exposed surface. Measurements were made using a scanning electron microscope from depths of 0 to 30 μm and using an optical microscope from 10 μm. As shown in figure 1, the density is $(1.14 \pm 0.06) \times 10^6$ tracks/cm² at 3.8 ± 1 μm from the surface and drops off rapidly with depth.

The remaining samples of the top surface were given varying treatments in H_2SO_4 and HNO_3 solutions to etch the MgF_2 coating. There are two layers of material covering the glass surface: the MgF_2 coating (about 1400 Å thick) and a second layer (about 0.4 μm thick), with a composition somewhat enriched in Si with respect to the glass. Possibly it is a silane coating used to enhance the adhesion of the MgF_2.

No tracks were found in the MgF_2, although the right etchant may not have been used. When the MgF_2 was removed by etching in HNO_3 at 75°C for 1 hr, the second film was left intact. Subsequent etching in dilute hydrofluoric acid gave a high density of shallow pits about $3 \times 10^8/\text{cm}^2$. Shallow pits also could be seen on the glass substrate in areas where the film was broken. However, similar pits were observed on the bottom, unirradiated portions of the glass and in some areas of a control glass provided by the Jet Propulsion Laboratory, Pasadena. Thus, it is not likely that they are nuclear particle tracks.

Additional etching in hydrofluoric acid removed the thick film and gave deep, characteristic track etch pits in the exposed glass surface. These tracks are oriented toward the camera opening. (See fig. 2.) The density of these pits is $(8.3 \pm 0.5) \times 10^5$ tracks/cm². When corrected for the geometry, this corresponds to a density of $(1.7 \pm 0.1) \times 10^6$ tracks/cm² in the plane of section I. As in other silicates, it is assumed that tracks are produced only by ions of the VH group ($Z \geqslant 20$). They could have been registered either on the Moon or in passage through the radiation belts. By using the data on the α/p ratio (ref. 6) and the C,N,O/α ratio (ref. 7), it is estimated that less than 10^{-3} tracks are from the Earth's radiation belts. It is most likely that the tracks were registered during energetic solar events, specifically the solar flare events of November 1968 and April and November 1969.

In figure 1, the data are compared with theoretical curves for different energy spectra of the type $(dN/dE) = CE^{-\gamma}$. The density was calculated as follows (see ref. 8):

$$\rho(d) = \sum_{x,z} \left(\frac{dN}{dE}\right)_{x,z} \left(\frac{dE}{dR}\right)_{x,z} \exp(-\psi R_{x,z}) \Delta R_c \, d\omega_{x,z} \quad (1)$$

where $\psi = 0$ for $E < 10$ MeV/nucleon. Each set of x, z corresponds to a particular location on the previously defined reference plane. $R_{x,z}$ is the distance traveled through the glass from the point x, z to a point located a distance, d, from the surface along the plane of section I. The corresponding solid angle per unit area is given by $d\omega_{x,z}$. Equation (1) was summed over 136 sets of x and z. Range energy data were taken principally from the tables of Henke and Benton (ref. 9). The stopping power of Fe in Pb was estimated following Barkas and Berger (ref. 10).

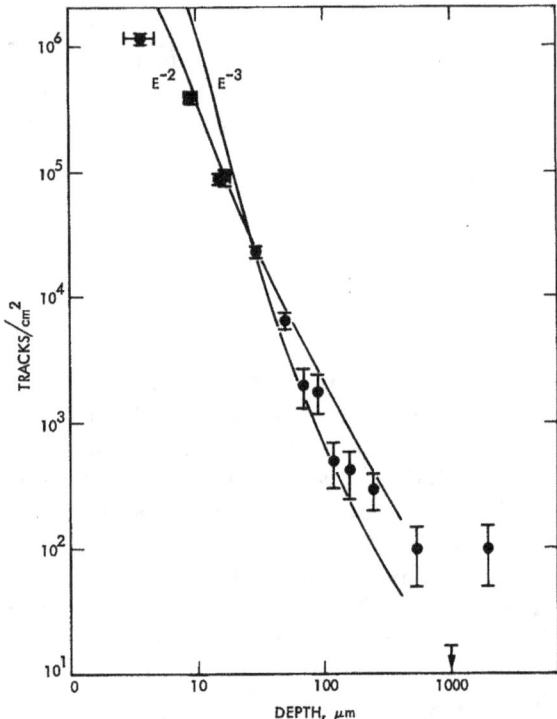

FIGURE 1.—Track density vs. depth for section I.

From 10 to 50 µm ($\sim$2 to 7 MeV/nucleon), our data are best fit with a power law spectrum of $dN/dE = 2.42 \times 10^6 \, E^{-2}$ (particles (p)/cm² sterad yr MeV per nucleon). At higher energies $dN/dE = 1.17 \times 10^7 \, E^{-3}$ (p/cm² sterad yr MeV per nucleon) seems a better fit, although the E^{-2} spectrum is possible.

The spectra are in accord with what is known about solar flares (refs. 11 and 12). Although the spectral index of the solar flare protons varies and even changes with time for a given flare, gamma values of about 3 are typical. The spectral index tends to be lower for alpha particles than for protons. The differential flux generally shows a sharp break to a lower spectral index between 4 and 8 MeV. J. Arnold et al. (ref. 13) also find that a spectral index of 2 gives the best fit to their radiochemical results on lunar rocks.

Two measurements of the absolute values of the flux are available for comparison: (1) a satellite estimate of about 3×10^2 p/cm² sec >20 MeV for the solar maximum of 1956 through 1960 (ref. 12) and (2) a radiochemically determined long-time average of about

25 p/cm² sec >10 MeV (ref. 13). As the solar flare flux varies by a factor of 10^5 from solar minimum to maximum, it is difficult to know, a priori, which value to choose. However, the radiochemical comparison of different isotopes indicates that the flux for the period April 20, 1967, to November 19, 1969, was no more than about two times the long-term average (ref. 13). Thus, a value of about 50 p/cm² sec >10 MeV could be expected.

Our data are not in good agreement with this estimate. Assuming a pure E^{-2} spectrum, we find values of 800 p/cm² sec >10 MeV and 400 p/cm² sec >20 MeV. Taking an E^{-3} spectrum starting at a depth of 30 µm, the corresponding numbers are 200 p/cm² sec >10 MeV and 50 p/cm³ sec >20 MeV. Based on this comparison, the combination of an E^{-2} with an E^{-3} spectrum is preferred.

Neither spectrum fits the data for a depth of (3.8 ± 1) µm (~ 1 MeV/nucleon); it is likely that the differential flux falls off below this value. Another possible interpretation we cannot ignore is that low-energy iron particles, having passed the maximum in their rate of energy loss, are no longer capable of producing tracks. Essentially the same track density is found on the external surface as at the 3.8-µm level. To register in the glass surface, the particles would have had to penetrate the two surface films encompassing about 0.5 µm of material. From this, we conclude that there are relatively few particles in the interval of about 30 keV/nucleon to 1 MeV/nucleon. Recently, Borg et al. (ref. 14) reported extremely high track densities ($>10^{11}$/cm²) in small (<1 µm) particles removed from the lunar soil. One explanation for the origin of these tracks is the 5- to 50-keV solar particle components observed in space probe experiments (ref. 15). However, no evidence for such particles is found here and the high track densities remain unexplained.

We showed (ref. 1) that the track data in lunar rocks 10017 and 10058 could be fit with either an E^{-2} or an E^{-3} spectrum, provided that an erosion rate of about 10^{-7} cm/yr was taken. Unpublished data on other lunar rocks confirm this. In these rocks, the track density vs. depth lies on a straight line on a log-log plot. However, the slope is gentler than in the Surveyor glass.

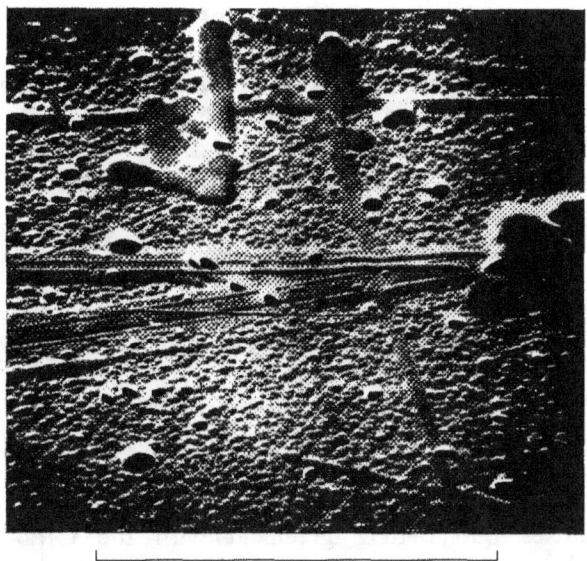

FIGURE 2.—Etched tracks on the top surface after removal of surface films. The tracks are the dark oval objects about 1 µm in size. They clearly point to the camera opening. The shallow pits also are found on the unirradiated part of the glass and are not produced by nuclear particles.

For example, in rock 10057 a tenfold change in depth produces only a sevenfold change in track density. The corresponding change in track density in figure 1 is a factor of 200 for a similar tenfold variation in depth. It is possible to show that this difference can be explained by an erosion rate of 10^{-7} cm/yr and a long-term spectral index of $\gamma = 2$.

Taking into account the solid angle, the stay time, and allowing a factor of 2 for the increase in solar activity over a long-time average, the track densities in figure 1 correspond approximately to those expected for a 2π irradiation in 1 year. Typical crystals from the rock surfaces have densities of about 10^9 tracks/cm² at a distance of 20 µm from the edge. Because of the geometry, this corresponds to about 10 µm in figure 1. The level of 10^9 tracks/cm² would be achieved in an irradiation time of about 2×10^3 years. As this is short compared to the cosmic-ray exposure age of the rocks, it is likely that the 10^9 tracks/cm² is an equilibrium value reflecting a balance between erosion and irradiation. Although the estimate of about 20 µm removed in 2×10^3 years is a factor of 10 greater than the limit of 10^{-7} cm/yr set by us from other considerations (see ref. 1), we do not consider this a serious disagreement in view of the various uncertainties. Our main point is that a considerable erosion rate is required.

The results also suggest that erosion occurs by a flaking of small thicknesses of material, possibly caused by solar wind irradiation. If whole crystals, from 100 to 300 μm in size, had to be removed every 2×10^3 years, the erosion rates would become unreasonably high.

One of the important points that needs to be examined in future work is the sensitivity of the Surveyor glass for particle track registration. If the glass were to register lighter particles of the C,N,O group, this would bring our absolute fluxes into better agreement with the radiochemical data and would modify somewhat (though not completely) our conclusions on lunar rock erosion.

References

1. Crozaz, G.; Haack, U.; Hair, M.; Maurette, M.; Walker, R.; and Woolum, D.: "Nuclear Track Studies of Ancient Solar Radiations and Dynamic Lunar Surface Processes." *Proceedings of the Apollo 11 Lunar Science Conference*, vol. 3, 1970, pp. 2051–2080.
2. Fleischer, R. L.; Haines, E. L.; Hart, H. R., Jr.; Woods, R. T.; and Comstock, G. M.: "The Particle Track Record of the Sea of Tranquillity." *Proceedings of the Apollo 11 Lunar Science Conference*, vol. 3, 1970, pp. 2103–2120.
3. Price, P. B.; and O'Sullivan, D.: "Lunar Erosion Rate and Solar Flare Paleontology." *Proceedings of the Apollo 11 Lunar Science Conference*, vol. 3, 1970, pp. 2351–2359.
4. Lal, D.; MacDougall, D.; Wilkening, L.; and Arrhenius, G.: "Mixing of the Lunar Regolith and Cosmic Ray Spectra: Evidence From Particle Track Studies." *Proceedings of the Apollo 11 Lunar Science Conference*, vol. 3, 1970, pp. 2295–2303.
5. Morey, G.: *The Properties of Glass*, Reinhold Publications, 1954.
6. Krimigis, S. M.; Verzariu, P.; Van Allen, J. A.; Randall, B. A.; Armstrong, T. P.; and Fritz, T. A.: "Trapped Energetic Nuclei Z Greater Than or Equal to 3 in the Earth's Outer Radiation Zone." *J. Geophys. Res.*, vol. 75, 1970, pp. 4210–4215.
7. Van Allen, J. A.; Randall, B. A.; and Krimigis, S. M.: "Energetic Carbon, Nitrogen, and Oxygen Nuclei in the Earth's Outer Radiation Zone." *J. Geophys. Res.*, vol. 75, 1970, pp. 6085–6091.
8. Fleischer, R. L.; Price, P. B.; Walker, R. M.; and Maurette, M.: "Origins of Fossil Charged-Particle Tracks in Meteorites." *J. Geophys. Res.*, vol. 72, 1967, pp. 331–353.
9. Henke, R. P.; and Benton, E. V.: U.S. Naval Radiological Defense Laboratory TR-1, 1966, p. 1102.
10. Barkas, W. H.; and Berger, M. J.: *Tables of Energy Losses and Ranges of Heavy Charged Particles*," National Academy of Sciences, National Research Council Publication 1133, 1964, pp. 103–172.
11. Fichtel, C. E.; and McDonald, F. B.: "Energetic Particles From the Sun." *Ann. Rev. Astron. Astrophys.*, vol. 5, 1967, pp. 351–398.
12. McDonald, F. B.: "Satellite Observations of Solar Cosmic Rays." *Intercorrelated Satellite Observations Related to Solar Events*, Reidel Publications, Holland, 1970, pp. 34–52.
13. S. H. R. E. L. L. D. A. L. F. F.: "Pattern of Bombardment-Produced Radionuclides in Rock 10017 and in Lunar Soil." *Proceedings of the Apollo 11 Lunar Science Conference*, vol. 2, 1970, pp. 1503–1532.
14. Borg, J.; Dran, J. C.; Durrieu, L.; Jouret, C.; and Maurette, M.: "High Voltage Electron Microscope Studies of Fossil Nuclear Particle Tracks in Extraterrestrial Matter." *Earth and Planetary Science Letters*, vol. 8, 1970, pp. 379–386.
15. Hundhausen, A. J.: "Composition and Dynamics of the Solar Wind Plasma." *Rev. of Geophys. and Space Phys.*, vol. 8, 1970, p. 729.

ACKNOWLEDGMENTS

We thank Neil L. Nickle of the Jet Propulsion Laboratory for his help in providing the glass and measurement of the reference geometry. We also thank the University of Chicago Cosmic Ray Group, in particular J. A. Simpson, K. C. Hsieh, and B. McKibben, for sharing their unpublished satellite flare data and for extensive discussions. We are grateful to J. Arnold for discussions concerning the most recent results of the radio chemistry work on lunar rocks.

We acknowledge many of our colleagues, including P. Fedders, M. Israel, D. Yuhas, J. Heymann, M. Harris, and C. Drabes, for their help in various aspects of this work; and P. Swan, who performed much of the experimental work.

PART B

VERY HEAVY SOLAR COSMIC RAYS: ENERGY SPECTRUM AND IMPLICATIONS FOR LUNAR EROSION

R. L. Fleischer, H. R. Hart, Jr., and G. M. Comstock

In November 1969, Apollo 12 astronauts removed and returned to Earth the Surveyor 3 television camera that had rested on the Moon for a period of 31 months at a time of maximum solar activity. Housed in the camera, but exposed directly to space, was a neutral density (clear flint) optical filter in the form of a glass plate highly suitable for particle track registration. This solid-state nuclear track detector has been used to study heavy solar nuclei and the effects of other cosmic rays from outside the solar system. From the number of nuclei stopping as a function of depth in the detector, the energy spectrum of the solar particles has been measured. The difference between the track density vs. depth relation found in the Surveyor glass and those previously observed in lunar samples allows an estimate of the rate at which erosion exposes new rock surfaces on the Moon. Although the data presented can be refined in order to yield more extensive and precise information, certain of the results are sufficiently clear and useful as to justify this account.

Glass as a detector material (refs. 1 through 3) ignores lightly ionizing particles. This high detection threshold, together with what is known of solar abundances (ref. 4), means that more than 90 percent of the observed solar particles will be the iron group nuclei Cr, Mn, Fe, Co, and Ni. Therefore, if from the energy spectrum it is clear that solar particles are present, it is also clear that they are dominantly iron group particles.

The filter glass used was a 3-mm-thick flint glass of density 3.60 and index of refraction 1.612 (± 0.002), with detection properties somewhat similar to the tektite glass reported in reference 3. The temperature near the glass, just behind the mirror, never exceeded 82°C. For comparison, less than 8 percent fading of ^{252}Cf fission tracks is observed after 1 hr at 125°C, giving considerable confidence that fading of similar tracks at the Moon was negligible. The minimum cone angle for ^{252}Cf fission fragments is 30° with an average etchable range of 11 μm; ^{20}Ne nuclei from a heavy ion irradiation were detectable, while ^{16}O nuclei were not. We estimate, by comparison with earlier calibrations (ref. 5), that the minimum cone angle for an iron nucleus is 30° to 35°, with an etchable range of 20 to 25 μm. By measuring individual tracks, we find cone angles ranging from 35° to 75° and an etchable length of 28 μm for the most abundant tracks, which we assume to be due to the Fe nuclei. Occasional tracks of length up to 55 μm indicate the presence of nuclei heavier than iron in less than 15 percent abundance. The cone angle is of great importance because it equals the minimum angle of inclination to a surface at which a track is etched (see refs. 1 through 3), and thus determines the solid angle through which tracks of incident particles are revealed by etching.

The specific geometry of the housing over the filter (ref. 6) was such as to shield out particles over most of the upward-facing hemisphere, allowing Fe nuclei of energy less than 22 MeV/nucleon entrance only over a solid angle factor estimated to be 1.3 sterad, centered around a line inclined at about 30° to the glass surface. Our measurements of individual tracks show them to be grouped within 23° or 24° of this line. Consequently, we may assume for simplicity that the particles detected arrive as a collimated beam at 30° incidence with an effective solid angle factor of 0.5. To observe the particles with maximum efficiency, the glass was cut and polished with a surface normal to this direction. In this geometry, a particle traveling through the center of the opening in the housing must traverse a distance through the glass equal to $\sqrt{3}$ times the distance from its intersection with

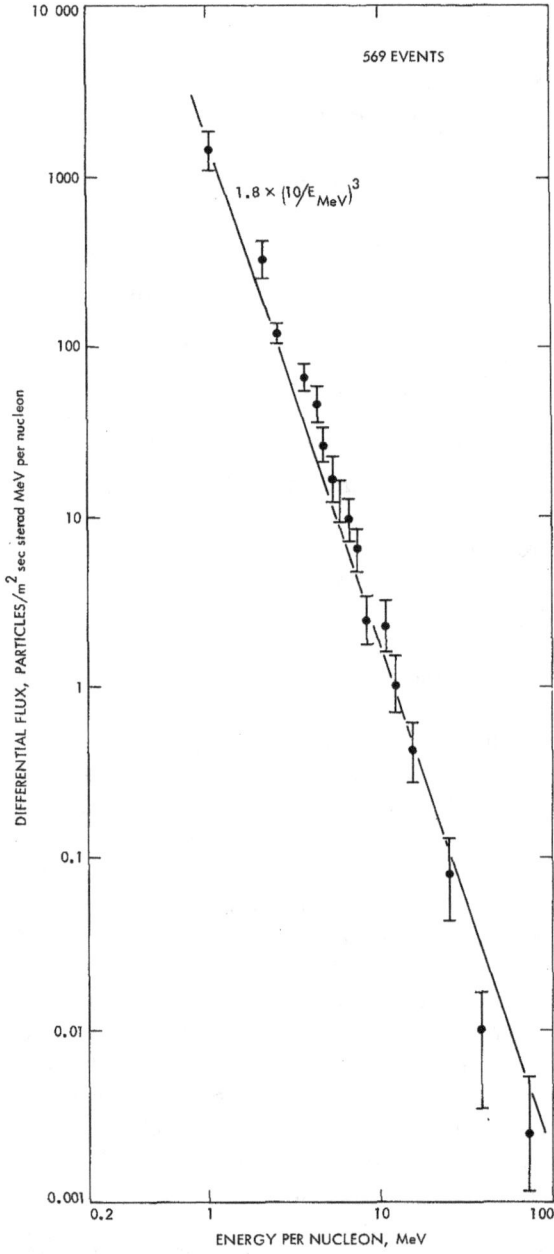

FIGURE 1.—Differential energy spectrum for iron group solar cosmic rays. The absolute position along the vertical scale may be in error by about 50 percent, primarily due to uncertainties in the solid angle factor and in the etchable range of an iron track (measured as 0.5 and 28 μm, respectively).

the polished surface up to the exposed original top surface of the glass. Tracks were etched using 5.7 percent hydrofluoric acid for 3 min to remove a 7-μm layer from the surface, revealing a track density vs. depth that descends sharply from 2.6 (± 0.2) $\times 10^6$ tracks/cm² at a position corresponding to a depth of 3.6 mg/cm² to 35 (± 20)/cm² at 700 mg/cm² and more.

These results, which are given in figure 1, show a flux, Φ, that is well fitted up to 100 MeV/nucleon by a relation $\Phi = 1.8 \times 10^3/E^3$ particles/m² sec sterad MeV per nucleon, where E is the energy in MeV/nucleon. Energy was deduced from range using curves calculated for olivine (ref. 7) and corrected (ref. 8) for the particular glass composition involved. (See ref. 9.) The energies for a given gram per square centimeter of stopping material were reduced by 20 percent, which should be within 5 percent of the true correction (ref. 8). The E^{-3} descent of the spectrum identifies it as solar in origin, since this is the behavior that is often observed for alpha particles from individual solar flares (ref. 10) and inferred less directly for iron group nuclei over a limited energy range (ref. 11). Although our curve gives no clear evidence of curvature, a flattening of the spectrum at still lower energies is likely. At higher energies (greater depths in glass), the track density levels off above 100 MeV/nucleon because of fission events produced by the penetrating background of galactic cosmic-ray protons and alpha particles.

Several particle tracks were of distinctive V-shapes that are characteristic of high-energy induced fission (ref. 12) such as has been seen after accelerator irradiations, but not previously from cosmic-ray irradiation. Because of the observed uniform distribution with depth at which the fission occurred, it is most likely caused by Pb in the glass fissioning in response to penetrating primary cosmic-ray particles (primarily galactic protons and alpha particles). We can estimate how frequent the formation of recognizable V-events should be and test this hypothesis by taking the following six-term product: the proton plus alpha particle fluence over 31 months ($2 \times 2.5 \times 10^7$/cm), the cross section (see ref. 13) for fission (0.8 to 1.2×10^{-25}/cm²) of lead, the number of lead atoms/volume in the glass (4.4×10^{21}/cm³), the fission fragment etchable range (10 μm), and the etching efficiency (ref. 12) for V-events (0.10 to 0.15). This product gives 10 ± 5/cm² as the expected track density. We observed eight definite V-tracks in 0.50 cm² or 16 (±6)/cm², in agreement with the calculated number.

The energy spectrum in figure 1 is a key to measuring erosion on the Moon. Figure 2 contains a shaded band giving the track densities vs. depth measured by four groups (refs. 11, and 14 through 16) in lunar samples 10017 and 10003. For Moon rocks, the exposures to low-energy cosmic rays occur typically over many millions of years (refs. 11, 14, 15, and 16), so that there is ample time for erosion processes to wear away surfaces. For such a situation, points at a final depth, R, in the rock have accumulated tracks when the actual depth ranged from R to $R + vt$, where v is the rate of erosion and t the time over which the sample served as a detector of heavy cosmic rays. If the erosion is uniform on a scale small compared with vt, and if the range spectrum of cosmic rays is proportional to $R^{-2.6}$ (which fig. 2 shows to be a good approximation), then the ratio of the observed to the uneroded track density will be $\{1 - 1/[1 + (vt/r)^{1.6}]\}$ $(R/1.6vt)$. Hence, for small R, the track density varies as $R^{-1.6}$, with the actual magnitude depending on v; at large R, the $R^{-2.6}$ variation occurs. The break from slope 1.6 to 2.6 depends on the product vt, so that, in principle, two measurements along the curve determine both v and t.

Figure 2 shows the track densities observed in two of the most extensively studied Apollo 11 Moon samples (refs. 11 and 13 through 16) with those calculated to result from the observed range spectrum if a constant rate of erosion takes place at 10^{-6}, 10^{-7}, 10^{-8}, or 10^{-9} cm/yr. The fluence observed here for 31 months is assumed to constitute half that of an average 11-year solar cycle. It is clear that a range of values from 0 to 2×10^{-8} cm/yr is in agreement with the data. Such values are consistent with the upper limits of 10^{-7} cm/yr set for a number of lunar samples (refs. 11 and 14 through 16) by other lines of reasoning, but are much less than the $\sim 2 \times 10^{-7}$ cm/yr implied by Shoemaker et al. (ref. 17) by extrapolation of observed impact crater frequencies into the unexplored region of small craters. This discrepancy suggests that the micrometeorite flux that is responsible for subcentimeter craters may be much less than has previously been thought. The observed erosion rate of an atomic layer per year (or per few years) suggests also the possibility that an atomic process may be responsible, rather than a macroscopic cratering phenomenon. For example, if one atom of the exposed surface were removed for each incident solar wind ion, it would give the observed erosion rate.

Although we imply a low rate of erosion here, it is worth noting that a number of effects could cause greater rates to be derived from track counts in Moon samples: statistical fluctuations

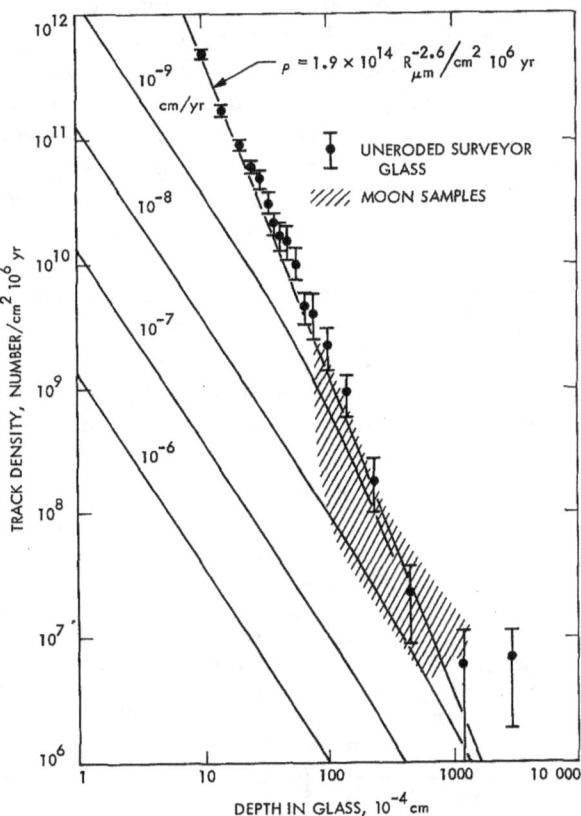

FIGURE 2.—Effect of erosion on the track gradients compared with those observed in lunar rocks. An erosion rate of 0 to 2×10^{-8} cm/yr is consistent with the track densities observed (see refs. 11 and 14 through 16) in lunar samples 10017 and 10003. The track densities and depths of the Moon samples have been recalculated for direct comparison with the Surveyor glass using a solid angle factor of $\pi/2$, and etchable range of Fe of 10 μm, and a 5-million-year exposure for one side of the rock. Only observations at depths greater than 50 μm are included. For the Surveyor glass, a solid angle factor of 0.5 and an etchable range of 28 μm are used.

in the actual erosion due to micrometeorites or secondary ejecta,[1] a thin covering layer of dust, or track counts not taken along the steepest track gradient. Nevertheless, our results show that higher erosion rates are by no means universal and that the Moon is a somewhat calmer place than previously thought to be.

We note also that there exists appreciable uncertainty in the long-term, average flux of solar particles. If the present solar cycle has furnished fewer track-forming flare particles than the long-time average (for example, by a factor of 2), then approximately twice the erosion must have occurred. It should be emphasized that we are implicitly assuming also that our functional relation of (energy)$^{-3}$ observed over 31 months of solar activity is an adequate representation over many millions of years, an assumption we cannot at present establish with certainty. However, one test of this assumption will be from repeated evaluations of t from the place of break in the curve from $R^{-1.6}$ to $R^{-2.6}$. If this age agrees consistently with values that are independently derived from the galactic cosmic rays (refs. 11 and 14 through 16), it will be strong support for the long-time applicability of an E^{-3} relation.

References

1. Fleischer, R. L.; and Price, P. B.: "Charged Particle Tracks in Glass." *J. Appl. Phys.*, vol. 34, 1963, pp. 2903–2904.
2. Fleischer, R. L.; Price, P. B.; and Walker, R. M.: "Solid-State Track Detectors: Applications to Nuclear Science and Geophysics." *Annual Review of Nuclear Science*, vol. 15, 1965, pp. 1–28.
3. Fleischer, R. L.; Price, P. B.; and Woods, R. T.: "Nuclear Particle-Track Identification in Inorganic Solids." *Phys. Rev.*, vol. 188, 1969, pp. 563–567.
4. Aller, L. H.: *The Abundance of the Elements*, Interscience Publishers, Inc., New York, 1961.
5. Price, P. B.; Fleischer, R. L.; and Moke, C. D.: "Identification of Very Heavy Cosmic Ray Tracks in Meteorites." *Phys. Rev.*, vol. 167, 1968, pp. 277–282.

[1] A recent flaking of a 100-μm chip would leave a surface with a lower track density characteristic of a greater (i.e., 100 μm) depth. Thus, for example, the large discrepancy between the track-density profiles of rock 10058 and of 10017 or 10003 would be resolved if a 75- to 300-μm chip had recently been removed from 10058, either by natural processes or in handling during recovery.

6. *Surveyor III Program Results*, NASA SP-184, Washington, D.C., 1969.
7. Henke, R. P.; and Benton, E. V.: *Range and Energy Loss Tables*, USNRDL-TR-1102, Nov. 21, 1966.
8. Norcliffe, L. C.; and Schilling, R. L.: "Range and Stopping Power Tables for Heavy Ions." *Nuclear Data*, to be published.
9. Morey, G. W.: *The Properties of Glass*, second edition, Reinhold Publishing Co., New York, 1954.
10. Anderson, K. A.: *Proceedings of the 11th International Conference on Cosmic Rays*, Budapest, 1969.
11. Fleischer, R. L.; Haines, E. L.; Hart, H. R., Jr.; Woods, R. T.; and Comstock, G. M.: "The Particle Track Record of the Sea of Tranquillity." *Proceedings of the Apollo 11 Lunar Science Conference*, vol. 3, 1970, pp. 2103–2150.
12. Fleischer, R. L.; Naeser, C. W.; Price, P. B.; Walker, R. M.; and Maurette, M.: "Cosmic Ray Exposure Ages of Tektites by the Fission-Track Technique." *J. Geophys. Res.*, vol. 70, 1965, pp. 1491–1495.
13. Perfilov, N. A.: "Fissionability of Nuclei by High Energy Protons." *JETP*, vol. 14, 1962, pp. 623–624.
14. Crozaz, G.; Haack, U.; Hair, M.; Maurette, M.; Walker, R.; and Woolum, D.: "Nuclear Track Studies of Ancient Solar Radiations and Dynamic Lunar Surface Processes." *Proceedings of the Apollo 11 Lunar Science Conference*, vol. 3, 1970, pp. 2051–2080.
15. Price, P. B.; and O'Sullivan, D.: "Lunar Erosion Rate and Solar Flare Paleontology." *Proceedings of the Apollo 11 Lunar Science Conference*, vol. 3, 1970, pp. 2351–2359.
16. Lal, D.; MacDougall, D.; Wilkening, L.; and Arrhenius, G.: "Mixing of the Lunar Regolith and Cosmic Ray Spectra: Evidence from Particle Track Studies." *Proceedings of the Apollo 11 Lunar Science Conference*, vol. 3, 1970, pp. 2295–2303.
17. Shoemaker, E. M.; Hait, M. H.; Swann, G. A.; Schleicher, D. L.; Schaber, G. G.; Sutton, R. L.; Dahlem, D. H.; Goddard, E. N.; and Waters, E. C.: "Origin of the Lunar Regolith at Tranquillity Base." *Proceedings of the Apollo 11 Lunar Science Conference*, vol. 3, 1970, pp. 2399–2412.

ACKNOWLEDGMENTS

We thank E. L. Haines for calling to our attention the fact that Surveyor 3 parts were to be returned to Earth; W. R. Giard and E. Stella for experimental assistance; E. W. Balis, L. B. Bronk, and D. H. Wilkins for prompt chemical analysis of the glass; R. C. DeVries for measuring the index of refraction; and N. Nickle for numerous communications relative to the exposure of the glass. We are indebted to NASA for supplying the Surveyor glass.

PART C

ENHANCED EMISSION OF IRON NUCLEI IN SOLAR FLARES

P. B. Price, I. D. Hutcheon, R. Cowsik, and D. J. Barber

From an analysis of tracks in a window of the Apollo 12 spacecraft and in a glass filter from the Surveyor 3 camera brought back from the Moon, we have determined the spectrum of Fe nuclei from about 1 to 30 MeV/nucleon in interplanetary space during the period from April 24, 1967, to November 24, 1969, and in the last 10 days of this period. The intensity and spectral slope were higher than expected on the basis of studies of alpha particles by other investigators (see ref. 1) during that period and assuming an Fe:He ratio equal to that in the solar photosphere (ref. 2). In addition to their relevance for solar physics, our results may have important consequences for galactic cosmic-ray processes. They also contribute significantly to the extremely high track densities observed in the lunar soil (ref. 3) and allow us to estimate the rate of erosion of lunar rocks.

The silica glass windows on the Apollo 12 Command Module were exposed to space with an effective recording solid angle of about 1 sterad from November 14 to 24, 1969. A clear flint glass filter over the lens system on the Surveyor 3 camera had approximately an 0.7-sterad effective view of space during the 31 months it resided on the lunar surface.

We received one Apollo 12 window and a small piece of the Surveyor camera filter for study. In both types of glass, tracks of heavily ionizing particles can be revealed by chemical etching. (See refs. 4 and 5.) The visibility of etched tracks depends on ionization rate and increases rapidly with atomic number. From bombardments of glasses with heavy ions, we conclude that ions with $Z \lesssim 16$ record with low efficiency and leave tracks that etch into pits with a low visibility when viewed in an optical microscope. For this reason, and because the solar abundance of ions with $Z > 16$ is strongly peaked at Fe, glass detectors discriminate strongly in favor of Fe. Ions of Fe with energy below about 6 MeV/nucleon, i.e., a range less than about 40 μm, have a sufficiently high ionization rate that they will leave tracks that can be etched into easily recognizable conical pits. Ions of Fe of higher energy have too low an ionization rate to leave etchable tracks at the surface, but the lower energy portions of their trajectories can be exposed by grinding or chemically etching away some of the glass, and these portions can then be detected by additional etching. The depth in the glass at which an Fe track is recorded is thus a measure of its energy.

By means of a sequence of etching (with dilute hydrofluoric acid) and grinding operations, densities of etch pits were measured throughout the entire 3-mm Surveyor glass thickness, corresponding to Fe energies from about 1 to 100 MeV/nucleon, and at the top surface of the Apollo 12 window. Figure 1 summarizes the measurements; the figure shows the etch pit distribution we would expect if Fe and He were emitted from the Sun in the ratio of their photospheric abundances. That distribution was calculated using the alpha-particle energy spectrum measured during the same 31-month period by solid-state detectors on IMP 4 and 5 by Lanzerotti[1] and by Hsieh and Simpson (ref. 1). Seven major solar flares contributed most of the flux. The alpha-particle spectrum scaled down by the recently redetermined (ref. 2) solar ratio $(Fe:He)_\odot = 4 \times 10^{-4}$ was used as the input for the calculation.

The large difference between the observed and predicted track densities was completely unexpected. After converting the observed track-density distribution to a rigidity (or energy) distribution, we obtain the important result that, at low rigidity (or energy), the solar particle Fe:He ratio is much higher than the photospheric abundance ratio, but decreases with

---

[1] L. T. Lanzerotti, unpublished data.

increasing rigidity until it approaches the photospheric value at a rigidity of about 500 MV (~25 MeV/nucleon for Fe). In the only previous observations of solar particles with $Z > 20$, Bertsch et al. (ref. 6) found 23 tracks of Fe-group nuclei in nuclear emulsions exposed in a 5-min rocket flight during the flare of September 2, 1966. They found Fe:He $\approx 2 \times 10^{-4}$ at $E > 24.5$ MeV/nucleon, which is not inconsistent with our results. However, one should keep in mind the possibility that, over a 31-month period, there may be a significant contribution by galactic cosmic rays with a much higher Fe:He ratio at energies beyond about 25 MeV/nucleon.

During the period November 14 to 24, 1969, in which the Apollo 12 windows were exposed, a small interplanetary enhancement occurred, contributing a flux of alpha particles only about 10^{-4} times the total contribution over the previous 31-month period.[2] The track counts corresponding to low-energy Fe nuclei in the Apollo 12 window indicate that the low-energy Fe flux during those 10 days was about 7×10^{-5} times the total over 31 months, in good agreement with the relative alpha-particle contribution. This result supports the assertion that the low-energy Fe tracks are of solar origin and are not an accumulated background of low-energy galactic Fe nuclei.

Our results represent the first evidence that heavy nuclei can be preferentially emitted from a source of energetic particles. Previously, Fichtel and co-workers (refs. 7 and 8) had found such a striking similarity between the abundances of energetic solar particles and of the photosphere that the earlier suggestion by Korchak and Syrovatskii (ref. 9) that heavy nuclei may be preferentially accelerated has largely been forgotten. Admittedly their mechanism, which applies when the acceleration rate is small, does not account for the strong enhancement of Fe observed by us, because acceleration of particles in solar flares takes place so rapidly that the energy loss suffered through ionization by the ions during the acceleration phase is negligible. Instead, we attribute the enhancement to preferential leakage of incom-

[2] L. T. Lanzerotti, private communication.

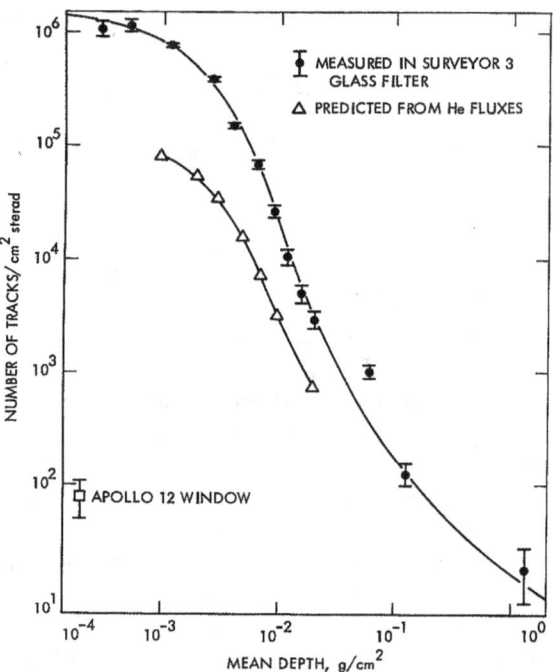

FIGURE 1.—Observed densities of Fe tracks penetrating to a given depth of Surveyor 3 glass and Apollo 12 glass compared with densities predicted assuming that the Fe:He solar particle ratio is the same as the photospheric ratio. The tracks are made visible by chemical etching if their residual range at a glass surface is less than about 40 μm.

pletely ionized heavy nuclei from the accelerating region.

The effective charge of an ion depends on its velocity as $Z^* = Z[1 - \exp(-125\beta/Z^{2/3})]$. From this, it can be seen that H and He are completely stripped of their electrons even at an energy of about 1 MeV/nucleon; Fe ions have an effective charge of only about 13 at 1 MeV/nucleon, increasing to about 24 at 15 MeV/nucleon and becoming very nearly equal to the nuclear charge, 26, only at energies above about 40 MeV/nucleon. Thus, heavy ions have rigidities higher by a factor $\sim Z/Z^*$ than that of an alpha particle at the same energy per nucleon.

If the probability of escape of the accelerated particles is a strong function of their rigidity, one can understand the enhanced Fe fluxes. It appears reasonable that heavy ions, which have a higher rigidity because of their smaller effective charge, should leak preferentially from the flare regions relative to alpha particles and protons of the same energy per nucleon. This preferential escape, which is a consequence of retention of some electrons around a heavy nucleus, should vanish at those high energies at which all the nuclei of interest are completely ionized. All previous observations (refs. 6 through 8) of solar particle composition have been made at sufficiently high energies that no enhancement would be expected. It is interesting to speculate on the possibility, as previously suggested by Davis (ref. 10), that this process of enhancement of heavy nuclei may operate as the injection mechanism for the galactic cosmic rays which are later accelerated to high energies. Thus, the overabundance of heavy nuclei in the cosmic rays may not be entirely indicative of source composition, but may be partly a consequence of preferential leakage from the source.

It also should be interesting to solar particle physicists to mention that the solar alpha particle and proton intensities summed over the 31-month period, when plotted as differential rigidity spectra, have the form $\sim A_i \exp(-R/R_0)$, with the same value of R_0 for protons and alpha particles. Their relative intensities, A_i, scale by a factor of about 12, consistent with model calculations of their photospheric abundance ratio. This agreement is consistent with our model of the enhanced Fe emission because both H and He should be completely stripped at energies above about 1 MeV/nucleon.

Turning now to some different implications of our results, we can use the track-density measurements in figure 1 to draw interesting conclusions about events in the distant past, assuming that the average level of solar activity has remained approximately constant over geologic time:

(1) Rocks exposed undisturbed on the lunar surface for 10^7 yr would accumulate about 6×10^{12} tracks/cm in the top 10 μm of their thickness. Accelerator bombardments of certain minerals with neon and argon ions (ref. 11) show that extensive strains and fractures occur at track densities of about 10^{12}/cm². Summing the contributions of all solar particles with $Z \gtrsim 10$, we conclude that the rate of radiation-induced erosion by fracturing of surface grains is likely to be about 10^{-9} cm/yr. In current unpublished electron microscope studies of Fe track densities as a function of depth in rocks exposed on the lunar surface for about 10^7 yr, we find a maximum track density of about 10^{10}/cm² at the surface. The difference between the observed gradient of track density and the gradient to be expected from figure 1 is attributed to various erosion processes including atomic sputtering by solar wind ions, cratering by micrometeorite bombardment, and flaking of radiation-damaged grains. We conclude that the total erosion rate of rocks that survive for about 10^7 yr on the lunar surface cannot exceed about 10^{-8} cm/yr. This limit is incompatible with the present estimated erosion rate by micrometeorites, about 10^{-7} cm/yr,[a] and allows us to conclude that the present micrometeorite flux measured on satellite detectors is about 10 times higher than the long-term average value.

(2) The lunar soil should contain heavily irradiated small grains, some with track densities of about 10^{12}/cm² that have flaked from radiation-damaged rock surfaces and some that were irradiated while residing at the top of the soil layer. Given a soil about 5 m deep that has accumulated over about 3.5×10^9 yr and has been frequently stirred by meteoritic impacts, we expect an average track density of about 10^{10}/cm² in grains of diameter less than about 10^{-3} cm. Because of the steepness of the solar Fe energy spectrum, larger grains should show visible gradients. We have previously reported all of these features (ref. 3), but were unable to account satisfactorily for the extremely high track densities without knowing the solar flare Fe spectrum. The steep track-density gradient in figure 1 now provides a reasonable, quantitative explanation for most of the observations.

References

1. Hsieh, K. C.; and Simpson, J. A.: "The Relative Abundances and Energy Spectra of ³He and ⁴He From Solar Flares." *Astrophys. J.*, vol. 162, 1970, pp. L191–L196.
2. Ross, J. E.: "Abundance of Iron in the Solar Photosphere." *Nature*, vol. 225, 1970, pp. 610–611.
3. Barber, D. J.; Hutcheon, I.; and Price, P. B.:

---

[a] F. Hörz, J. B. Hartung, and D. E. Gault, Lunar Science Institute Contribution 09, unpublished.

Extralunar Dust in Apollo Cores?" *Science,* vol. 171, 1971, pp. 372–374.
4. FLEISCHER, R. L.; AND PRICE, P. B.: "Charged Particle Tracks in Glass." *J. Appl. Phys.,* vol. 34, 1963, pp. 2903–2904.
5. FLEISCHER, R. L.; PRICE, P. B.; AND WOODS, R. T.: "Nuclear-Particle-Track Identification in Inorganic Solids." *Phys. Rev.,* vol. 188, 1969, pp. 563–567.
6. BERTSCH, D. L.; FICHTEL, C. E.; AND REAMES, D. V.: "Relative Abundance of Iron-Group Nuclei in Solar Cosmic Rays." *Astrophys. J.,* vol. 157, 1969, pp. L53–L56.
7. BISWAS, S.; AND FICHTEL, C. E.: "Composition of Solar Cosmic Rays." *Space Science Reviews,* vol. 4, 1965, pp. 709–736.
8. DURGAPRASAD, N.; FICHTEL, C. E.; GUSS, D. E.; AND REAMES, D. E.: "Nuclear-Charge Spectra and Energy Spectra in the September 2, 1966, Solar-Particle Event." *Astrophys. J.,* vol. 154, 1968, pp. 307–315.
9. KORCHAK, A. A.; AND SYROVATSKII, S. I.: "On the Possibility of a Preferential Acceleration of Heavy Elements in Cosmic-Ray Sources." *Dokl.—Soviet Phys.,* vol. 3, 1958, pp. 983–985.
10. DAVIS, L., JR.: *Space Sciences,* ch. 12, John Wiley & Sons, New York, 1963.
11. SEITZ, M.; WITTELS, M. C.; MAURETTE, M.; AND WALKER, R. M.: *Radiation Effects.* To be published.

ACKNOWLEDGMENTS

We are indebted to R. Baldwin of NASA for arranging the loan of an Apollo 12 window, to N. Nickle of JPL for supplying the Surveyor 3 glass filter, and to L. Lanzerotti and K. Hsieh for supplying us with their unpublished data.

PART D

SOLAR FLARES, THE LUNAR SURFACE, AND GAS-RICH METEORITES

D. J. Barber, R. Cowsik, I. D. Hutcheon, P. B. Price, and R. S. Rajan

High track densities and steep track-density gradients have been observed in interior grains of certain gas-rich meteorites (refs. 1 and 2), in the top millimeters of lunar rocks, and in crystals and glass from the lunar soil (refs. 3 through 8). The tracks were almost certainly produced by heavy nuclei ($Z \approx 26$) emitted in solar flares with a steeply falling energy spectrum. Heavy nuclei in the galactic cosmic rays have an energy spectrum that rises less steeply at low energies than does the solar particle spectrum, but that penetrates much more deeply, down to several centimeters. The presence of tracks of solar origin in isolated grains that were later compacted into meteorites indicate that the peak shock pressure during compaction did not exceed about 100 kilobars, the value below which tracks made visible by chemical etching are not erased (ref. 9). The presence of solar tracks in subsurface lunar soil shows that those layers were once exposed at the surface.

If the rock surface were being eroded during its irradiation or if it were separated from the source of energetic particles by either solid or gaseous matter, the observed track-density gradient would be lower than the predicted gradient (ref. 10). Until now, the use of this concept to infer erosion rates and irradiation history has been impeded by ignorance of the average interplanetary energy spectrum of Fe-group nuclei ($Z \approx 26$).

Three recent developments make it profitable to re-examine lunar erosion, ancient solar flares, and the history of the lunar soil and gas-rich meteorites:

(1) Techniques for observing track densities up to about $5 \times 10^{11}/cm^2$ with high-voltage electron microscopy.

(2) Direct measurement of the gradient of Fe tracks in glass from the Surveyor 3 camera after a 31-month exposure to solar flares during 1967 through 1969.

(3) Calibration of the glass and of lunar minerals with beams of 10.3 MeV/N ^{40}Ar and ^{84}Kr ions.

Energy Spectrum of Interplanetary Fe Nuclei

Three groups have measured the track density from interplanetary Fe-group nuclei as a function of depth in portions of the clear flint filter that was exposed on the lunar surface from May 20, 1967, until the Apollo 12 astronauts brought the Surveyor 3 camera back to Earth in November 1969 (refs. 11 through 13). We have critically compared the available data; in figure 1, we present a revised differential rigidity spectrum for the 31-month period that takes into account several factors not previously considered by all three groups: (1) Using beams of ^{40}Ar and ^{84}Kr ions, we have determined the dependence of cone angles of etched tracks in the glass on ionization rate. On the basis of this, we estimate that the track of an Fe ion can be recognized between 2 and 35 μm of its residual range. (2) We have calculated an accurate range/energy relation for the composition of the flint glass. (3) Fleischer et al. (ref. 13) have shown that there is a uniform background of cosmic-ray-induced fission of lead atoms in the glass that distorts the highest energy part of the spectrum. (4) Nickle[1] has provided us with a recently measured size distribution of lunar dust particles on the glass surface that distorts the lowest energy part of the spectrum. We have taken into account the degrading effect of these dust grains, which are discontinuously distributed and allow some of the lowest energy particles to reach the glass without energy loss. The remaining, primary source of uncertainty in the new spectrum is the human efficiency for observing etched tracks with various cone angles inclined at various angles to the glass surface. We measured, using optical and scanning electron microscopy, a track density of 1.5×10^6/cm^2 sterad very close to the filter's surface. The values obtained by the two methods were self-consistent and agree with the value by Crozaz et al. (ref. 14). Deeper in the glass, we use the

[1] N. Nickle, JPL, private communication.

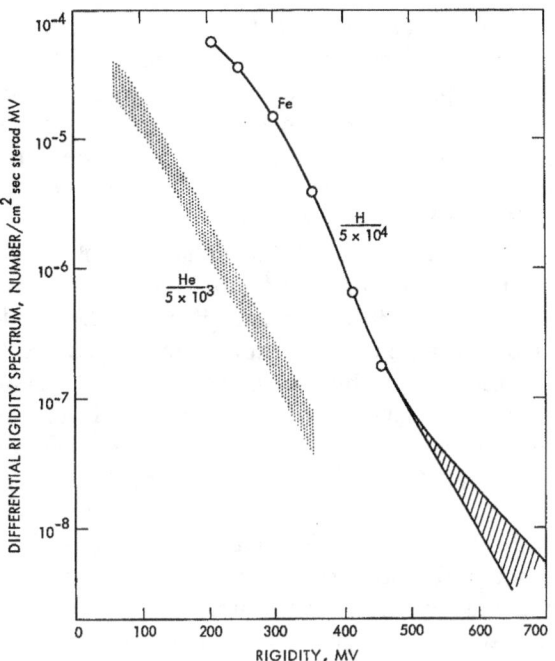

FIGURE 1.—The observed tracks in Surveyor glass are assumed to be due entirely to Fe ions, and the rigidity spectrum is derived, using observed variation with depth. The hatched region at large rigidities represents uncertainty due to a background of fission tracks. The expected curve, with an indicated uncertainty factor of 2, was computed on the basis of satellite measurements of solar protons and alphas, using the proper photospheric abundance ratios. This predicted curve falls much below the observed curve, indicating enhanced emission of Fe-group nuclei.

data of Fleischer et al. (ref. 13), who have a high efficiency for the observation of tracks, since they scan a surface that is normal to the mean direction of the flux of the Fe ions.

Recently, we pointed out (ref. 10) that the flux of Fe nuclei in figure 1 is far greater than expected on the assumption that the Sun emits energetic particles in the ratio of their photospheric abundances. In drawing this conclusion, we have used the published satellite data for solar alpha particles and protons of Lanzerotti[2] and the proton data of Bostrom et al.[3] and of Hsieh and Simpson.[4] In the energy level at overlap, the spectra of these three groups agree

[2] L. Lanzerotti, unpublished data.

[3] C. Bostrom, D. Williams, and J. Areno, unpublished data.

[4] K. Hsieh and J. Simpson, unpublished data.

reasonably well for most flares, with the data of Hsieh and Simpson [4] low by as much as a factor of 2 for the April 12, 1969, flare. The agreement with the solar proton spectra inferred by Finkel et al. (ref. 15) from radiochemical measurements on ^{56}Co and ^{54}Mn in rock 12002 is fair, tending to be low [5] by up to a factor of 2.

As an illustration of the enhancement of Fe flux, the broad curve in figure 1 shows the predicted Fe rigidity spectrum given by the product of the alpha-particle spectrum of Lanzerotti and the solar abundance ratio (Fe:He) $\approx 2 \times 10^{-4}$ calculated by Ross (ref. 16) from the recent solar Fe abundance deduced by Wolnik et al. (ref. 17). We have attributed the difference between the observed and predicted flux to the preferential leakage of low-energy Fe nuclei from the accelerating region because of their incomplete ionization and consequent high magnetic rigidity (ref. 11). Our present analysis places this observation of a heavy ion enhancement on an even firmer basis than before. The enhancement is far greater than the maximum uncertainty in solar proton spectra based on satellite and radiochemical data.

In the remainder of this paper, we assume that the spectrum in figure 1 represents the flux level during the active half of an 11-yr solar cycle and that the flux drops to zero during the inactive half, so that in 10^6 yr the accumulated number of Fe tracks would be ($10^6 \times 3.2 \times 10^7$ sec/yr), 5.5 times higher, assuming that the intensity of the present solar cycle is equal to the average intensity over millions of years.

Lunar Erosion Rate

The method of determining rock erosion rate depends on a comparison of track-density gradients in a rock and in the Surveyor glass and is completely independent of the ratio of Fe ions to protons in solar flares.

The track-density gradient in a lunar rock is most reliably obtained from measurements on an etched, polished section rather than on individual grains removed from various locations. Figure 2 shows the track-density profile taken in a region of rock 12022 that contains no impact pits. This is a particularly valuable rock because

[5] J. R. Arnold, private communication.

of its simple history; in contrast to 10017 (refs. 6, 3, and 4) and 12063 (ref. 14), rock 12022 was irradiated only from one direction and appears not to have received a subsurface exposure. Our measurements, which combine transmission electron microscopy and optical microscopy, cover four orders of magnitude of depth and extend from track densities of about 3×10^6/cm² to about 6×10^9/cm² at a depth of about 3 µm below the surface.

From the profile deep inside the rock, because of galactic Fe-group nuclei, we infer a surface residence time of about 10^7 yr. In figure 2, we show the expected track densities at various depths deduced (making the mentioned assumptions) from the measurements on the Surveyor glass filter. This dashed curve was derived assuming a mean erosion rate of 3 Å/yr and is in excellent agreement with the measured track

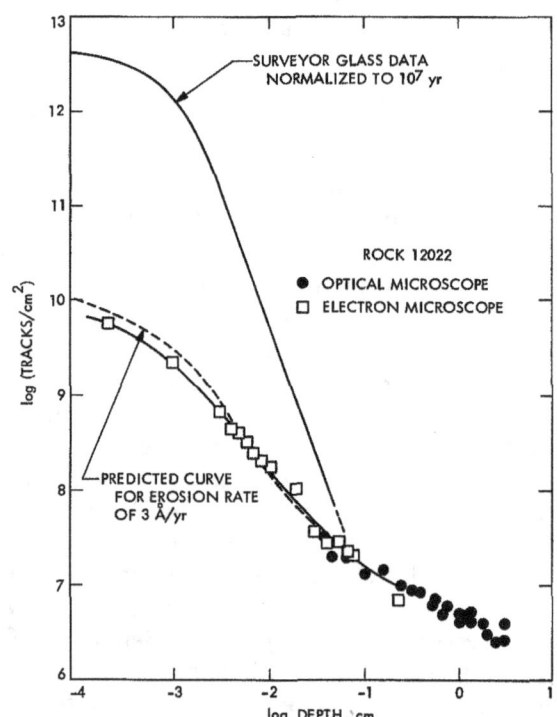

FIGURE 2.—The solar contribution to track densities during 10^7 years is estimated by assuming that the observed track densities in Surveyor glass represented the contribution over one-half of an 11-year solar cycle. On the basis of this, and assuming a uniform erosion rate of 3 Å/yr, the track-density gradient predicted for a lunar rock is in good agreement with that measured for rock 12022.

densities in rock 12022. The erosion rate of about 3 Å/yr is reliable only to the extent that the spectrum in figure 1 truly represents the average solar flare spectrum over 10^7 yr. From their radiochemical study of 12002, Finkel et al. (ref. 15) conclude that the proton intensity in the present solar cycle is representative of the average over about 10^6 yr. Thus, about 3 Å/yr seems to be a reasonable average value for recent lunar history, but may not be representative of erosion processes at an earlier epoch. For example, to build a regolith of about 7 m thick at Oceanus Procellarum (ref. 18) in 3.3×10^9 yr (ref. 19) would necessitate an *average* erosion rate of about 20 Å/yr if the regolith were derived from comminution of local rock. We emphasize that these two rates are not incompatible if they apply to quite different epochs, or if the regolith is built dominantly by large meteorite collisions that would destroy rocks of the size brought back to Earth.

At least three processes are responsible for the erosion of the rocks:

(1) Sputtering of individual atoms by the solar wind (mainly hydrogen) may remove as much as 0.4 Å/yr (ref. 20), depending on the average angle of inclination of the rock surface to the Sun.

(2) The flux of heavy nuclei emitted in solar flares is sufficiently great (fig. 1) that, in the absence of other erosion processes, the outer 10 μm of rock would accumulate about 10^{12} Fe-group tracks in a million years, as well as a considerably larger density of more lightly damaged regions produced by ions of abundant elements like C, O, Ne, Mg, Si, and S. At a dose of about 10^{13}/cm² Ar ions, certain minerals develop extensive strains and fractures in regions where the ions have stopped (ref. 21), so that excessively radiation-damaged layers might flake and contribute to the regolith. We estimate that the erosion rate by this mechanism might reach 0.1 or 0.2 Å/yr for feldspars and certain other minerals that are especially susceptible to radiation damage. If, however, sputtering removes as much as 0.4 Å/yr, the density of solar flare tracks would be limited to about 10^{11}/cm²; flaking probably would not occur.

(3) Micrometeorites contribute to rock erosion. The magnitude of their contribution is uncertain because of uncertainty in the present-day flux of micrometeorites (known to within no better than $\pm 3\times$) and in the long-time constancy of the flux. In a very careful study of microcraters on lunar rocks, Hörz et al. (ref. 22) arrive at an average erosion rate of from 1 to 2 Å/yr and a surface lifetime of about 10^7 yr before destruction by a large micrometeorite, subject to the above uncertainties. It is not clear from microscopic observations alone whether the crater distributions on Apollo 12 rocks have reached a steady state or not. To account for the 3 Å/yr inferred for rock 12022, microcratering must be a more important mechanism than sputtering, which cannot remove more than about 0.4 Å/yr.

At the Apollo 12 Lunar Conference, we estimated the erosion rate of 12022 to be no more than 1 Å/yr; Fleischer et al. (ref. 13) quoted a rate of 0 to 2 Å/yr in their analysis of track gradients reported by several groups at the Apollo Conference. Crozaz and Walker (ref. 12) quoted a value of about 10 Å/yr, based on an apparent erosion equilibrium for the track gradient in 12063. The agreement between our revised value of about 3 Å/yr for 12022 and the rate of 1 to 2 Å/yr by Hörz et al. (ref. 22) is sufficiently close that one can conclude that the micrometeorite flux over the last 10^7 yr must have been fairly similar to the present-day rate, on which they based their calculations.

The effects of erosion undoubtedly depend to some extent on the size of the body being eroded and must be taken into account in attempting to understand the origin of the lunar fines and of the highly irradiated grains in gas-rich meteorites. Atomic sputtering should be essentially independent of the size of the body. Micrometeorite bombardments will not affect track gradients in submillimeter particles because none of these particles will survive a single collision. The only fine particles available for study are those that have avoided collision.

To summarize, track gradients in small particles are less steep than that predicted from the energy spectrum observed in the Surveyor glass. This discrepancy could arise from sputtering-type erosion or from coverage by a layer of matter, but probably not from erosion by radiation-induced cracking. Micrometeorite bombardment erodes large particles and rocks, but destroys small particles. This discussion is pertinent to the section that follows.

Highly Irradiated Grains in the Regolith and in Gas-Rich Meteorites

Any features common to both the Moon and the meteorites contribute to our understanding of the origin of both. For example, the chemical composition and mineralogy of many of the lunar rocks are similar to those of eucrites (ref. 23). Ganapathy et al. (refs. 24 and 25) conclude from an analysis of enriched concentrations of certain trace elements that 2 percent of the lunar soil is of carbonaceous chondritic origin, presumably from accumulated infall. The enrichment is most pronounced in the small grain size fraction.

Using a 1-MeV electron microscope, Borg et al. (ref. 7) and Dran et al. (ref. 26) have found extremely high track densities ($>10^{11}/cm^2$) in a large fraction of the finest grains in the lunar soil, but have failed to find any tracks in grains taken from gas-rich meteorites. They have emphasized the difference in habit and texture features of lunar and meteoritic grains. Their inability to etch the tracks in the lunar grains and the predominance of high track densities in the *smallest* grains led them to suggest that solar suprathermal heavy ions, with damage rates below the threshold for etching, were responsible for the tracks. Suprathermal protons at high flux levels have been observed on several occasions by Frank (ref. 27).

With a 650-keV electron microscope, we have found extremely high densities of etchable tracks at all depths down to 60 cm in fines from Apollo cores (ref. 8) as well as in thin sections of the Pesjanoe, Pantar, and Fayetteville gas-rich meteorites. From both etching and dark-field work, we deduce that about 20 percent of fines less than 5 μm in diameter have track densities greater than $10^{10}/cm^2$. In Fayetteville, about 5 to 10 percent of the smaller grains have about 10^{10} tracks/cm^2. Micrographs of tracks in a particle of lunar soil and in the Fayetteville meteorite are presented in figure 3. Although the meteorite studies are still at a preliminary stage, our observations of track densities comparable

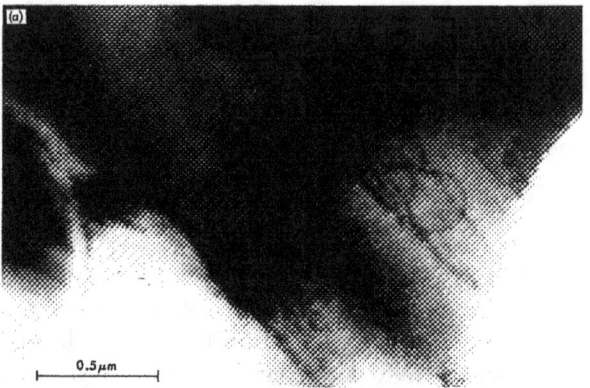
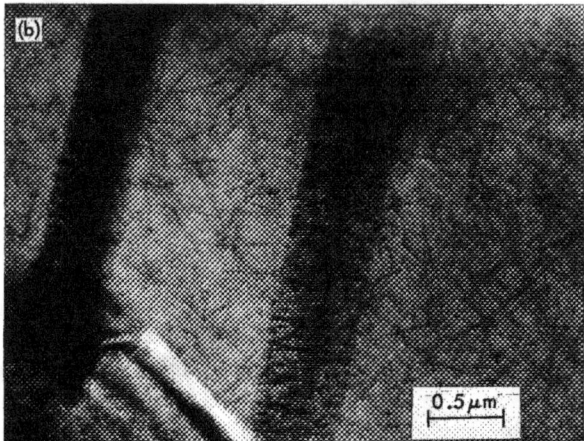

FIGURE 3.—Dark-field electron micrographs (650 keV). (a) Tracks of solar flare particles in a grain from the Apollo 12 lunar fines. (b) Fossil particle tracks in a section of the Fayetteville meteorite thinned by sputter etching with 5-keV argon ions.

to those in lunar grains and at least 20 times greater than had been originally reported in track-rich meteorite grains (refs. 1 and 2) are highly significant because they remove one of the previous major distinctions (ref. 7) between lunar grains and meteoritic grains and suggest the possibility of a similar origin. Track densities of $2 \times 10^{10}/cm^2$ or higher are present in the interior of crystals more than 10 μm in diameter within sections of the Fayetteville meteorite which were thinned by ion-beam machining. Similar high track densities also exist in smaller ($\gtrsim 0.5$ μm in diameter) euhedral crystals. In previous optical microscope and scanning electron microscope studies of etched grains of gas-rich meteorites, those with track densities exceeding $10^{10}/cm^2$ were never noticed because they completely dissolved in the standard etching process. The tracks in meteorites have been studied previously by diffraction contrast imaging (i.e., without etching); the absence of the amorphous layer present on lunar grains makes etching less beneficial (ref. 8) in seeing tracks. Etching the meteorites is disadvantageous because it can cause grains, which are barely held within the thinned and weakened fabric, to drop out. We have established, however, that the tracks in most lunar and meteoritic minerals are etchable, under suitable conditions.

If radiation-induced flaking is a more important erosion process than is sputtering, we could attribute the track-rich grains in the lunar soil to flaked-off surface layers of rocks. We believe, however, that there are some difficulties. The surfaces of rocks show no evidence of extreme stress, nor do they contain track densities as high as $10^{10}/cm^2$. Most of the track-rich grains, both in the lunar soil and in the meteorites (fig. 3), exhibit electron diffraction patterns that argue against extreme radiation damage. Admittedly, many of the fines have amorphous outer layers (about 500 Å thick) attributed to accumulated damage by solar wind bombardment, but the interiors are still crystalline. (See fig. 3(a).) These findings are in agreement with the work of Dran et al. (ref. 26) and Borg et al. (ref. 7), who emphasized that the grains were not disordered. The observed euhedral habits of some of the meteoritic track-rich grains also argue against radiation stress-induced fracture. In the Kapoeta meteorite, however, electron microscopy reveals that many of the small grains contain minute cracks, and microstructural features are severely distorted. The electron diffraction patterns correspondingly exhibit arcs and extended spots. So far, we have failed to see tracks in the carbonaceous chondrites, Murray and Orgeuil, and other observations we have made suggest that tracks will not be found.

We have recently suggested that some of the highly irradiated lunar grains are fragments of infallen extra-lunar dust (ref. 8). It has been suggested previously that some of the gas-rich meteorites were assembled by sintering of circumsolar grains (refs. 1 and 2). Continuing observations of ion-beam-thinned sections of gas-rich meteorites should provide severe constraints on their mode of origin. It would be especially useful to find a large grain with a high track density and a gradient that could be related to an erosion process.

We regard it as highly unlikely that suprathermal heavy ions were responsible for the observed high track densities. A suprathermal ion energy spectrum should continue to rise to a peak at low energy, so that the track length distribution should be peaked at short lengths. Our electron microscope observations of both etched and unetched tracks show that the track length distribution on lunar grains ≥ 10 μm, thinned by sputter-etching, is not peaked at short length and is typical of randomly oriented tracks that penetrate the entire grain.

References

1. LAL, D.; AND RAJAN, R. S.: "Observations Relating to Space Irradiation of Individual Crystals of Gas-Rich Meteorites." *Nature*, vol. 223, 1969, pp. 269–271.
2. PELLAS, P.; POUPEAU, G.; LORIN, J. C.; REEVES, H.; AND AUDOUZE, J.: "Primitive Low-Energy Particle Irradiation of Meteoritic Crystals." *Nature*, vol. 223, 1969, pp. 272–274.
3. CROZAZ, G.; HAACK, U.; HAIR, M.; MAURETTE, M.; WALKER, R.; AND WOOLUM, D.: "Nuclear Track Studies of Ancient Solar Radiations and Dynamic Lunar Processes." *Proceedings of the Apollo 11 Lunar Science Conference*, vol. 3, 1970, pp. 2051–2070.
4. FLEISCHER, R. L.; HAINES, E. L.; HART, H. R., JR.; WOODS, R. T.; AND COMSTOCK, G. M.: "The Particle Track Record of the Sea of Tranquillity." *Proceedings of the Apollo 11 Lunar Science Conference*, vol. 3, 1970, pp. 2103–2120.
5. LAL, D.; MACDOUGALL, D.; WILKENING, L.; AND ARRHENIUS, G.: "Mixing of the Lunar Regolith and Cosmic Ray Spectra: Evidence From Particle-Track Studies." *Proceedings of the Apollo 11 Lunar Science Conference*, vol. 3, 1970, pp. 2295–2303.
6. PRICE, P. B.; AND O'SULLIVAN, D.: "Lunar Erosion Rate and Solar Flare Paleontology." *Proceedings of the Apollo 11 Lunar Science Conference*, vol. 3, 1970, pp. 2351–2359.
7. BORG, J.; DRAN, J. C.; DURRIEU, L.; JOURET, C.; AND MAURETTE, M.: "High Voltage Electron Microscope Studies of Fossil Nuclear Particle Tracks in Extraterrestrial Matter." *Earth and Planetary Science Letters*, vol. 8, 1970, pp. 379–386.

8. Barber, D. J.; Hutcheon, I.; and Price, P. B.: "Extralunar Dust in Apollo Cores?" *Science*, vol. 171, 1971, pp. 372–374.
9. Ahrens, T. J.; Fleischer, R. L.; Price, P. B.; and Woods, R. T.: "Erasure of Fission Tracks in Glasses and Silicates by Shock Waves." *Earth and Planetary Science Letters*, vol. 8, 1970, pp. 420–426.
10. Price, P. B.; Rajan, R. S.; and Tamhane, A. S.: "On the Preatmospheric Size and Maximum Space Erosion Rate of the Patwar Stony-Iron Meteorite." *J. Geophys. Res.*, vol. 72, 1967, pp. 1377–1388.
11. Price, P. B.; Hutcheon, I.; Cowsik, R.; and Barber, D. J.: "Enhanced Emission of Iron Nuclei in Solar Flares." *Phys. Rev. Letters*, to be published.
12. Crozaz, G.; and Walker, R. M.: "Solar Particle Tracks in Glass From the Surveyor 3 Spacecraft." *Science*, to be published.
13. Fleischer, R. L.; Hart, H. R., Jr.; and Comstock, G. M.: "Very Heavy Solar Cosmic Rays: Energy Spectrum and Implications for Lunar Erosion." *Science*, to be published.
14. Crozaz, G.; Walker, R.; and Woolum, D.: "Cosmic Ray Studies of 'Recent' Dynamic Processes." *Proceedings of the Apollo 12 Lunar Science Conference*. To be published.
15. Finkel, R. C.; Arnold, J. R.; Reedy, R. C.; Fruchter, J. S.; Loosli, H. H.; Evans, J. C.; Shedlovsky, J. P.; Imamura, M.; and Delany, A. C.: "Depth Variation of Cosmogenic Nuclides in a Lunar Surface Rock." *Proceedings of the Apollo 12 Lunar Science Conference*. To be published.
16. Ross, J. E.: "Abundance of Iron in the Solar Photosphere." *Nature*, vol. 225, 1970, pp. 610–611.
17. Wolnik, S. J.; Berthel, R. O.; and Wares, G. W.: "Shock-Tube Measurements of Absolute gf-Values for FeI." *Astrophys. J.*, vol. 162, 1970, pp. 1037–1047.
18. Shoemaker, E. M.; and Hait, M. H.: "The Bombardment of the Lunar Maria." *Proceedings of the Apollo 12 Lunar Science Conference*. To be published.
19. Albee, A. L.; Burnett, D. S.; Chodos, A. A.; Haines, E. L.; Huneke, J. C.; Papanastassiou, D. A.; Podosek, F. A.; Russ, G. P.; Tera, F.; and Wasserburg, G. J.: "Rb-Sr Ages, Chemical Abundance Patterns and History of Lunar Rocks." *Proceedings of the Apollo 12 Lunar Science Conference*. To be published.
20. Wehner, G. K.; KenKnight, C.; and Rosenberg, D. L.: "Sputtering Rates Under Solar-Wind Bombardment." *Planet. Space Sci.*, vol. 11, 1963, pp. 885–895.
21. Seitz, M.; Wittels, M. C.; Maurette, M.; and Walker, R. M.: "Accelerator Irradiations of Minerals: Implications for Track Formation Mechanisms and for Studies of Lunar and Meteoritic Minerals." *Rad. Effects*, vol. 5, 1970, pp. 143–148.
22. Hörz, F.; Hartung, J. B.; and Gault, D. E.: "Lunar Microcraters." *Proceedings of the Apollo 12 Lunar Science Conference*. To be published.
23. Wänke, H.; Rieder, R.; Baddenhausen, H.; Spettle, B.; Teschke, F.; Quijano-Rico, M.; and Balacescu, A.: "Major and Trace Elements in Lunar Material." *Proceedings of the Apollo 11 Lunar Science Conference*, vol. 2, 1970, pp. 1719–1727.
24. Ganapathy, R.; Keays, R. R.; Laul, J. C.; and Anders, E.: "Trace Elements in Apollo 11 Lunar Rocks: Implications for Meteorite Influx and Origin of Moon." *Proceedings of the Apollo 11 Lunar Science Conference*, vol. 2, 1970, pp. 1117–1142.
25. Ganapathy, R.; Keays, R. R.; and Anders, E.: "Apollo 12 Lunar Samples: Trace Element Analysis of a Core and the Uniformity of the Regolith." *Science*, to be published.
26. Dran, J. C.; Durrieu, L.; Jouret, C.; and Maurette, M.: "Habit and Texture Studies of Lunar and Meteoritic Materials With a 1 MeV Electron Microscope." *Earth and Planetary Science Letters*, vol. 9, 1970, pp. 391–400.
27. Frank, L. A.: "On the Presence of Low Energy Protons ($5 \leq E \leq 50$ keV) in the Interplanetary Medium." *J. Geophys. Res.*, vol. 75, 1970, pp. 707–716.

ACKNOWLEDGMENTS

We thank A. Ghiorso for the Ar and Kr ion irradiations. We also thank N. Nickle and D. Robertson for help with the Surveyor glass; K. C. Hsieh, J. D. O'Sullivan, C. O. Bostrom, L. O. Lanzerotti, J. R. Arnold, and R. C. Reedy for helpful discussions and data on solar protons and alpha particles.

X. Soil Property Analyses

PART A

BEARING STRENGTH OF THE LUNAR SOIL

L. D. Jaffe

Before lunar soil samples were returned to Earth, a number of measurements of the mechanical properties of lunar soil were made from spacecraft. (See refs. 1 through 22.) No equipment specifically designed for such measurements was carried on spacecraft, except the soil penetrometer on Luna 13 (ref. 16). The soil mechanics experiment on the Surveyor spacecraft utilized a device designed primarily to sample the soil (ref. 23). In general, the soil mechanical properties were determined by using imaging and other equipment that was aboard the spacecraft for other purposes.

The problem of measuring surface mechanical properties, without returned samples, will probably arise for other planets. As a guide in evaluating probable techniques, it seems worthwhile to compare measurements of soil mechanical properties made on the Moon, as mentioned above, with mechanical property measurements on lunar soil returned to Earth.

A unique opportunity for comparative measurements was provided by the return to Earth of 6.5 g of lunar soil contained in the scoop of the Surveyor 3 surface sampler, together with the scoop itself. This scoop had been used to measure soil properties on the Moon during Surveyor 3 operations (ref. 12). Other soil property measurements had been made within about 1 m of the same spot using other equipment on Surveyor 3 (ref. 10). The scoop and the soil within it were removed and returned to Earth by Apollo 12 astronauts Conrad and Bean. This soil sample had been used for mechanical property measurements on the Moon and could be used again for such measurements on Earth.

This article presents a discussion of one aspect of the on-Earth laboratory measurements: bearing strength and bearing load-penetration relations, measured in air as a function of bulk density.

Material

After the scoop of the Surveyor 3 surface sampler was returned to the Lunar Receiving Laboratory in Houston, Tex., it was placed in a polyethylene bag. During subsequent handling, some of the lunar soil in the scoop fell into the bag. This soil was recovered, and 1.3 g of lunar soil were provided by NASA for this and related investigations.

The few particles larger than about 1 mm had been removed by hand, but the soil had not been sieved or otherwise intentionally fractionated. Particle size distributions, measured on part of the 1.3-g sample, are reported in reference 24.

The material was stored in air during and after its transfer to Earth.

Equipment

A commercial vertical, screw-driven, tension/compression testing machine equipped for recording load vs. deformation was used. Full-scale load-recording ranges extended from 2 g upward. As the lower ranges could only be used in tension, the test fixture was designed accordingly. The cup that contained the soil under test was made of poly(methyl methacrylate) and had an inside diameter of 1.0 cm and a depth of 1.1 cm. (For the first tests, the inside diameter was 0.6 cm.) The bearing load was applied by a vertical rod, 2.0 mm in diameter. The rod tip tapered inward about 0.35 mm on the diameter in the 5 to 9 mm above the end to provide friction relief on the sides of the rod as it penetrated the soil. The rod was integral with a cylindrical brass weight suspended by a thin wire from the load cell at the top of the test machine.

Procedure

For the low bulk densities, soil was gently brushed into the cup from its top, or spooned in with a spatula. For high packing densities, the cup was tapped or, in a few cases, vibrated. Density was determined by weighing on an analytical balance and measuring the depth optically or on radiographic prints. Radiography was used in many runs to check freedom from voids larger than the particle size. Bulk densities obtained ranged from 1.15 to 1.93 g/cm³.

Tests were made in air at 70°C; relative humidities were recorded as 40 to 50 percent. To test, the cup containing the soil was driven upward against the rod tip at the rate of 0.0021 cm/sec (0.05 in./min). Motion was measured as travel of the lower cross head, load as reduction of the weight suspended from the upper cross head. Runs generally were started with the recording system at high sensitivity. If the load went off scale, cross-head motion was stopped, the load recorder was switched to lower sensitivity, and cross-head motion was resumed.

After test, the surface of the material was observed and changes were noted. Some specimens were reradiographed after test to provide additional information on the nature of the deformation.

Results

No voids were visible in radiographs made before test, although some small denser clumps were noted in one run at a bulk density of 1.26 g/cm³. The other specimens radiographed appeared to be uniform before test.

Figure 1 shows bearing stress vs. penetration curves for four runs. At low penetrations, the relation was about linear, with some tendency to curve toward higher force as the penetration increased. In most of the runs at medium and high bulk densities, the slope of the stress vs. penetration curves suddenly increased sharply,

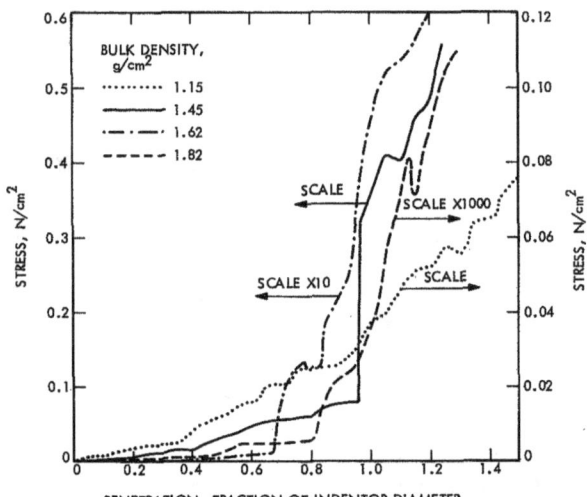

FIGURE 1.—Bearing stress vs. penetration. Four individual test runs, at various bulk densities, are plotted. Note different vertical scales. Indentor tip diameter = 2 mm.

leading to a rapid increase in stress, often amounting to an order of magnitude or more. (See fig. 1.) A few of the runs at high bulk density showed one or more decreases in load with increasing penetration; these load decreases generally were accompanied by visible local bulging of the top surface of the material.

The top surfaces after test showed bulging and cracking for all runs at bulk densities above 1.62 g/cm³. No cracking or bulging was observed for any run below 1.61 g/cm³ (except for a small amount of bulging in one run at 1.42 g/cm³). Radiographs after test were in complete agreement with these visual observations. For material of low bulk density, it was usually possible to see in these radiographs a cylindrical plug of denser material directly below the indentor hole. The holes retained their vertical sides after the indentor was withdrawn, displaying the soil cohesion.

Discussion

The shape of the stress vs. penetration curves agreed with those ordinarily found for terrestrial particulate materials with corresponding bulk densities and relatively low cohesion, except for the initial low stress level, followed by the sudden slope increase. To elucidate these characteristics, bearing tests were made on crushed terrestrial basalt, with a particle size distribution and mechanical properties resembling the lunar material. In these tests, larger indentors (6-mm diameter) and larger cups (150-mm diameter, 75-mm depth) were used, as well as the small ones used for the lunar material. Lunar material could not be tested with the larger cups and indentors because the sample was too small. Sudden slope increases were found with the terrestrial basalt tested with the 2-mm-diameter indentor. With the larger indentor, in the large cup, the initial low stress level and sudden slope increase were never found. Instead, the stress level immediately rose to levels corresponding to those encountered after a sudden increase. Tests using the 2-mm-diameter indentor in the large cup showed that the initial low stress, followed by the sudden increase, was characteristic of packing procedures in which a thin, loose layer of material was placed on a well-compacted substrate and the cup then tapped to compact the material further. It appears, therefore, that the initial low stress level was due to a surface layer of lower density than that below. Placement of particulate material in the small cup, followed by tapping, is apparently likely to lead to this condition.

Accordingly, the stress levels before the sudden slope increase are probably not representative of the overall bulk density. In most runs where such an increase occurred with lunar material, the increase took place before penetration reached 1 indentor diameter. The stress at penetration equal to 1 indentor diameter was taken as the bearing capacity. (See table 1.) In a few cases in which a sudden increase occurred at high penetration, or the indentor tilted before penetration equaled 1 diameter, the curve was extrapolated to this penetration.

In general, when motion of the testing machine head was stopped to permit switching the range of the load sensor, the load promptly fell to zero or almost zero. When indentor motion

TABLE 1.—*Bearing capacity and density of lunar soil*

Bulk density, g/cm^3	Bearing capacity at penetration = indentor diameter, N/cm^2	Remarks	Cracking and bulging
1.15	0.038		No
1.18	.027		No
1.22	.038	6-mm cup diameter	
1.26	.021		No
1.42	.048		No cracking; minor bulging
1.45	.35		No
1.46	1.9		No
1.48	.82		No
1.54	1.4		No
1.60	5.6	Extrapolation of stress vs. penetration curve; 6-mm cup.	
1.61	8.2		Yes
1.62	4.4		No
1.70	6.2		Yes
1.70	10		Yes
1.76	12.5	Extrapolation	Yes
1.79	>6.2	Extrapolation	Yes
1.80	16		Yes
1.82	11	Extrapolation	Yes
1.83	100	At yield	Yes
1.84	33		Yes
1.84	36		Yes
1.86	>6.2	Extrapolation	Yes
1.90	32		Yes
1.93	>6.2	Extrapolation	Yes

was resumed, the load rose rapidly to its previous value, but a detectable penetration occurred during the load increase. This penetration was deducted in the analysis of stress vs. penetration curves.

The bearing capacity is plotted vs. bulk density in figure 2. Despite the scatter, the trend is obvious. Drawn in the figure is a linear least-squares fit for log of the bearing capacity, p, vs. bulk density, d, corresponding to the relation

$$\log_{10} p = -6.94 + 4.62d \qquad (1)$$

where p is in newtons per square centimeter and d in grams per cubic centimeter. The standard deviation is equivalent to a difference of 0.06 g/cm³ in bulk density. A slightly better fit was obtained with a quadratic least-squares, but the improvement was not statistically significant.

Comparison With Lunar Results

To compare the laboratory results with lunar measurements, it is necessary to know the bulk density in-situ on the Moon. Unfortunately, no reliable measurements of lunar regolith density have been published. The in-situ measurements from Luna 13 (see refs. 15 and 25) are ambiguous and questionable (ref. 26). Measurements on cores returned by Apollo 11 and 12 undoubtedly reflect significant disturbances on packing caused by insertion of the core tubes themselves. (See refs. 27 and 28.) Indeed, the author attempted to calculate density from the in-situ bearing strength measurements, obtaining 1.1 g/cm³ at the surface and 1.6 g/cm³ at a depth of 5 cm (ref. 29). Perhaps the best results at the moment are those from Apollo 12 core tubes, indicating that the bulk density probably averages about 1.8 g/cm³ for the top 30 cm of material (ref. 28). The corresponding bearing capacity shown by figure 2 is about 20 N/cm².

The in-situ bearing data providing the most direct comparison with the present measurements are those of Scott and Roberson (ref. 12), using the same Surveyor 3 soil mechanics surface sampler, with its scoop closed, at positions including that from which the soil sample used in the present work was obtained, and all within 1.5 m of it. Scott and Roberson obtained a bearing pressure of 2 N/cm² at a depth of 2.5

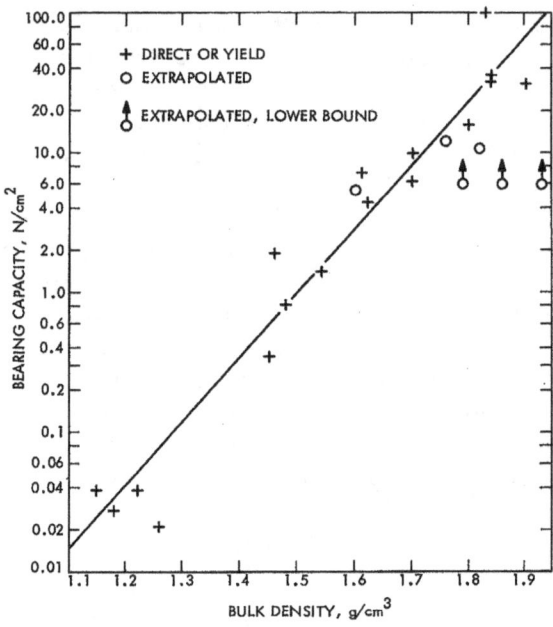

FIGURE 2.—Bearing capacity vs. bulk density. Line indicates least-squares fit. Bearing capacity was taken at a penetration equal to indentor diameter.

cm and bearing plate width of 2.5 cm. Other Surveyor 3 bearing stress measurements for nearby soil included 10 N/cm², for depths of 4 to 5 cm and bearing plate width of 0.32 cm, from the surface sampler with scoop open (ref. 28); and 4 N/cm² for a depth of about 2.5 cm and a bearing diameter of about 25 cm, from a footpad indentation (ref. 10).

The closed-scoop surface sampler value of 2 N/cm² was obtained at a (penetration depth): (bearing plate width) ratio of 1, corresponding to the condition used for figure 2. Match to the curve of figure 2 occurs not at a bulk density of 1.8 g/cm³, but at about 1.6 g/cm³. This tends to suggest that the bulk density of the lunar soil at Surveyor 3, and a depth of 2.5 cm, is about 1.6 and not 1.8 g/cm³. Corrections to the bearing strengths should, in principle, be made for differences in scale, geometry, gravity, and perhaps atmosphere. It seems best, however, to await results of other tests, including shear tests, planned for the same sample of lunar soil before attempting those corrections.

An almost linear stress vs. penetration curve for linear soil was found in in-situ measurements at the Surveyor 7 site, near Tycho, using a soil mechanics surface sampler (ref. 13). The bearing capacity observed with the scoop was essentially the same as at Surveyor 3, described above (ref. 13).

Other Surveyor and Lunar Orbiter results have been summarized by the author (ref. 29). The indicated bearing capacity was about 0.1 N/cm² at 0.1-cm depth and 1.7 N/cm² at 2-cm depth. Whether this variation is due to change of bulk density with depth remains to be determined.

Observations on the lunar surface by Apollo 11 astronauts gave stresses of 0.5 to 1.5 N/cm² for penetration:diameter or penetration:width ratios $\ll 1$, and depths of 1 to 8 cm (ref. 30). The present laboratory results seem consistent with these observations.

Tests by Costes et al. (ref. 30) on lunar soil returned by Apollo 11, in which a penetrometer was inserted to the depth necessary to reach a fixed load, gave, at a penetration:diameter ratio near 1, a bearing stress of about 1 N/cm² at a bulk density of 1.14 g/cm³, <5 to 14 N/cm² at 1.77 g/cm³, and 30 N/cm² at 1.80 g/cm³. The results at 1.77 to 1.80 g/cm³ are consistent with those found in this work; that at 1.36 g/cm³ is five times higher than the value indicated by figure 2. The failure modes in the laboratory tests of Apollo 11 material were the same as in the present tests.

Conclusions

(1) Bearing capacities of lunar soil returned from Surveyor 3 vary from 0.02 to 0.04 N/cm² at a bulk density of 1.15 g/cm³ to 30 to 100 N/cm² at 1.9 g/cm³. The relation between bulk density and logarithm of the bearing capacity is about linear. These results are for measurements with an indentor of 2-mm diameter, in air, on Earth, and at a penetration equal to the diameter of the indentor.

(2) Shapes of the load vs. penetration curves are similar to those obtained with particulate material of terrestrial origin.

(3) At bulk densities below 1.61 g/cm³, deformation was by compression of the material below the indentor ("local shear," "compressible failure"). At bulk densities above 1.62 g/cm³, deformation was, by outward displacement of the material ("general shear," "incompressible failure").

(4) Preliminary comparison with bearing measurements made in-situ on the Moon by remote-control techniques, before return of samples from the Moon, suggests good agreement if the lunar material has a bulk density of about 1.6 g/cm³ at a depth of 2.5 cm. Definitive comparison is dependent upon the availability of better data on bulk densities of the lunar soil and other tests of mechanical properties of returned materials, as well as additional analysis.

References

1. KUIPER, G. P.; LePOOLE, R. S.; AND STROM, R. G.: "Interpretation of the Ranger Records." *Ranger VIII and IX. Part II—Experimenters' Analyses and Interpretations*, TR 32–800, Jet Propulsion Laboratory, Pasadena, Calif., 1966, pp. 35–248.
2. MOORE, H. J.: "Cohesion of Material on the Lunar Surface." *Ranger VIII and IX. Part II—Experimenters' Analyses and Interpretations*, TR 32–800, Jet Propulsion Laboratory, Pasadena, Calif., 1966, pp. 263–270.
3. JAFFE, L. D.: "Strength of the Lunar Dust." *J. Geophys. Res.*, vol. 70, 1965, pp. 6139–6146.
4. JAFFE, L. D.: "Lunar Surface Strength." *Icarus*, vol. 6, 1967, pp. 75–91.
5. JAFFE, L. D.: "Surface Structure and Mechanical Properties of the Lunar Maria." *J. Geophys. Res.*, vol. 72, 1967, pp. 1727–1731.
6. JAFFE, L. D.: "Surveyor 6 Lunar Mission." *J. Geophys. Res.*, vol. 73, 1968, pp. 5297–5300.
7. JAFFE, L. D.; AND SCOTT, R. F.: "Lunar Surface Strength: Implications of Luna 9 Landing." *Science*, vol. 153, 1966, pp. 407–408.
8. CHRISTENSEN, E. M.; BATTERSON, S. A.; BENSON, H. E.; CHANDLER, C. E.; JONES, R. H.; SCOTT, R. F.; SHIPLEY, E. N.; SPERLING, F. B.; AND SUTTON, G. H.: "Lunar Surface Mechanical Properties—Surveyor I." *J. Geophys. Res.*, vol. 72, 1967, pp. 801–813.
9. CHRISTENSEN, E. M.; CHOATE, R.; JAFFE, L. D.; SPENCER, R. L.; SPERLING, F. B.; BATTERSON, S. A.; BENSON, H. E.; HUTTON, R. E.; KO, H. Y.; SCHMIDT, F. N.; SCOTT, R. F.; AND SUTTON, G. H.: "Surveyor V—Lunar Surface Mechanical Properties." *Science*, vol. 158, 1967, pp. 637–640.
10. CHRISTENSEN, E. M.; BATTERSON, S. A.; BENSON, H. E.; CHOATE, R.; JAFFE, L. D.; JONES, R. H.; KO, H. Y.; SPENCER, R. L.; SPERLING, F. B.; AND SUTTON, G. H.: "Lunar Surface Mechanical Properties at the Landing Site of Surveyor 3." *J. Geophys. Res.*, vol. 73, 1968, pp. 4081–4094.
11. CHRISTENSEN, E. M.; BATTERSON, S. A.; BENSON, H. E.; CHOATE, R.; HUTTON, R. E.; JAFFE, L. D.;

Jones, R. H.; Ko, H. Y.; Schmidt, F. N.; Scott, R. F.; Spencer, R. L.; and Sutton, G. H.: "Lunar Surface Mechanical Properties." *J. Geophys. Res.*, vol. 73, 1968, pp. 7169–7192.
12. Scott, R. F.; and Roberson, F. I.: "Soil Mechanics Surface Sampler: Lunar Surface Tests, Results, and Analyses." *J. Geophys. Res.*, vol. 73, 1968, pp. 4045–4080.
13. Scott, R. F.; and Roberson, F. I.: "Soil Mechanics Surface Sampler." *J. Geophys. Res.*, vol. 74, 1969, pp. 6175–6214.
14. Choate, R.; Batterson, S. A.; Christensen, E. M.; Hutton, R. E.; Jaffe, L. D.; Jones, R. H.; Ko, H. Y.; Spencer, R. L.; and Sperling, F. B.: "Lunar Surface Mechanical Properties." *J. Geophys. Res.*, vol. 74, 1969, pp. 6149–6174.
15. Cherkasov, I. I.; Kemurjian, A. L.; Mikhailov, L. N.; Mikheyev, V. V.; Morozov, A. A.; Mosatov, A. A.; Savenko, I. A.; Smorodinov, M. I.; and Shvarev, V. V.: "Determination of the Density and Mechanical Strength of the Surface Layer of the Lunar Soil at the Landing Site of the Luna-13 Probe." *Kosm. Issled.*, vol. 5, 1967, pp. 746–757. Translated in *Cosmic Res.*, vol. 5, 1967, pp. 636–645.
16. Cherkasov, I. I.; Gromov, V. V.; Zobachev, N. M.; Musatov, A. A.; Mikheyev, V. V.; Petrukhin, V. P.; and Shvarev, V. V.: "Soil-Density Meter Penetrometer of the Automatic Lunar Station Luna 13." *Dokl. Akad. Nauk SSSR*, vol. 179, 1968, pp. 829–831. Translated in *Dokl.–Soviet Phys.*, vol. 13, 1968, pp. 336–338.
17. Cherkasov, I. I.; Mikhailov, L. N.; Morozov, A. A.; Petrukhin, V. P.; Shvarev, V. V.; Mikheyev, V. V.; Smorodinov, M. I.; and Zobachev, N. M.: "Determination of the Structural-Mechanical Properties of the Lunar Soil With the Aid of the Automatic Lunar Station Luna 13." *Inzh. Fiz. Zh.*, vol. 14, 1968, pp. 581–585.
18. Cherkasov, I. I.; and Shvarev, V. V.: "First Results of the Close Investigations of the Lunar Soil." *Zemlya i Vselennaya*, vol. 2, 1968, pp. 15–24.
19. Filice, A. L.: "Lunar Surface Strength Estimate From Orbiter II Photograph." *Science*, vol. 156, 1967, pp. 1486–1487.
20. Eggleston, J. M.; Patterson, A. W.; Throop, J. E.; Arant, W. H.; and Spooner, D. L.: "Lunar 'Rolling Stones.'" *Photogrammetric Eng.*, vol. 34, 1968, pp. 246–255.
21. Halajian, J. D.: "Mechanical, Optical, Thermal and Electrical Properties of the Surveyor I Landing Site." *J. Astronaut. Sci.*, vol. 14, 1967, pp. 270–281.
22. Karafiath, L. L.; and Nowatzki, E. A.: "Surveyor V Landing—The Effect of Slope on Bearing Capacity." *Science*, vol. 161, 1968, pp. 601–602.
23. Scott, R. F.: "Soil Mechanics Surface Sampler Experiment for Surveyor." *J. Geophys. Res.*, vol. 72, 1967, pp. 827–830.
24. Jaffe, L. D.; Strand, J.; and Scott, R. F.: *Particle Size Distribution of Lunar Soil From the Surveyor 3 Surface Sampler*. To be published.
25. Morozov, A. A.; Smorodinov, M. I.; Shvarev, V. V.; and Cherkasov, I. I.: "Measurement of Lunar Surface Density by the Automatic Station 'Luna 13.'" *Dokl. Akad. Nauk SSSR*, vol. 179, 1968, pp. 1087–1090. Translated in *Dokl.–Soviet Phys.*, vol. 13, 1968, pp. 348–350.
26. Scott, R. F.: "The Density of the Lunar Surface Soil." *J. Geophys. Res.*, vol. 73, 1968, pp. 5469–5471.
27. Lunar Sample Preliminary Examination Team: "Preliminary Examination of Lunar Samples from Apollo 11." *Science*, vol. 165, 1969, pp. 1211–1227.
28. Scott, R. F.; Carrier, W. D.; Costes, N. C.; and Mitchell, J. K.: "Mechanical Properties of the Lunar Regolith." *Apollo 12 Preliminary Science Report*, NASA SP-235, Washington, D.C., 1970, pp. 161–182.
29. Jaffe, L. D.: "Lunar Surface Material: Spacecraft Measurements of Density and Strength." *Science*, vol. 164, 1969, pp. 1514–1516.
30. Costes, N. C.; Carrier, W. D.; Mitchell, J. K.; and Scott, R. F.: "Apollo 11 Soil Mechanics Investigation." *Science*, vol. 167, 1970, pp. 739–741.

ACKNOWLEDGMENTS

I thank A. Sorkin for suggesting major features of the indenting device and J. Greenleaf for his help in the testing.

PART B

CRACKING OF THE LUNAR SOIL

L. D. Jaffe

The tendency of lunar soil to break into clods when disturbed was recognized on examination of the first Surveyor 1 imagery (refs. 1 and 2). Pictures of disturbances produced in bearing tests with the Surveyor 3 soil mechanics surface sampler and shown by Scott and Roberson (ref. 3) suggested to some observers that the soil layer in Oceanus Procellarum tends to crack into thin flat "tiles" (figs. 1(a) and 2(a)) and, therefore, that it consists of a thin, rather rigid crust over a softer substrate. Pictures of areas disturbed by the Apollo 11 closeup camera in Mare Tranquillitatis (fig. 3(a)) and by the Apollo 12 Lunar Module (LM) descent engine in Oceanus Procellarum conveyed a similar impression, as mentioned by Gold (ref. 4) and by Gold, Pearce, and Jones (ref. 5). None of the authors mentioned were misled, and, indeed, they cautioned against this interpretation (refs. 3 through 5). Nevertheless, photographs made by the Apollo 12 astronauts of the areas disturbed by Surveyor 3 help to clarify the matter.

Figures 1(b) and 2(b) are portions of Apollo 12 photographs showing the same bearing test imprints as figures 1(a) and 2(a). The appearance of the areas inboard (toward the Surveyor) is quite different; rather than a pattern of cracks in an apparently flat surface, these Apollo photographs suggest an irregular surface that consists of clods or fragments. The difference in appearance arises from differences in lighting angle. The pictures in figures 1(a) and 2(a) were made at Sun angles of 80° and 40°, respectively, above the horizon to the west. Those in figures 1(b) and 2(b) were taken with the Sun 23° above the eastern horizon. The 13° slope, downward to the west, of the crater wall on which Surveyor 3 rests, makes the Sun angles to the surface about 87°, 53°, 10°, and 10°, respectively. The "flat" appearance of figures 1(a) and 2(a) is due to the high Sun angle; the low angle lighting of figures 2(a) and 2(b) shows the topography more clearly. The high angle lighting is useful, however, in showing the existence and extent of the pattern along which the soil broke into the clods evident at low Sun (ref. 3).

Figure 1(c) adds additional information. It was taken a few seconds after figure 1(b), from a slightly different angle. Figures 1(b) and 1(c) form a stereopair. Viewed stereoscopically, they show clearly the three-dimensional character of the disturbed material; instead of flat "tiles," one sees a pile of roughly equiaxed clods.

The Apollo closeup photographs, such as figure 3(a), were taken with artificial illumination at a high angle above the horizontal (ref. 4); no comparable pictures of the same areas were made with low lighting. However, a picture forming a stereopair with figure 3(a) is available (fig. 3(b)); stereoviewing of the pair reveals strong vertical displacements across the "cracks." At normal stereoviewing separation angles, the vertical displacements seem greater than the horizontal size of the blocks. This exaggeration effect arises from the high stereoseparation angle of the camera used (ref. 4); the blocks are probably nearly equiaxed. The impression of flat "tiles" and "crusting," obtained by monoscopic viewing, is an illusion. Rather, the lunar soil deforms and cracks in the same manner as homogeneous isotropic terrestrial soils of moderate bulk density, having a small amount of cohesion. (See refs. 1 through 4.)

References

1. SURVEYOR SCIENTIFIC EVALUATION AND ADVISORY TEAM: "Surveyor I: Preliminary Results." *Science*, vol. 152, 1966, pp. 1737–1750.
2. CHRISTENSEN, E. M.; BATTERSON, S. A.; BENSON, H. E.; CHANDLER, C. E.; JONES, R. H.; SCOTT, R. F.; SHIPLEY, E. N.; SPERLING, F. B.; AND SUTTON, G. H.: "Lunar Surface Mechanical

Properties—Surveyor I." *J. Geophys. Res.*, vol. 72, 1967, pp. 801–813.
3. SCOTT, R. F.; AND ROBERSON, F. I.: "Soil Mechanics Surface Sampler: Lunar Surface Tests, Results, and Analysis." *J. Geophys. Res.*, vol. 73, 1968, pp. 4045–4080.
4. GOLD, T.: "Apollo 11 and Apollo 12 Close-Up Photograph." *Icarus*, vol. 12, 1970, pp. 360–375.
5. GOLD, T.; PEARCE, F.; AND JONES, R.: "Lunar Surface Closeup Stereoscopic Photography." *Apollo 12 Preliminary Science Report*, NASA SP 26–35, Washington, D.C., 1970, pp. 183–188.

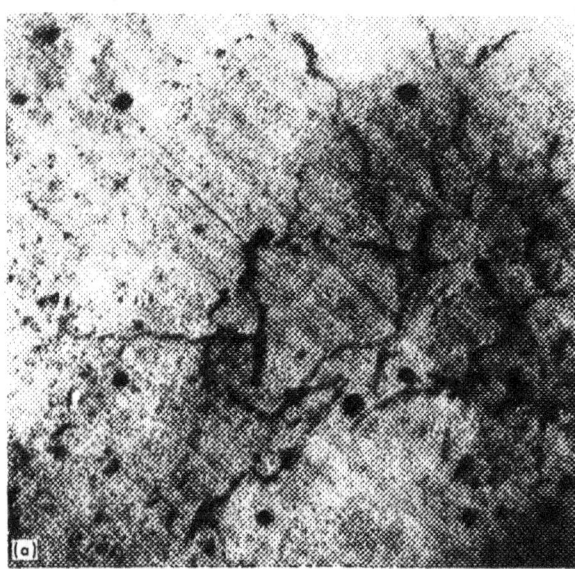

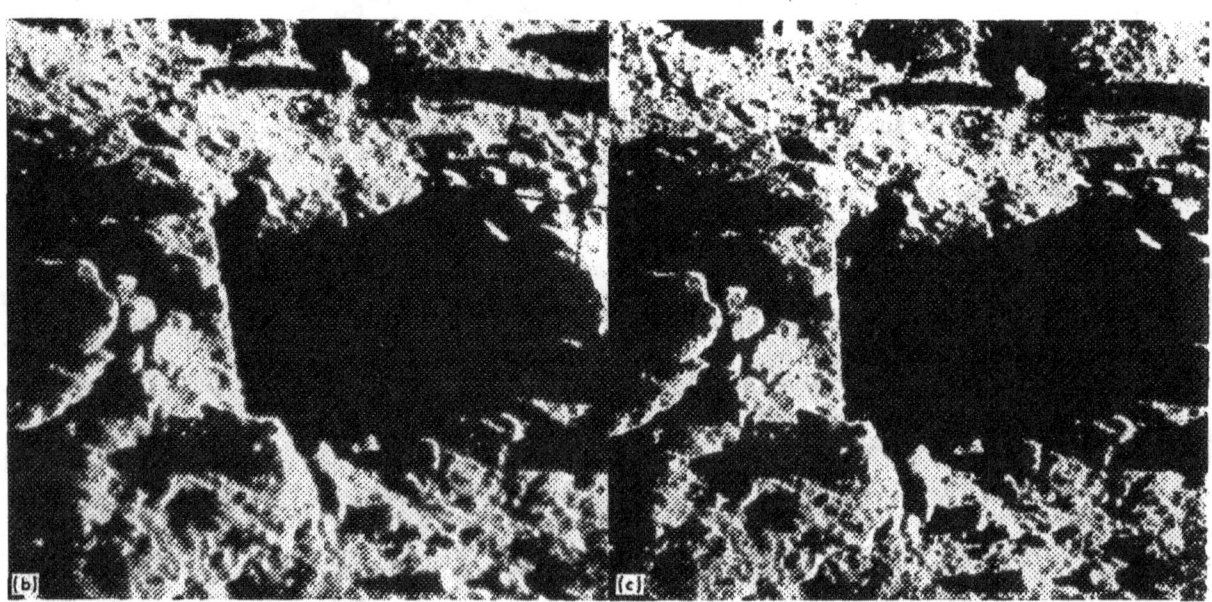

FIGURE 1.—Lunar surface disturbed by bearing test (bearing test 2) made with Surveyor 3 surface sampler. (a) Portion of picture made with Surveyor 3 television camera on April 27, 1967, at 08:56:45 GMT. View is from the west, rotated to match figures 1(b) and 1(c). (b), (c) Portions of photographs made with Apollo 12 hand-held camera on November 20, 1969, at about 06:30 GMT. View is from the south. Figures 1(b) and (c) form a stereopair (from NASA photographs AS12–48–7106 and AS12–48–7107).

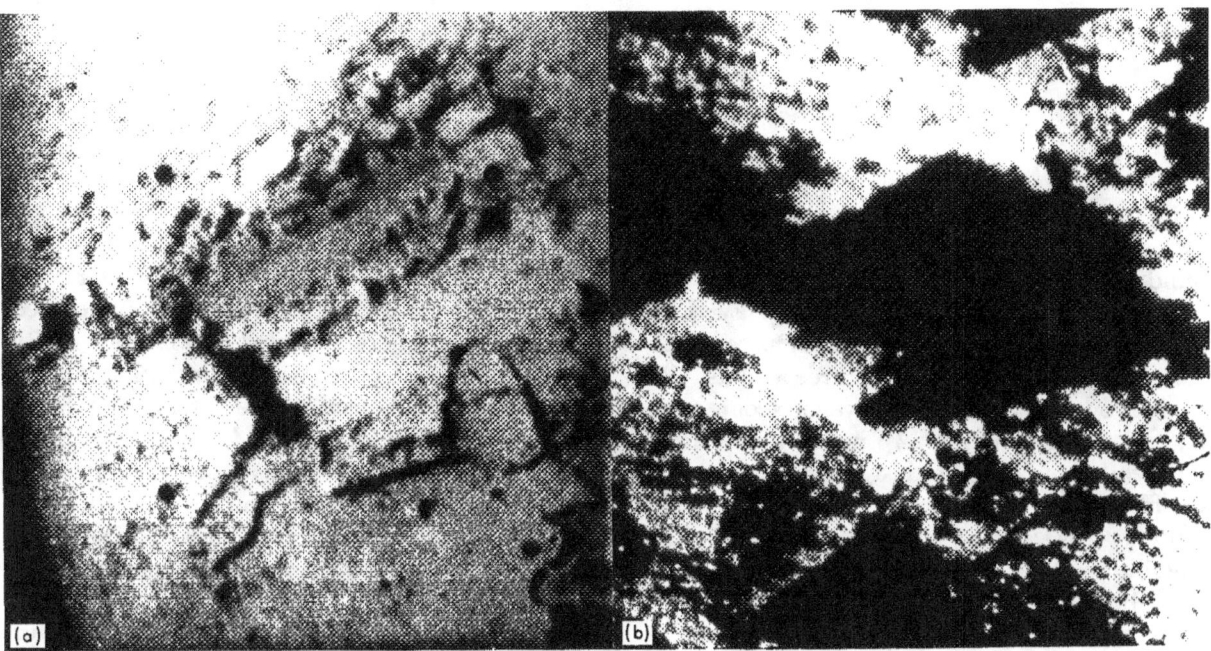

FIGURE 2.—Lunar surface disturbed by bearing test (bearing test 5) made with Surveyor 3 surface sampler. (a) Portion of picture made with Surveyor 3 television camera on April 30, 1967, at 15:39:30 GMT. View is from the south. (b) Portion of photograph made with Apollo 12 hand-held camera on November 20, 1969, at about 06:35 GMT. View is from the north, rotated to match figure 2(a) (from NASA photograph AS12–48–7126).

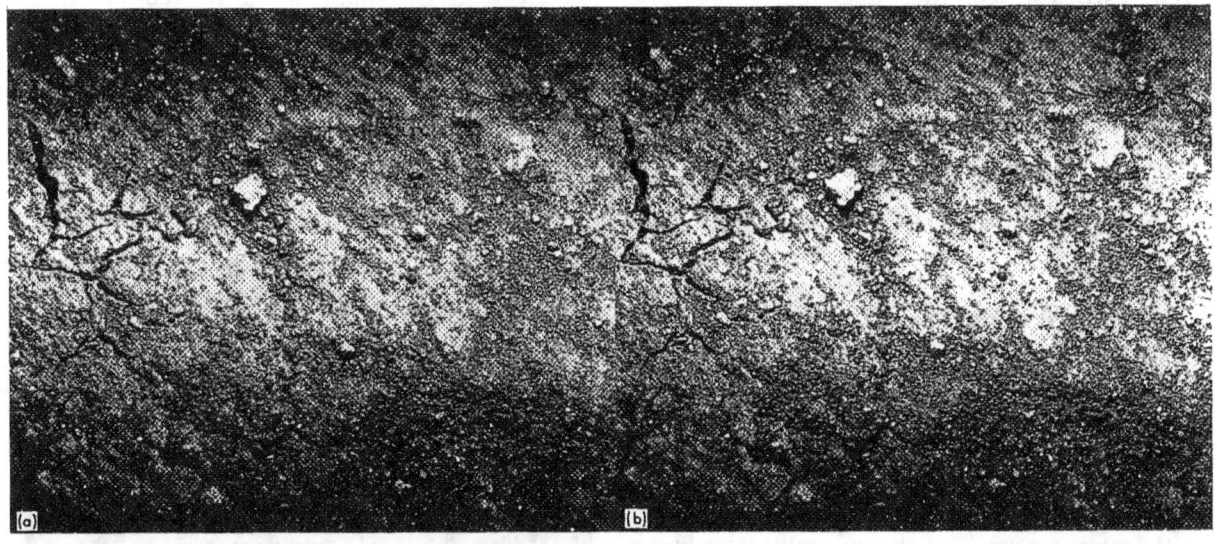

FIGURE 3.—Lunar surface disturbed by pressure from the hood of Apollo 11 closeup camera. Illuminated by flashbulb. Figures 3(a) and (b) form a stereopair (from NASA photograph AS11–45–6702–1).

PART C

WHISKERS ON THE MOON

D. Brownlee, W. Bucher, and P. Hodge

A considerable amount of lunar dust was found on the optical filters of the Surveyor television camera, which was returned to Earth by the Apollo 12 astronauts. This dust presumably was deposited as a result of three events: (1) the Surveyor landing, (2) Surveyor scoop activity, and (3) the Apollo 12 Lunar Module landing. While studying the dust distribution on the red filter, we noticed an unusual particle. This particle was a 6-μm nontransparent, rough spheroid with many small transparent filamentary fibers protruding from it. (See fig. 1.) All fibers were straight and appeared transparent; they ranged in length up to about 10 μm and were less than 0.5 μm in diameter. The particle had the general appearance of a sea urchin, except that none of the fibers were tapered. This particle was discovered using an optical microscope at 100× magnification with upper darkfield illumination. The rest of the slide was scanned at 100× magnification and, although the filter was covered with about 10^6 micrometer-sized particles, no other fiber particles were found.

A large fraction of the slide then was scanned at 500× magnification; 20 particles with fibers were found at this higher power. The largest particle with fibers was 20 μm long, but most of them were about 1 μm. The number of fibers on these particles varied; on the average, a fiber particle had only a few fibers. Some had only one; others, like the first particle encountered, had many. All of the fibers were straight and of a very small diameter. The diameters of most of the fibers were unresolvable optically; one of the particles had fibers that showed interference lines, indicating that the fibers were transparent and of fairly large diameter (about 0.5 μm). The average length of the fibers was about the size of the parent particle. Two unusually long fibers were found; the longest, from a 2-μm particle, was 50 μm and was above the plane of the filter at an angle of about 45°.

The argument against these fiber particles being contaminants, other than the existence of no plausible source, is that the fibers occurred on a wide variety of particle types. Particles with fibers were in no other way distinguishable from the other lunar dust on the filters. The fibers occurred on angular, rounded, transparent, nontransparent, large, and small particles. Perhaps the most convincing argument for a lunar origin for the particles is that three fibers were found on a reddish-brown transparent lunar spherule (4 μm). If the fiber particles were contaminants, then presumably they would have a common origin and similar structure. The fibers on the spherule seem to exclude all probable origins except lunar.

The red filter was shadowed with two 40-Å aluminum coatings, evaporated from 20° above the filter plane and 180° apart, for analysis in a scanning electron microscope (SEM). Four of the fiber particles are shown in figures 2 through 5. Because of the small size of the fibers, the latent heat of vaporization in the aluminum apparently partly melted many of the fibers. (Compare the SEM picture of the fiber particles in fig. 2 with its appearance before evaporation in fig. 1.) It must be remembered that the SEM pictures show the appearance of the fibers

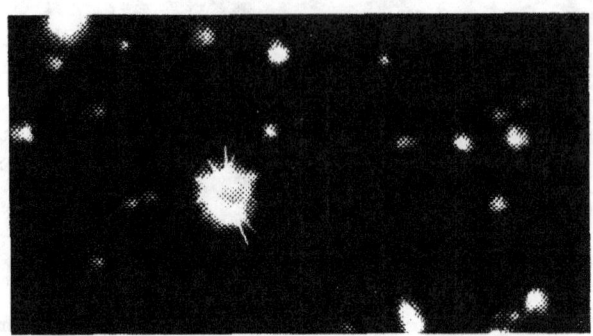

FIGURE 1.—Optical micrograph of a 6-μm nontransparent particle with whiskers.

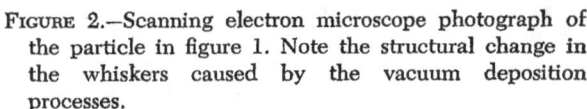

FIGURE 2.—Scanning electron microscope photograph of the particle in figure 1. Note the structural change in the whiskers caused by the vacuum deposition processes.

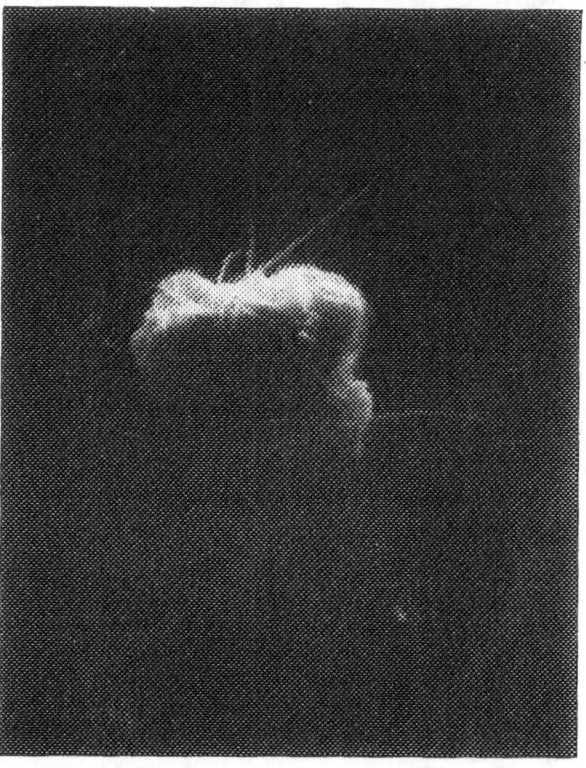

FIGURE 4.—Scanning electron microscope photograph of two whiskers on a 5-μm particle.

FIGURE 3.—Scanning electron microscope photograph of an 0.8-μm particle with a single long whisker.

FIGURE 5.—Scanning electron microscope photograph of a 5-μm particle with many small whiskers.

after an alteration by the evaporation process, and that originally the fibers were straight and not tapered or fused together.

We hypothesize that the fibers on these particles are whiskers that grew on the particles from clouds of vaporized lunar rock during macro-sized cratering events. At the time of whisker formation, the particles were on or slightly above the lunar surface. In a cratering event, transient vapor clouds are formed, and it seems entirely plausible that whiskers would grow on particles in contact with such clouds. The contact time between the vapor and the particles is short, but Berg and McDonnell (ref. 1) have reported similar whisker growth from transient vapor clouds produced by exploding wires and by hypervelocity impact of small particles on thin foils.

To our knowledge, the existence of whiskers on lunar soil grains has not been observed by other groups. This can be explained by any of the following reasons:

(1) The whiskers were produced by some unknown circumstance unique to particles on Surveyor 3.

(2) Any whiskers were broken off of Apollo 11 and 12 lunar fines by abrasion during transportation from the Moon to the laboratory.

(3) Any whiskers on Apollo 11 and 12 lunar fines were not detected because the technique used was not sufficient to detect them.

We believe that reason (2) is the most probable. The particles on the Surveyor optical filters are unique because, unlike the soil samples, they did not undergo mechanical abrasion during transport from the Moon to the laboratory. Reason (3) is also possible in view of the difficulties in locating whisker particles. Most of the whiskers are on the order of 1000 to 2000 Å in diameter and are extremely difficult to locate. They also are very rare. Less than one particle in 10 000 on the Surveyor filter had a whisker. Detection of the whisker particles requires optical scanning with high magnification, proper illumination, and adequate spacing of particles.

We conjecture that whisker growth on lunar dust particles may be a relatively common event. Few particles can maintain whiskers because of mechanical abrasion during natural lunar processes. If this explanation is correct, then measurement of the fraction of lunar particles that contain whiskers may allow setting limits to theories that predict migration of dust over the lunar surface by various processes.

Reference

1. BERG, O. E.; AND McDONNELL, J. A. M.: "Filamentary Crystal Growth Associated with Impact Craters from Hypervelocity Microparticles." *Science*, vol. 168, 1970, pp. 820–822.

XI. Microbe Survival Analyses

PART A

SURVEYOR 3: BACTERIUM ISOLATED FROM LUNAR-RETRIEVED TELEVISION CAMERA

F. J. Mitchell and W. L. Ellis

On April 20, 1967, the unmanned Surveyor 3 spacecraft successfully landed on the lunar surface near the eastern shore of Oceanus Procellarum. On November 20, 1969, two Apollo 12 crew members walked from the Lunar Module to inspect and photograph Surveyor 3. The entire television camera and other selected components were then retrieved for return to Earth (fig. 1). Upon return to Earth, the camera and lunar soil samples were placed in quarantine in the Lunar Receiving Laboratory (LRL) at the NASA Manned Spacecraft Center (MSC) at Houston, Tex. The quarantine was lifted on January 7, 1970, and inspection and disassembly of the returned television camera began the next day.

Microbial analysis was the first of several studies of the retrieved camera and was performed immediately after the camera was opened. A serious constraint placed upon this analysis was the need to obtain samples without compromising any planned subsequent studies. As a consequence, not all desired microbial samples could be obtained. The emphasis of the microbial analysis was placed, therefore, upon isolating microorganisms that could be potentially pathogenic for man.

Decontamination measures taken before the Surveyor 3 launch did not eliminate the possibility that the spacecraft carried microorganisms to the Moon. The following statement reflects the decontamination guidelines which were current at the time of the Surveyor spacecraft launches (ref. 1):

The precautions against the contamination of the Moon, once strict have now been relaxed in view of our

FIGURE 1.—Surveyor 3 spacecraft on its lunar landing site with astronaut Conrad inspecting the television camera. The Lunar Module Intrepid appears in the background.

FIGURE 2.—The retrieved Surveyor 3 television camera, complete with shroud, original collar and cables, as it appeared in the laminar-flow hood of the Lunar Receiving Laboratory.

developing knowledge of the inhospitable environment for terrestrial life that exists on the lunar surface and the belief that landed contamination, if it survives, will remain localized. For these reasons, lunar landing spacecraft may have on board a low level of microbial life—they must be decontaminated, but not sterile.

The extensive experience gained from Apollos 11 and 12 indicated that extraterrestrial microorganisms would not be isolated from Surveyor 3 (refs. 2 through 5). The recovery of terrestrial microorganisms originally present in the camera would be possible if these microorganisms had been able to survive in the lunar environment. However, verifying the origin of any isolate would be complicated by the possibility of post-retrieval contamination.

Because it had not been anticipated at launch in 1967 that the television camera would be returned to Earth at a future time, no pre-launch microbial analysis of the camera interior was performed; therefore, no appropriate experimental control was available for comparison. However, substitute controls were available. Several identical backup Surveyor cameras had been held in bonded storage during the same time period that Surveyor 3 had remained on the lunar surface. One backup camera was used to refine techniques for disassembly and microbial sampling before performing any definitive procedures on the retrieved Surveyor 3 camera. A second backup camera, designated the type-approval test camera (TAT-1), was disassembled after the Surveyor 3 camera and sampled identically to the retrieved Surveyor 3 camera.

Disassembly and Sampling Procedures

The retrieved camera was placed in a laminar-outflow hood equipped with high-efficiency particulate air filters (fig. 2) in the LRL astronaut debriefing room, which has an air-conditioning system separate from the system used by the rest of the LRL. Each surface of the laminar-flow hood that would be exposed to the camera was thoroughly washed twice with isopropyl alcohol before the camera was placed into the hood. A sterile cloth was placed on the floor of the hood to retain any lunar material that might accumulate as a result of the disassembly procedures. Only the personnel di-

FIGURE 3.—Surveyor 3 camera interior with shroud and cables removed.

rectly responsible for disassembling and sampling the camera were permitted in the room. They were clothed in laboratory attire, including surgical caps, face masks, and sterile gloves. Other participating personnel observed and coordinated activities from behind a viewing window.

To prepare the camera for disassembly, the original collar of the camera was removed and replaced with a special tripod permitting easy manipulation of the camera in the laminar-flow hood. To remove the camera shroud, the outer

TABLE 1.—*Microbial sampling sites of the Surveyor 3 and TAT-1 television cameras* [a]

Sampling site	Tube number		
	TSB	THIO	YMB
1. Metal surface under front half of collar	1	11	21
2. Nylon ties, Teflon wrapping, cable connector surface	2	12	22
3. Surface area on support studs	3	13	23
4. Surface area on electronic conversion unit	4	14	24
5. Circuit-board support-plate edges and screw studs	5	15	25
6. Surface area of all three cable connectors inside camera	6	16	26
7. Nylon ties and cable wrappings	7	17	27
8. Debris in bottom of shroud	8	18	28
9. Large area on inside of shroud	9	19	29
10. Top surface of exposed circuit boards	10	20	30
11. Foam samples from between circuit boards	31	32	33

[a] Sampling sites 1 and 2 are exterior camera samples pertaining to the Surveyor 3 camera only. The TAT-1 camera had no collar or cables; consequently, no sample of site 1 was taken. Site 2 included all three exterior metal cable connector surfaces.

aluminum and inner clear Teflon wrappings were removed from the cable connectors. The cable connectors were sampled and then washed with isopropyl alcohol. Retaining screws on the shroud were removed, and the cable connectors were pushed inside the shroud. The shroud then was removed from the bottom of the camera and the biological samples were immediately taken. The shroud fit very tightly on the camera; although the camera was not hermetically sealed, the interior of the camera was extremely clean and no evidence of lunar material was observed within the television camera when the shroud was removed (fig. 3). The only evidence that the camera had been on the Moon and retrieved was a small number of particles (no larger than 1 mm^3) that had accumulated in the bottom of the shroud. These particles were determined to be bits of ceramic insulation which had shaken loose during the flight to the Moon or during the return flight.[1]

Identical procedures were used for sampling the Surveyor 3 and the TAT-1 cameras. Three sterile calcium alginate swabs were arranged with the swab heads in tandem, moistened with sterile phosphate-buffered saline (0.0003 M PO$_4^{3-}$, 0.147 M NaCl), and used to swab the maximum surface area of each site (table 1). The swabs were then separated. One each was placed into 5 ml of trypticase soy broth (TSB) for aerobic analysis, 5 ml of thioglycollate broth (THIO) for anaerobic analysis, and 5 ml of yeast malt broth (YMB) containing 33 units/ml of penicillin G and 62 μg/ml of streptomycin for mycological analysis. In confined areas where this method of swabbing could not be used, three sequential samples were taken and placed in the appropriate media. The first such sample was always placed into TSB, the second into THIO, and the third into YMB.

Dry swabs, arranged as described previously, were employed at three sampling sites because of the nature of the material to be sampled or the requirements of prescribed followup studies. These samples included the bits of ceramic debris in the camera shroud base, the cable surfaces in the camera interior, and the top surface of the circuit boards. Samples 31, 32, and 33 consisted of bits of polyurethane foam. This foam had been used as insulation between the two aluminum plates of the circuit boards. The space between the aluminum plates was about 4 mm. This thin layer of foam was accessible only where holes had been cut into the plates for the placement of electronic components. Only by using long, curved, needle-nosed forceps

---

[1] R. Riglin, Hughes Aircraft Co., Culver City, Calif., and W. Carroll, Jet Propulsion Laboratory, Pasadena, Calif., private communication, 1970.

could one reach through the hole and into the space between the aluminum plates to obtain bits of the foam. The largest bit of foam extracted was approximately 1 mm^3. Samples obtained with forceps or with dry swabs were cultured according to the same procedures and in the same media as prescribed for wet-swab samples.

The protocol established for the aerobic and anaerobic analyses (fig. 4) maximized the possibility of detecting and quantitating low numbers of microorganisms in a sample, while at the same time yielding valuable clues as to the source of any microorganism detected. The protocol inherently contained a system of redundancy and cross-checks designed to identify suspected laboratory contamination. For example, growth on any blood agar (BA) plate from the 10^2 dilution tube without simultaneous growth in the original tube, the two dilution tubes and on the BA plates from the 10^1 dilution tube and from the original tube containing the foam sample would be suspect. Growth on any BA plate without growth in the tube from which aliquots were taken to place on the BA plate would require extreme care in interpretation; in this case, one would probably suspect contaminated BA plates. Growth in either the 10^1 or the 10^2 dilution tube without growth in the original tube would require some logical reason why growth did not occur in the undiluted tube.

The replicate BA plates provide the requirement for consistent results, within experimental error, on each plate and provide a check on techniques used in making dilutions. Streaking fresh BA plates with aliquots from each tube after 24 hr of incubation was intended to provide an opportunity for early isolation and separation in case the sample contained more than one microorganism with the result that one specie might overgrow another specie if the tube were allowed to incubate to full turbidity without examination. Again this operation provided another opportunity to cross-check with the results from the previous dilutions and plating. With growth on the 10^2 dilution BA streak plate, one also would expect growth on the 10^1 dilution BA streak plate and on the BA streak plate from the original undiluted tube. Whatever the results, any observed growth would have to be consistent and logical in view of the redundancy and cross-checks built into the protocol. Obvious cases of laboratory contamination could easily be identified and reported as such.

The swab or sample was placed into 5 ml of the selected broth and vortexed. One milliliter of this broth was spread in 0.2-ml aliquots onto five BA plates. Aliquots of 0.1 ml were taken from 10^1 and 10^2 dilutions of the broth containing the selected sample and spread on BA plates in replicates of three. The tube containing the original sample, the two dilution tubes, and all plates were incubated for 24 hr. The TSB tubes and a set of BA plates containing 5 percent sheep (lamb) blood were incubated aerobically. The THIO tubes and a second set of BA plates were incubated anaerobically, using GasPak (BBL) systems in stainless-steel jars. Aliquots from each tube were then streaked onto fresh BA plates, and all plates and tubes were returned for incubation at 37°C for 30 days. Any observed growth on the plates was quantitated and identified. Growth in the incubated tubes was also identified. The YMB tubes were handled according to the established LRL procedures for mycological analysis.

Surveyor 3 Camera Results

The only sample to produce visible microbial growth was sample 32, a 1-mm^3 piece of foam incubated in undiluted thioglycollate broth. The initial growth was observed on the fourth day of incubation as a white "tail" of growth 2 to 3 mm in length, hanging from the piece of foam which was floating in the middle of the tube. No other growth was observed on that day. The next day this tube was turbid with growth and the 10^1 dilution tube exhibited approximately 100 foci of growth scattered predominantly at the top of the tube. No growth was observed in the 10^2 dilution tube or on any BA plate or BA streak plate for the remainder of the study.

In both tubes containing growth, only a single cellular morphology was observed, that of a gram-positive coccus in chains. Since the initially observed growth had required 4 days of incubation in THIO and since no growth was observed on the initial five anaerobic BA plates, these media were again inoculated with the isolate. In

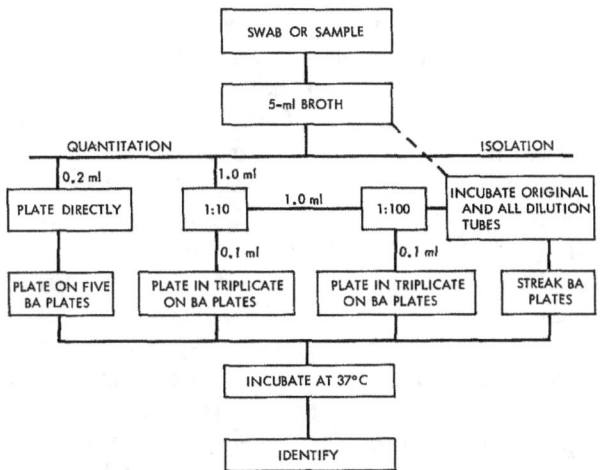

FIGURE 4.—Protocol established for the aerobic and anaerobic samples. This protocol was followed for both the retrieved Surveyor 3 camera and the backup TAT-1 camera.

addition, TSB and aerobic BA plates were inoculated with the isolate. Growth was observed in both THIO and TSB within 24 hr. Growth was observed on the aerobic BA plates within 24 hr and on the anaerobic BA plates within 72 hr (first examination). As a precaution, 1-ml aliquots containing, respectively, 10^3, 10^4, 10^5, and 10^6 viable cells of the isolate were injected intraperitoneally (in replicates of five) into 5-week-old white male CD-1 mice, with no observed effect.

The isolate was identified, with confirmation from the U.S. Public Health Service Center for Disease Control in Atlanta, Ga., as alpha hemolytic *Streptococcus mitis*.[2]

TAT-1 Television Camera Results

The results from the backup TAT-1 camera, sampled identically as the retrieved Surveyor 3 camera, provide observations on microbial survival at ambient atmospheric pressure and room temperature for the same period of time that the Surveyor 3 camera rested on the lunar surface. The TAT-1 camera was held undisturbed in bonded storage in its original shipping container for this time period. Terrestrial microorganisms were isolated in very low numbers from one ex-

[2] R. Facklam, Communicable Disease Center, Atlanta, private communication, 1970.

terior and five interior locations. One bacterial isolation and five mycological isolations were made after long incubation periods varying from 6 to 27 days. All six isolations were made from accessible metallic and nonmetallic sampling sites.

From a Teflon-covered cable within the TAT-1 camera, a *Bacillus* species was isolated in THIO after 6 days of incubation. Growth appeared only in the tube containing the undiluted sample.

An *Aureobasidium* species was isolated after a sampling of the TAT-1 exterior metal cable connectors was incubated in YMB for 14 days. The same species was also isolated after a sampling of the metal electronic conversion unit within the TAT-1 camera was incubated in THIO for 27 days.

Aspergillus pulvinus was isolated from three sites in the interior of the camera. This isolate was detected after a sampling taken from the top surface of the nonmetallic circuit board was incubated in THIO for 12 days. A second isolation was made from a sampling of the metal cable connectors after 14 days of incubation in TSB. The third isolation was made from a sampling of the metal electronic conversion unit after 21 days of incubation in YMB.

In the five fungal isolations, growth appeared only in the tube containing the undiluted sample, indicating very low numbers of microorganisms originally present on the sampled surfaces. To illustrate, when three swabs were used in tandem to sample one of the selected sites, *Aspergillus pulvinus* was isolated from only one of the swabs; an *Aureobasidium* species was isolated from a second swab; and the third swab was negative.

Discussion

Every step in the retrieval of the Surveyor 3 television camera was analyzed for possible contamination sources, including camera contact by the astronauts; ingassing in the Lunar Module and Command Module during the mission or at "splashdown"; and handling during quarantine, disassembly, and analysis at the LRL.

Contact by the astronauts during retrieval on the Moon was not considered a probable source of contamination. Microorganisms were undoubtedly present on exterior surfaces of the astronauts' space suits during each lunar landing and selenological sample collection excursion. However, no viable terrestrial microorganism has ever been detected in the selenological samples collected by the astronauts (refs. 2 through 5).

After the television camera was removed from the Surveyor 3 spacecraft, it was placed into a back pack carried by one of the astronauts. The pack was zipper-closed, although there was no capability for sealing it. The pack was placed in storage first aboard the Lunar Module and then the Command Module and finally was flown to the LRL by jet aircraft. At the LRL, the camera was removed from the pack and placed in a Teflon bag. The bag was heat sealed, and then the camera and first bag were placed into a second Teflon bag, which was also heat sealed. The double-bagged camera was then placed in bonded storage at room temperature until the lunar sample was released on January 7, 1970.

When the Apollo 12 Lunar Module landed on the Moon, lunar dust was disturbed with such force that it traveled approximately 155 m with a reported velocity of at least 70 m/sec, "sandblasting" the Surveyor 3 spacecraft (ref. 6). Shadows in the exterior paint of the Surveyor 3 camera were clearly visible wherever a strut or other part had shielded the camera from this hail of lunar particles caused by the rocket exhaust.

While the television camera was being disassembled, it was observed that barely visible particles of lunar dust had accumulated underneath the camera collar. The presence of this fine dust in this protected area is a reflection of the minute size of some lunar particles and the "sandblasting" force which caused the penetration. It has already been noted that no such presence or accumulation of lunar particles was found in the interior of the camera protected by the shroud despite the "sandblasting." This suggests that the camera shroud may have provided a formidable barrier to ingassing, carrying fine particles, perhaps even the size of a bacterium, from the environment into the camera interior.

The lunar material under the camera collar was sampled for viable microorganisms. None were recovered. As the two layers of Teflon wrappings were removed from the exterior of the metal cable connector, a sampling was made of both layers of the Teflon wrappings as well as the metal surface of the cable connector. Again, no viable microorganisms were detected. This was a deliberate attempt to detect any microorganisms which might have been available in the external environment and which might have entered the camera interior during ingassing.

The Apollo 12 astronauts, spacecraft, and space suits were sampled before launch and after recovery. All three astronauts carried species of a number of genera of microorganisms, including S. mitis.[3] As a result, the cabin air of both the Lunar and Command Modules undoubtedly contained a number of different bacteria as an aerosol load.

Assuming that microorganisms had entered the camera interior during ingassing, a representation of the entire microbial population available would be expected rather than a single species. This representative population of microorganisms would be expected to be randomly distributed in the camera. Therefore, if large surface areas of the camera interior were sampled, microbial contamination due to ingassing should be detected. Even if S. mitis were the only one of the population carried in by ingassing to survive, it should have been found randomly distributed over large surface areas instead of in the only relatively inaccessible location that was sampled.

On a unit area basis, at least 10 000 times the area in which the isolate was detected was sampled; this area represents large exposed surface areas of different types of materials throughout the camera interior. That S. *mitis* cells (alone from all the microorganisms available in the external environment) could enter the camera and find their way to the least accessible sampling site without being detected in 10 000 times that area of readily exposed surface is difficult to envision. In the absence of any other microorganisms isolated and in view of the large sampling area, it is considered improbable that ingassing at any point in the retrieval could

---

[a] J. Ferguson, Manned Spacecraft Center, Houston, private communication, 1970.

be responsible for depositing S. *mitis* in the relatively inaccessible location where it was isolated.

Extreme precautions were taken at all times during the analysis to prevent any handling errors that might have caused contamination. Experimental controls of the implements and media used in the analysis did not initiate microbial growth.

To determine whether low numbers of organisms alone could cause the delay in initial growth, a dilution series of THIO containing the isolate was prepared. From each dilution tube, 0.1 ml was transferred to a THIO tube and to five aerobic BA plates. Visible growth appeared within 24 hr, even in the dilution tube initially containing less than 10 viable cells as determined by the colony count on the five BA plates. Furthermore, the presence of the foam sample did not account for the initial delay in growth, since growth was not delayed when the isolate was cultured in a dilution series of THIO containing foam sections the same size and composition as the original samples.

The fact that no growth was observed until the fourth day of incubation in liquid broth indicated that the isolated bacterium required an adaptation period. Growth delays are not uncommon in bacteria recovering from lyophilization (ref. 7). No colonies were found on the first set of five anaerobic BA plates, indicating either that no viable cells were placed on the BA plates, or that the cells could not adapt and replicate on the solid agar surface as they had in the liquid broth media.

The "tail" of growth which streamed from the underside of the foam on the fourth day of incubation indicates a direct relationship between the organism and the foam sample and is an important observation. When a control dilution series of the broth containing the isolate was made with similarly sized foam sections, no such relationship (no "tail") was observed in any of the dilution tubes, indicating no spontaneous association of the bacteria with the foam.

The initial delay in growth of the isolate, the direct association of the bacterium with the foam sample from which it was isolated, the relatively inaccessible location from which the isolate was obtained, and the absence of any other isolates in the large sampling area are, in our opinion, not consistent with the hypothesis that the Surveyor 3 television camera was contaminated with the isolate during or after its retrieval.

It is inadequate to simply imply that the foam sample or the thioglycollate tube became contaminated and that this readily explains the growth in the original undiluted tube and the 10^1 dilution tube. That would not be examining all the data and it would require unsupported asumptions; for example, the assumption that somehow the contaminant came into intimate contact with and remained in association with the foam sample despite vortexing so that it eventually grew as a "tail" to the foam. It would have to assume that, for some reason, the S. *mitis* cells were damaged and growth was delayed 4 days; that, of all the tubes in the experiment, contamination occurred only in this particular tube despite the control data, or that contamination occurred in the sample taken from the most inaccessible of all the sampling sites. Still other such assumptions would be required for such a simple explanation. No one single observation is adequate. Every bit of data must be considered. In the opinion of the authors, the total data are consistent with the hypothesis that the isolated bacterium was in intimate association with and isolated from the piece of foam sample which was taken from the camera interior and processed in an aseptic manner under controlled conditions.

The isolated bacterium, S. *mitis*, is a spherical microorganism measuring from 0.5 to 1.0 μm in diameter and is a frequent, normal, benign inhabitant of the respiratory tract. Man constantly sheds microorganisms into the air, a large portion of which comes from the respiratory tract. Although normal talking drives out considerable numbers of organisms, a good healthy sneeze may dispense as many as 20 000 aerosol droplets, which may vary in diameter from 10 μm to 2 mm and the larger of which may travel about 5 m (15 ft) before reaching the ground. These larger droplets settle rapidly, adhering to particles of dirt, and dry leaving organisms attached to the particles (ref. 8).

A single aerosol droplet could contain large numbers of organisms. It has been estimated that saliva contains an average of 750 millions of organisms per milliliter (ref. 9). In addition, saliva contains many organic constituents, the major portion of which is protein and the principal salivary protein of which is mucin. "It seems that mucin exerts much of its effect on the oral microbiota by physical localization of bacterial growth. Mucin probably protects bacteria primarily by a coating effect with the formation of a temporary artificial capsule about the cell; this has been demonstrated with such oral microorganisms as staphylococci, streptococci and lactobacilli" (ref. 10).

Other organic constituents of saliva are carbohydrates, including hexosamine, methyl pentose, galactose, mannose, deoxyribose, and glucose (ref. 10). "The synthesis of intracellular glycogen in the presence of excess carbohydrate, and its rapid catabolism to lactate in the absence of exogenous carbohydrate, has been observed in S. mitis. The polysaccharide appears to function as the sole reserve of energy of this organism and may provide the cell with energy in a utilizable form. The conclusion seems to be justified that the possession of glycogen by S. mitis favors its survival during starvation" (ref. 11). In addition, when drying bacteria, the presence of glucose in the suspending fluid in concentrations of between 5 and 10 percent greatly increased the survival rate both immediately and after storage (ref. 12).

As noted in the Hughes Aircraft Co. report (ref. 13), "There were opportunities for contaminants to deposit on the camera prior to launch." A number of these opportunities came while the shroud of the camera was removed for prelaunch inspections or repairs. In addition, the pre-launch thermal vacuum testing of the camera provided conditions conducive to lyophilization. The Surveyor 3 and TAT-1 cameras were subjected to a series of thermal vacuum tests following inspections and repairs. Information provided by personnel of the Hughes Aircraft Co., El Segundo, Calif., where the Surveyor 3 television camera was tested before launch indicates that, before launch, the Surveyor 3 television camera was exposed, under a 10^{-5}-torr vacuum, at least 12 times to temperatures of $-29°C$ and at least three times each to temperatures of $-45°$ and $-118°C$. Exposure at these temperatures was for at least 1 hr and in many cases longer. The highest temperature attained during any testing cycle was $52°C$. The last thermal vacuum test of the camera before it was placed on the spacecraft occurred late in January 1967, leaving approximately 90 days before launch. After the Surveyor 3 camera was attached to the Surveyor 3 spacecraft, it was again exposed to extreme temperature and vacuum conditions in the course of spacecraft thermal vacuum testing.

If the bacterium was deposited in the camera before launch, one can only speculate as to how many of these lyophilizing cycles the bacterium experienced. In one report, a paracolon bacillus culture was subjected to repeated lyophilization and reconstitution without allowing for further growth. Approximately the same percentage of cells survived each cycle of lyophilization and reconstitution (ref. 12). It is certain that, if deposited in the camera, the bacterium would have experienced at least one cycle when the camera was attached to the Surveyor 3 spacecraft and the spacecraft underwent its thermal vacuum testing. In addition, since the television camera was not maintained under a continuous vacuum, ambient pressure returned to the camera for approximately 90 days while the spacecraft awaited its launch to the Moon. The survival of the bacteria inside an aerosol droplet in the foam would depend, it would seem, on the amount of protective substances which might surround the bacteria and the effect the lyophilizing conditions had on the dried droplet. Considering the fact that tubercle bacilli can survive in dried sputum for at least 8 months (ref. 8), it would seem possible that, if the bacteria were encapsulated in a protective coating and dried, they might survive until they experienced the continuous vacuum of space after launch. "The haemolytic streptococcus group B is very resistant to drying, and one strain, which shows a survival rate of 100% even in serum water, was, in another experiment, not entirely killed 18 months after drying in distilled water. It seems impossible to kill this strain by drying" (ref. 12).

It has been reported that when bacteria and viruses are dry they require, like isolated enzymes, a higher temperature for irreversible damage (ref. 14). Engineering estimates at MSC suggest the maximum temperature experienced inside the television camera while on the lunar surface was 70°C.[4] Perhaps in such a dried state and under the high continuous vacuum of space, survival of lyophilized bacteria is possible. It has been shown that several *Streptococcus* species have remained viable for at least 20 years after lyophilization under routine laboratory conditions (ref. 15). Finally, in dealing with large numbers of microorganisms, even the loss of 99.9+ percent of the original population can still leave considerable numbers of survivors. It is estimated that between 2 and 50 cells or clumps of cells (chains) of *S. mitis* were isolated from the foam sample.

It would be very desirable to be able to define the exact conditions under which the isolated bacterium may have been deposited on the foam, the amount of protection which may have been provided by its source in the respiratory tract, the tolerance of bacteria contained in an aerosol droplet to heat and high vacuum, and the initial concentration of bacteria. Although the literature contains many reports of experiments which at first appear to be applicable, they all seem to suffer from the same shortcomings: the test species were different, the vacuum or temperature was not high enough, and most common of all, the experiment did not last long enough.

The isolated bacterium was lyophilized upon its initial isolation and is available for further testing as time, money, and facilities are available. The bacterium will be submitted for addition to the American Type Culture Collection.

The available data indicate that *Streptococcus mitis* was isolated from the foam sample and suggest that the bacterium was deposited in the Surveyor 3 television camera before spacecraft launch. It is suggested that the bacterium may have been provided some protection from its source in the respiratory tract and that lyophilizing conditions to which the camera was subjected before launch and later while it was on the lunar surface may have been instrumental in the apparent survival of this terrestrial microorganism.

[4] R. Erb, Manned Spacecraft Center, Houston, private communication, 1970.

References

1. HALL, L. B.: *NASA Requirements for the Sterilization of Spacecraft*, NASA SP–108, Washington, D.C., 1966, pp. 25–36.
2. OYAMA, V. I.; MEREK, E. L.; AND SILVERMAN, M. P.: "A Search for Viable Organisms in a Lunar Sample." *Science*, vol. 167, 1970, pp. 773–775.
3. TAYLOR, G. R.; FERGUSON, J. K.; AND TRUBY, C. P.: "Methods Used to Monitor the Microbial Load of Returned Lunar Material." *Appl. Micro.*, vol. 20, 1970, pp. 271–272.
4. OYAMA, V. I.; MEREK, E. L.; SILVERMAN, M. P.; AND BOYLEN, C.: "Search for Viable Organisms in Lunar Samples: Further Biological Studies on Apollo 11 Core, 12 Bulk, and 12 Core Samples." *Proceedings of the Apollo 12 Lunar Science Conference.* To be published.
5. TAYLOR, G. R.: "Microbial Assay of Lunar Samples." *Proceedings of the Apollo 12 Lunar Science Conference.* To be published.
6. JAFFE, L. D.: "Blowing of Lunar Soil by Apollo 12: Surveyor III Evidence." *Science*, vol. 171, 1971, pp. 798–799.
7. SINSKY, T. J.; AND SILVERMAN, G. J.: "Characterization of Injury Incurred by Escherichia Coli Upon Freeze-Drying." *J. Bact.*, vol. 101, 1970, pp. 429–437.
8. SMITH, D. T.; CONANT, N. F.; AND OVERMAN, J. R.: "Microbiological Ecology and Flora of the Normal Human Body." *Zinsser Microbiology*, ch. 10, 1964, pp. 162–172.
9. ROSEBURY, T.: "The Indigenous Cocci." *Microorganisms Indigenous to Man*, ch. 2, The Blakiston Division, McGraw-Hill, 1962, pp. 9–47.
10. BURNETT, G. W.; AND SCHERP, H. W.: "Oral Microbiology and Infectious Disease." *The Microbial Flora of the Oral Cavity*, Williams & Wilkins Co., 1968, pp. 273–327.
11. VAN HOUTE, J.; AND JANSEN, H. M.: "Role of Glycogen in Survival of *Streptococcus mitis*." *J. Bact.*, vol. 101, 1970, pp. 1083–1085.
12. FRY, R. J.; AND GREAVES, R. I. N.: "The Survival of Bacteria During and After Drying." *J. Hyg.*, vol. 49, 1951, pp. 220–246.
13. *Test and Evaluation of the Surveyor III Television Camera Returned From the Moon by Apollo XII*, vols. I and II, SSD 00545R, Hughes Aircraft Co., Culver City, Calif., 1970.
14. DAVIS, B. D.; DULBECCO, R.; EISEN, H. N.; GINSBERG, H. S.; AND WOOD, W. B., JR.: "Sterilization and Disinfection." *Principles of Microbiology and Immunology*, ch. 11, Harper & Row, 1968, pp. 335–353.
15. RHOADES, H. E.: "Effects of 20 Years Storage on Lyophilized Cultures of Bacteria, Molds, Viruses, and Yeasts." *Am. J. Vet. Res.*, vol. 31, 1970, pp. 1867–1870.

ACKNOWLEDGMENTS

Grateful acknowledgment is extended to M. D. Knittel of the Jet Propulsion Laboratory, Pasadena, Calif., for his interest and cooperation, especially in supervising the selection of the sampling sites and assisting in the sampling of all cameras (except for foam sample extraction); R. G. Riglin of the Hughes Aircraft Co. for his thorough and intimate knowledge of the Surveyor television cameras and especially for his fortuitous choice and extraction of the foam sample containing the isolate; T. C. Molina of Brown & Root-Northrop for competently performing all laboratory microbiological analyses; E. I. Hawthorne, R. G. Riglin, and P. M. Blair of the Hughes Aircraft Co., as well as W. F. Carroll of the Jet Propulsion Laboratory, for their expert disassembling of the retrieved television camera.

Our sincere thanks are extended also for technical and administrative assistance to R. H. Green of the Jet Propulsion Laboratory; P. A. Volz of Eastern Michigan University; T. C. Allison of Pan American College; M. S. Favero of the U.S. Public Health Service, Phoenix; R. R. Facklam of the Communicable Disease Center, Atlanta; and the following personnel of the NASA Manned Spacecraft Center at Houston, Tex.: J. G. Zarcaro, A. D. Catterson, W. W. Kemmerer, Jr., J. A. Mason, R. B. Erb, J. L. McQueen, J. K. Ferguson, G. R. Taylor, B. C. Wooley, H. E. Eitzen, R. C. Graves, S. Jacobs, L. G. Leger, K. L. Suit, and W. C. Alexander.

PART B

MICROBIOLOGICAL SAMPLING OF RETURNED SURVEYOR 3 ELECTRICAL CABLING

M. D. Knittel, M. S. Favero, and R. H. Green

Some of the many studies regarding the survival of microorganisms in deep space have exposed selected bacterial cultures during actual space flights (ref. 1); some simulation studies of the space environment also have been made (ref. 2). Some long-term investigations have been made to study the effect of a single parameter such as vacuum (ref. 3).

The plan of Apollo 12 to land near the site of the Surveyor 3 spacecraft offered the unique opportunity for retrieval of selected parts for scientific and engineering studies. The microbiological examination of parts of the spacecraft could provide possible information concerning whether or not microorganisms could survive in the harsh environment of space. A piece of electrical wiring bundle running from the television camera to another part of the spacecraft was selected for microbiological examination. This part was selected because—

(1) Previous information obtained during the planetary quarantine monitoring of Mariners 6 and 7 showed a high level of bacterial contamination associated with wiring bundles.

(2) Surveyor 3 had not been sterilized before launch.

(3) The cable could easily be removed and packaged against contamination.

In this experiment, the kinds and number of microorganisms initially present on the cable were not known and there was no available "control" such as an identical Surveyor cable exposed to terrestrial environment for the same length of time as Surveyor 3 had been on the Moon.

Microbiological Materials and Methods

Media. Bacteriological media used in this study were Eugon and thioglycollate broths. The thioglycollate broth was chosen for isolation of any anaerobic bacteria, and the Eugon broth for the growth of aerobic bacteria.

Equipment. Sampling of the wiring bundle was conducted in a glove box manufactured by Blieckman Co. Forceps, wire strippers, scissors, wire cutters, and a vise were used to remove pieces of the cable.

Sterilization of Materials. The interior of the glove box and surfaces of containers entering the glove box were sterilized with 2 percent peracetic acid. Media to be used in the assay were placed in glass screw cap tubes and sterilized by autoclaving at 121°C for 20 min. Tools and other hardware were placed in metal cans and sterilized by dry heat at 180°C for 4 hr.

Electrical Wiring Bundle Sampling Methods

The electrical wiring bundle (see fig. 1) was dissected into its component parts and each piece placed into one of the bacteriological culture media. This culture enrichment method was chosen over a dilution and plating procedure because it was expected that the numbers of surviving microorganisms (if any) would be low and would be missed if a serial dilution and culturing technique were used.

Sampling was performed by first removing the outside nylon ties that held the wrappings, then removing pieces of the exterior wraps until the bundle of wires was completely exposed. The insulation was removed from individual wires with wire strippers; pieces of the exposed stranded wire were removed with wire cutters.

The procedures were performed inside a stainless-steel glove box in order to isolate the experiment from airborne bacterial contamination. During the sampling, a piece of similar sterilized control wiring bundle was sampled in the same way according to numbers taken from a random numbers table. These sterile control samples represented 10 percent of the total and were an internal standard for contamination monitoring.

The objective of this experiment was to determine whether terrestrial microorganisms present on Surveyor 3 when it was launched could survive 31 months of lunar exposure. If, during the actual sampling of the wires, a contaminant were accidentally induced, it would be impossible to separate it from a lunar survivor. Therefore, it was necessary to perform several simulated assays with a piece of sterile wiring bundle before the lunar sample assay in order to perfect technique. During these simulated assays, all procedures that were to be used in sampling of the Surveyor 3 cable were employed to determine if the sampling could be conducted without contamination. As can be seen in table 1, the procedures that were adopted provided the protection needed to sample three separate, sterile wiring bundles without contamination. These procedures insured that the Surveyor 3 cable could be examined without fear of contamination during assaying, which would negate the experiment.

The sealed environmental sample container (SESC) containing the Surveyor 3 cable and other parts was checked for leaks before opening; evidence of leakage was found. When the SESC was opened, it was found to contain a high concentration of oxygen. This led to modification from previous sampling plans to swab sample the outside and inside of each wrap to

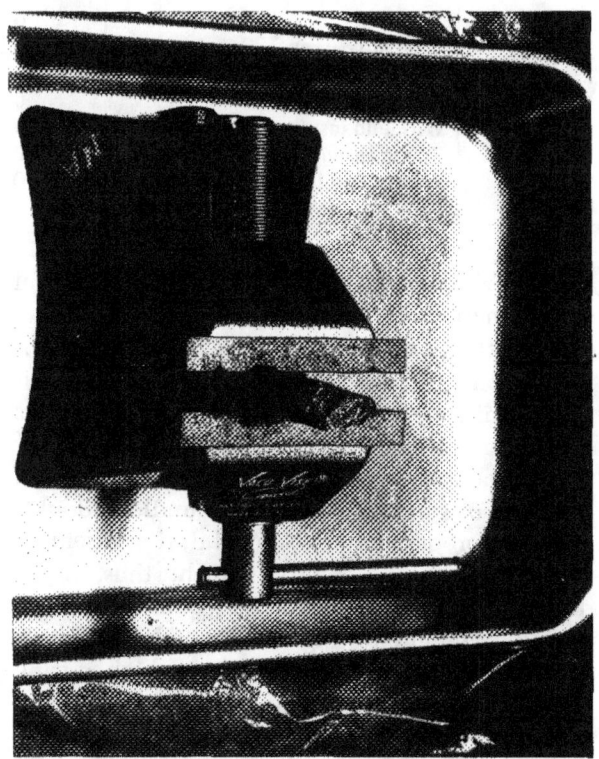

FIGURE 1.—Electrical wiring bundle.

TABLE 1. — *Results of dissection of sterilized cable during 3 separate simulated sampling runs*

Sampling no.	Number of samples		Number positive
	Eugon	Thioglycollate	
1.	20	20	0
2.	20	20	0
3.	20	20	0

determine if contamination of the wiring bundle had occurred through the leak and if the wrappings of the cable protected the underlying components from contamination. Table 2 shows the results of an experiment performed to find out if the wraps on the cable would protect against migration of the contaminants to the internal parts of the wiring bundle. As can be seen, if care is exercised during removal of the wraps, the contamination on the exterior surfaces remains on that surface.

The results of the dissection and culture enrichment of the pieces of the Surveyor 3 wiring bundle are presented in table 3. All samples taken were negative for bacterial growth after 6 weeks of incubation at 25°C. All random sterile controls were also negative. All agar plates exposed at various times in the glove box were negative.

In order to determine whether the procedure contained processes that would be inhibitory to the isolation of any survivors on the Surveyor cable, a piece of unsterilized control wiring bundle was sampled using the same procedures. From 40 samples taken, 30 positive samples were obtained: 21 samples yielded gram-positive cocci, 6 spore-forming rods, 2 gram-positive non-spore-forming rods, and 1 gram-negative rod. Random numbered sterile controls taken at the same time as the sampling were negative; agar plates exposed during various times of the sampling also were negative.

Results

The results that have been presented show that no viable microorganisms were recovered from that part of the Surveyor 3 cable sampled. Some factors that could have contributed to the sterility of the cable are thermal vacuum testing, natural dieoff, change in pressure during launch, and lunar vacuum and temperature.

Thermal vacuum testing of the Mariner 9 spacecraft has been found to reduce the number of viable microorganisms. A reduction of about 80 percent in the number of spore-forming bacteria and a more than 90 percent reduction in the number of viable non-spore-forming bacteria occurred as a result of this testing. The thermal vacuum testing of Surveyor 3 could have been responsible for a major reduction in the bacterial contamination.

Recently, in the laboratory, it has been shown that when a surface is protected from redeposition of microorganisms, such as within the layers of a thermal blanket, the initial population of microorganisms is reduced to near zero during 100 days of storage. The wiring bundle undoubtedly was prepared and wrapped during assembly and not reopened before launch; thus, a redeposition of microorganisms could not occur during pre-flight testing. The initial population of microorganisms may have been high during assembly of the wiring bundle but, because of natural dieoff, this number may have been reduced significantly.

A microbiological sampling was conducted of

TABLE 2. — *Experimentally inoculated surface wraps*

Test cable	Wrapping area sampled	Colonies (*Bacillus subtilis*)
1	Exterior wrap:	
	Outside	2.6×10^3
	Inside	7.7×10^3
	Interior wrap:	
	Outside	0
	Inside	0
2	Exterior wrap:	
	Outside	9×10^2
	Inside	0
	Interior wrap:	
	Outside	0
	Inside	0

NOTE: During the unwrapping manipulation, the wrap slipped from the forceps and curled back upon itself.

TABLE 3. — *Results of culture enrichment of Surveyor 3 wiring bundle*

Number of samples	Sample description	Results
6	Nylon ties	0
7	Protective wrap	0
23	Insulation from wires	0
13	Wire	0
17	Wire and insulation	0
3	Teflon sleeving	0
9	Wire or insulation sterile controls	0

spare flight Surveyor television cameras that had been in storage for 1 yr or more. All internal surfaces that were sampled, including pieces of polyurethane under the circuit boards, were sterile, demonstrating that if a surface were protected from redeposition by airborne microorganisms, the natural dieoff would reduce the surface population.

The remaining population of microorganisms on the wiring bundle would have been reduced by the change in atmospheric pressure during the launch of Surveyor 3. Our research has shown that when the dried cells of a bacterium are subjected to a rapid change in pressure from 760 torr to 1×10^{-5} torr within a 12-min period, loss of viability does occur. This can cause a decrease in population of 10 percent with spore-forming bacteria and up to 50 percent with non-spore-forming bacteria.

The exposure of bacteria to high vacuum (10^{-10} torr) has shown that vacuum itself is not sterilizing, even though a reduction in the number of viable bacteria does occur. However, if during the vacuum exposure, the cells also are heated to 60°C and above, death of the bacteria is accelerated. For instance, *Bacillus subtilis* var. *niger* spores, when exposed to 10^{-10}-torr vacuum and heated to 60°C for 14 days, lost 69 percent of viability. *Staphylococcus epidermiditis* exposed to the same conditions lost 99 percent of viability. During the 31 months of lunar exposure of Surveyor 3, the spacecraft was exposed to vacuum and temperature cycling.

Reduction in numbers of bacteria caused by pre-launch environmental effects (i.e., from the time the cable was wrapped to launch), coupled with the effects of launch and lunar environment exposure, could easily have reduced the numbers of contaminating microorganisms to such a low number that portions of the sampled wiring bundle were sterile.

References

1. deSenes, F. J.: "Effects of Radiation During Space Flight on Microorganisms and Plants on the Biosatellite II and Gemini Missions." *Life Science and Space Research*, vol. 7, 1969, pp. 62–66.
2. Horneck, G.; Bucher, G. H.; and Wollenhaupt, H.: "Survival of Bacterial Spores Under Some Simulated Lunar Surface Conditions." *COSPAR Space Research XI*, Akademie-Verlag, Berlin, 1971.
3. Cameron, R. E.; Morelli, F. A.; and Conrow, H. A.: *Survival of Microorganisms in Desert Soil Exposed to Five Years of Continuous Very High Vacuum*, TR 32–1454, Jet Propulsion Laboratory, Pasadena, Calif., 1970.

ACKNOWLEDGMENTS

We thank the crew of Apollo 12 for their successful flight and return of material used in this study. We also thank G. M. Renninger, M. Adams, J. H. Stevens, and D. C. Schneider for their assistance in the experiment's early planning stages, and D. M. Taylor for his helpful discussions during this experiment.

Appendix A

Spacecraft Orientation and Exposure to Environment

N. L. Nickle

Subsequent to the compilation of solar exposure data presented below, the writer became aware of a change in the spacecraft's orientation since the end of the Surveyor 3 mission. Measurements made from Apollo 12 photographs of the Surveyor spacecraft indicate that footpad 1 (downhill) was anchored, while footpads 2 and 3 rotated from 7 to 8 cm counterclockwise about footpad 1 (fig. A–1). Shock absorbers attached to legs 1 and 3 collapsed, causing the spacecraft to tilt an additional 5° (ref. A–1; also see ch. IV, pt. L, of this document).

The effect of the change in orientation on the exposure values depends on when the change occurred. If the change occurred early, the exposure values will vary up or down depending on the spatial orientation of a given surface. If the change occurred late in the 31-month residence time for the returned parts, then the values are considered valid.

R. Scott informed the writer of the possibility that the change in orientation occurred late due to evidence observed in an Apollo 12 photograph (AS12–48–7124) of footpad 3. He contends that clumps of dirt emplaced on the white upper surface of the footpad during landing were moved and partially removed by an episodic event, such as a failure of the shock absorbers. The time at which the movement occurred can be estimated only from the comparison of the shielded and unshielded portions of footpad 3 and a knowledge of the mechanism and rate of the process that tans the painted surface.

An evaluation of the change in exposure to solar radiation due to later spacecraft movement is not planned unless an estimate of the time of movement can be established. It is judged that the change in total exposure will be negligible except for those surfaces that received little radiation at grazing angles of incidence.

Exposure to Solar Radiation

The Surveyor 3 spacecraft landed halfway down the eastern slope of a 200-m-diameter

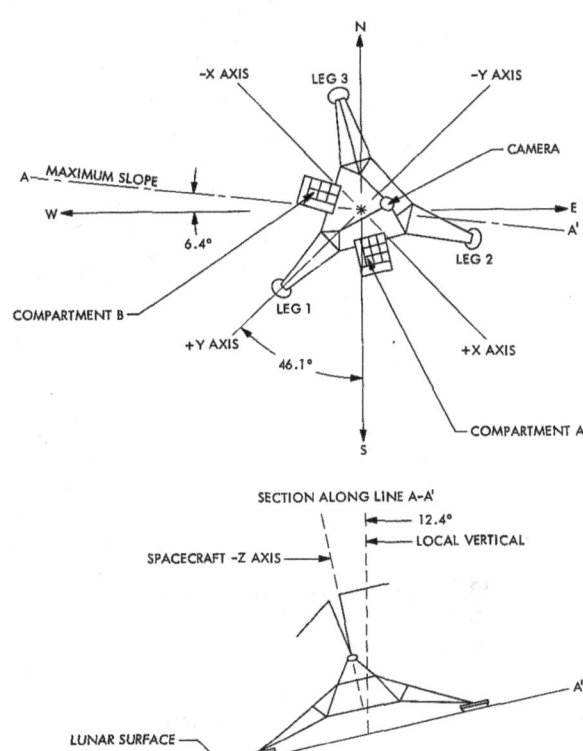

FIGURE A–1.—Geometrical configuration of Surveyor 3 as it existed at the end of the 1967 mission. The planar array antenna and solar panel shown in section view are not displayed in plan view.

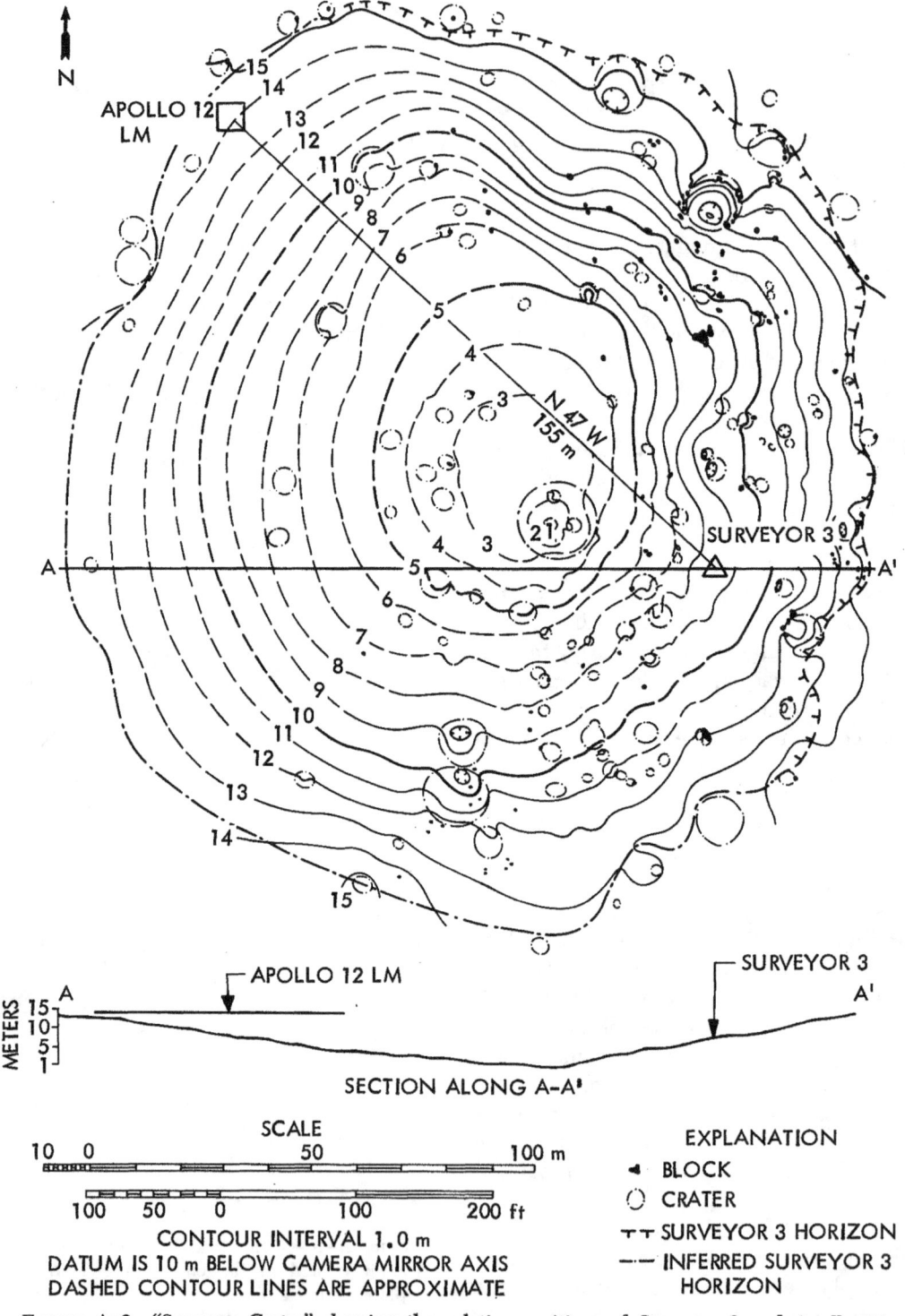

FIGURE A-2.—"Surveyor Crater" showing the relative positions of Surveyor 3 and Apollo 12 in plan and cross-sectional views. The Lunar Module was situated 155 m away from N 47° W of, and at a ground level of 4.3 m higher than the Surveyor television camera. (See ch. IV, pt. I, of this document.)

TABLE A-1.—*Data used to evaluate solar exposure of various parts of the Surveyor 3 spacecraft*

Synodical month	29.5 days
Angular velocity of Sun from Moon	0.51°/hr
Sunrise on the spacecraft	7°
Sunset on the spacecraft	178°
Maximum exposure to sunlight per lunation	171°; 335 hr
Duration of lunar stay of the returned parts	942 days; 31.9 lunations

crater at 2.99° S latitude and 23.34° W longitude (ref. A-2). The horizon visible to the spacecraft permitted an exposure to solar radiation between 7° and 178° to the local horizontal (see fig. A-2). Therefore, the 171° of arc in the plane of the ecliptic is equivalent to 335 hr of exposure to solar radiation for each lunar day. Table A-1 lists the values used in calculating the exposures found in tables A-2 and A-3.

Material removed from the spacecraft and returned to Earth remained on the lunar surface for 31.9 lunations. None of the returned parts received the maximum 10 686 hr of exposure because of shadowing by the planar array antenna, solar panel, thermal-control compartments, or other parts of the spacecraft.

To determine the actual exposure of specific parts to sunlight, six series of photographs were taken at the science and engineering testing laboratory (SETL) at JPL. A one-fifth-scale model spacecraft was oriented to a collimated light source simulating the orientation of Surveyor with the Sun (fig. A-3). Three cameras were set up to view different parts of the spacecraft; photographs were taken at the minimum illumination angle (2°), at each 10° interval through 170°, and at the maximum angle possible (178°). The data obtained from these photographs permitted an evaluation of the effects of exposure to solar radiation on the camera and its parts, the surface sampler scoop, and the strut from the radar altimeter and doppler velocity sensor (RADVS).

Television Camera

The Z-axis of the camera was tilted 23.5° from the local vertical in a direction N 43° W during the Surveyor 3 mission (ref. A-2). The upper

FIGURE A-3.—A one-fifth-scale model spacecraft shown in the orientation of Surveyor 3 at the end of the mission. The axes of the mirrors represent the plane of the ecliptic. This configuration was used to evaluate the exposure of various parts of the spacecraft to solar radiation.

shroud of the camera and the normal to the plane of the television camera mirror faced N 83° E. The pivot axis of the mirror is estimated to have been 1.5 m above the lunar surface. The surface of the lower shroud facing northeast was oriented parallel to the X-coordinate of the spacecraft. (See fig. A-1.)

Various surfaces of the camera have been evaluated for their solar exposure and are presented in table A-2. Various external features of interest are defined in figures A-4 to A-6.

The camera was equipped with four optical filters. (See table A-4 for specifications.) The clear filter was situated over the lens of the camera at the termination of the mission; because of its relative position in the hood assembly, it received longer exposures than did the other filters. The exposure values listed in table

TABLE A-2.—*Exposure of selected parts of the camera to solar radiation*

Surface [a]	Exposure/ lunation,[b] hr	Total exposure, hr	Sun angle at first exposure	Angle of incidence [c] at first exposure	Sun angle at last exposure	Angle of incidence [c] at last exposure
Lower shroud							
A [d]	137	4383	....	45	77	10
B	26	814	122	2	135	11
	45	1440	155	25	178	39
C	167	5322	7	43	92	20
D [e]	20	626	125	8 (avg)	135	16 (avg)
	26	814	165	39 (avg)	178	64 (avg)
E	22	689	7	10	18	0
Elevation drive housing [f]							
F	133	4256	7	80	75	30
G	12	376	172	5	178	6
H	33	1064	125	19	142	32
	14	438	171	64	178	73
I	120	3819	17	0	78	58
	35	1126	125	69	143	53
	14	438	171	26	178	19
J	20	626	7	11	17	0
Vidicon thermal radiator							
Top [g]	145	4633	18	0	92	71
Bottom [h]	22	689	7	10	18	0
Optical filters							
Clear	141	[i] 4180	18	0	90	69
Red	122	[i] 3554	18	0	80	61
Green	108	[i] 3115	30	76	85	64
Blue	[i] 3	[i] 3	Not exposed			

[a] See figs. A-5 and A-6 for identification of surfaces.

[b] Assuming non-moving surfaces, as opposed to the upper shroud, elevation drive mechanism, mirror, etc.

[c] The angle of incidence is measured from the plane of the surface.

[d] This surface was oriented within 1° of being tangent to the Apollo 12 Lunar Module. (See text footnote 1.)

[e] This surface has a radius of curvature of 6.95 cm. Ninety-five percent of the area was continuously shaded by the mast and other supporting structures.

[f] This part rotated with the camera head assembly. The values listed in the "exposure/lunation" column are too large for the first lunar day. It is estimated, therefore, that the values for the total exposure would lie within 10 percent of the values listed in that column.

[g] Painted surface.

[h] Unpainted aluminum.

[i] It is difficult to assess the exposure during the first lunar day while the mission was in progress. The camera head assembly was oriented in all azimuth directions for various periods of time. The additional exposure experienced during this period, which can be obtained by the tedious reduction of the mission command tapes, must be added to these values. Of the 331 hr of daylight on the first lunar day, it is estimated that sunlight would have been incident upon one or more filters less than 1 percent of the time (≤ 3 hr). Therefore, 3 hr have been added to the "total exposure" column for 31 lunations instead of 32.

TABLE A-3.—*Exposure of the surface sampler scoop to solar radiation*

Surface	Exposure/ lunation, hr	Total exposure, hr	Sun angle at first exposure	Angle of incidence at first exposure	Sun angle at last exposure	Angle of incidence at last exposure
Total.	284	9078	7	Variable	152	Variable
	6	188	163	Variable	166	Variable

TABLE A-4.—*Camera filter specifications*

Filter	Manufacturer	Remarks
Clear.	Bell & Howell.	Clear flint, $\rho = 360$ g/cm³, $n = 1.612$;[a] MgF$_2$ coating on both sides.
Red.	Corning.	3–76.
Green.	Schott.	OG4 (top; light yellow, bottom coated with Inconel [b]).
	Chance.	OGR3 (bottom; uncoated).
Blue.	Schott.	GG15 (top; bottom coated with Inconel).
	Schott.	BG1 (bottom; uncoated).

[a] R. Fleischer, personal communication, 1971.
[b] Inconel is 78.5 percent Ni, 14 percent Cr, 6.5 percent Fe.

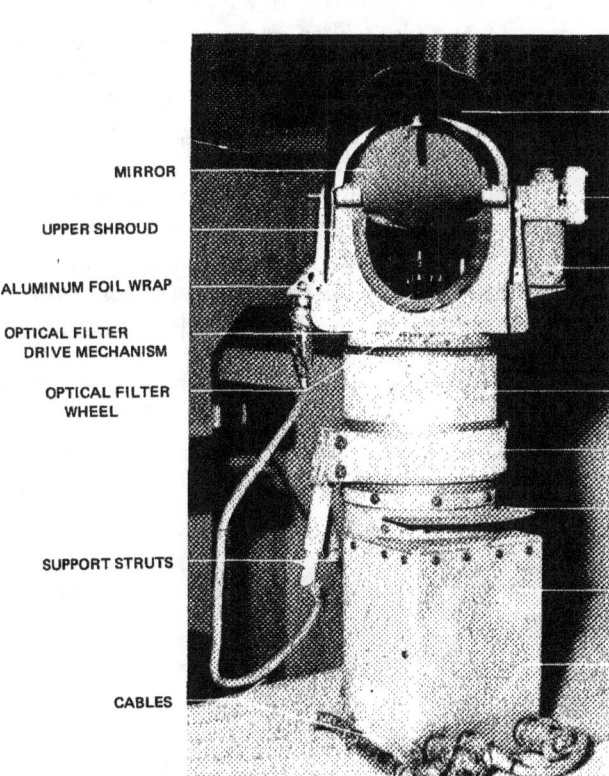

FIGURE A-4.—Surveyor 3 camera as it was unbagged at the Lunar Receiving Laboratory. Dents in the visor occurred during transport from the lunar surface.

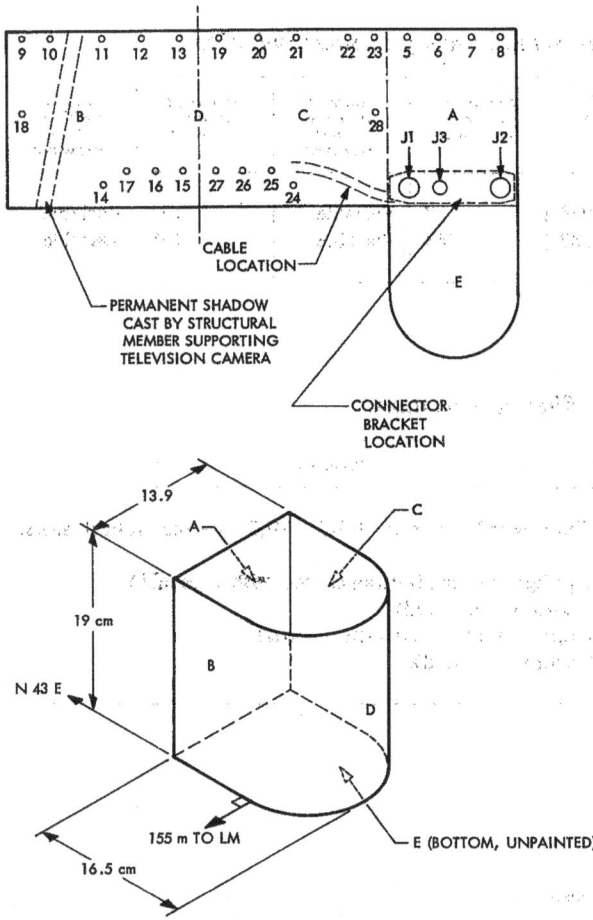

FIGURE A-5.—Lower shroud of the television camera identifying the various surfaces as used in table A-2 for the calculation of exposures to solar radiation. The numbered circles refer to the positions of numbered screws and washers removed during disassembly.

A-2 are not valid for the entire surface of the filters because, at some time, part of each filter was shaded by either the front opening of the camera, the filter-wheel drive mechanism, and/ or the mirror. The values listed in table A-2, therefore, represent the maximum exposure experienced by selected parts of each filter.

The upper portions of all filters were covered with varying amounts of lunar dust. Preliminary data from peels taken from the clear filter (see ch. IV, pt. B, of this document) indicate the median grain size is 0.8 μm, with particles ranging in size from less than 1 to greater than 15 μm. Fifty percent of all particles are under 1 μm in size. Particle density averages 0.18 particle/μm²; the surface area covered by particulate material entrapped in this peel, therefore, is 25.0 percent. This value is in agreement with the value calculated by Carroll[1] and Rennilson et al. (see ch. IV, pt. E, of this document) from spectral transmission data taken from the clear filter. They determined that about 25 percent of the surface area was covered with particles by comparing data taken before and after cleaning the filter of its particulate material.

The effective shadowing of the upper surface of the filters by adhering particles is a function of exposure geometry. If 25 percent of the surface of the clear filter is shadowed by particles from an incident beam oriented 90° to the plane of the filter, then the effective shadowing will be considerably greater at angles approaching that plane.

Figure A-7 shows the effective shadowing for incidence angles ranging from 0° to 90°. The data were compiled by calculating areas of elliptical shadows cast by hypothetical spheres from the particle characteristic data compiled by Robertson et al. (See ch. IV, pt. B.)

The visibility, or solid angle of view, that each

[1] W. Carroll, JPL, private communication, 1970.

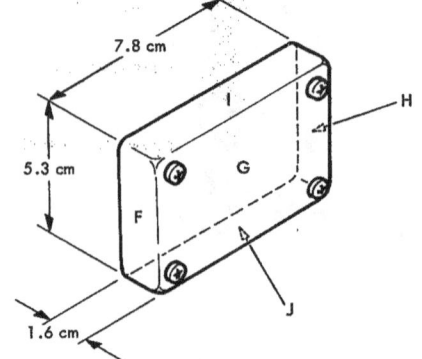
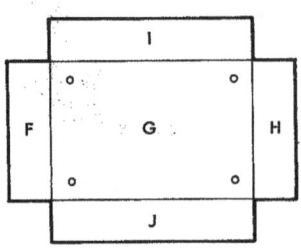

FIGURE A-6.—Elevation drive housing of the television camera indicating the various surfaces as used in table A-2 for the calculation of exposures to solar radiation.

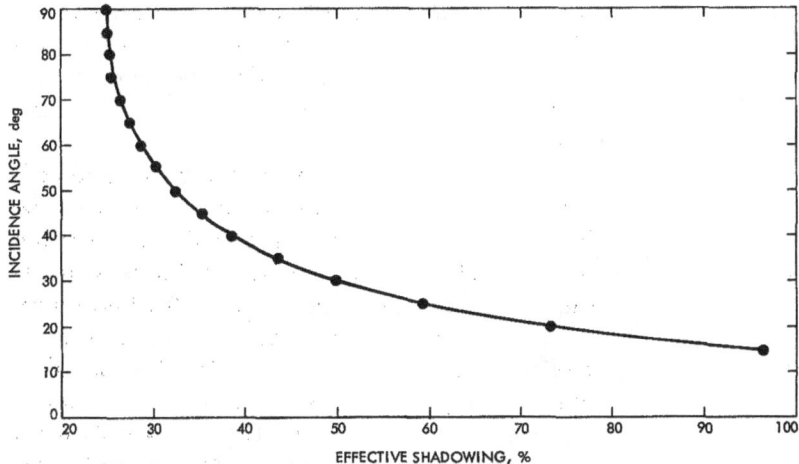

Figure A-7.—Effective shadowing created by particles adhering to the upper surface of the clear flint filter by an incident beam ranging from 0° to 90°. Data are not available for the other filters.

filter had through the front opening of the camera was determined to evaluate the results of particle track studies, alpha radioactive fallout, and micrometeorite impact flux measurements. The visibility was determined at the center of the filters, except for the blue filter which had no direct view from the camera. Figure A-8 shows the relative position of the red (R), clear (ND), and green (G) filters; their visibility was 0.49, 2.19, and 0.73 sterad, respectively. These values vary most with position on the red and green filters where parts of the filter drive mechanism partially covered them.

Surface Sampler Scoop

The arm of the scoop was left fully extended and at maximum elevation at the end of the

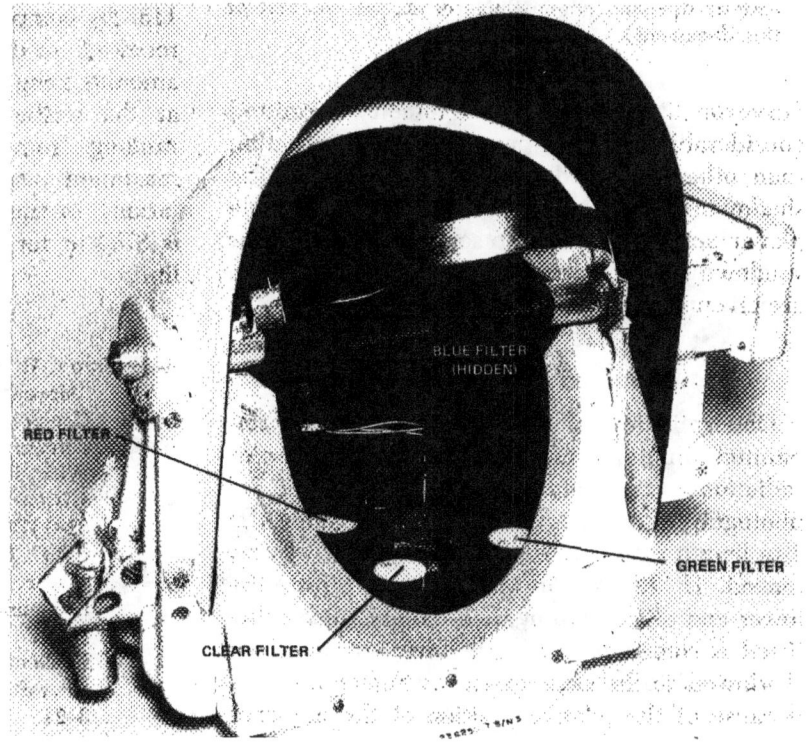

Figure A-8.—Upper part of a duplicate Surveyor television camera head showing the relative positions of the optical filters and the filter-wheel drive assembly. The labels and crosses show the central position from which visibility measurements were made.

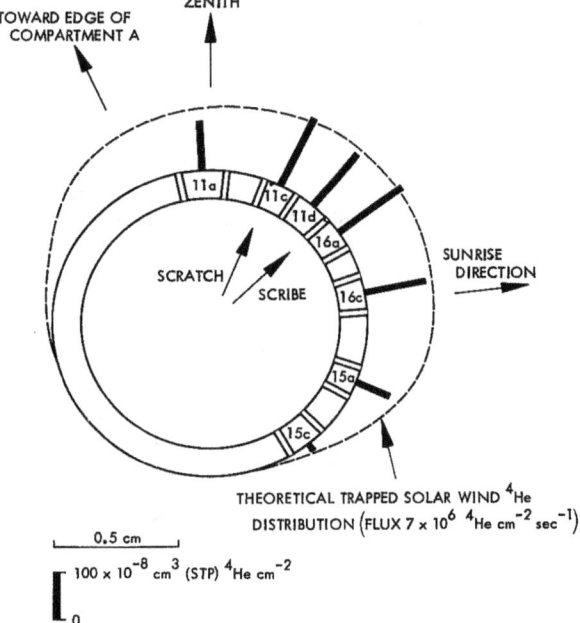

FIGURE A-9.—Cross section of an annulus of the polished aluminum tube showing the results of a preliminary examination designed to verify the rotational orientation of the tube on the lunar surface. The scribe line is coincident with the theoretical maxima and is situated 42° above the local horizontal, facing an easterly direction (from Buhler et al.; see ch. VIII of this document).

Surveyor 3 mission. This geometry permitted considerably more exposure to solar radiation than other returned parts. The relatively little shadowing the scoop received came from the planar array antenna, the solar panel, and, near sundown, by the spacecraft itself. Exposure data are given in table A-3.

Unpainted Aluminum Tube

Determination of the exposure of the unpainted aluminum tube (RADVS strut) to solar radiation was hampered by the absence of photography to verify precisely where along the 1-m length of tubing the 19.7-cm section was removed. If the section was obtained from the lower end of the tubing, then the exposure value listed is somewhat low. The lower end was not shadowed to the same extent as higher portions because of the relative position of the adjacent thermal-control compartment (compartment A). There is reason to believe that the section was removed from approximately the center (see footnote 1), however, and the exposure value reflects that assumption.

At the Lunar Receiving Laboratory, a scratch was made by hand along the length of the tube for orientation purposes (ref. A-3). The scratch was too light, so a heavier scribe line was subsequently made in the same area. This line represents the surface of maximum exposure to solar radiation, assuming that the direction of maximum implantation of solar wind rare gases is coincident with sunlight. Figure A-9 shows a cross section of the tube and a trapped solar wind helium envelope along with the direction of sunrise, the zenith, and shadowing. (See ch. VIII.) The scribe line is believed to represent an orientation on the Moon that is 42° above the local horizontal, and facing an easterly direction.

A bearing of N 15° E, plunge 18° S was determined for the axis of the tube by means of a Brunton compass on a one-fifth-scale model spacecraft (fig. A-3). Sunrise occurred on the tube at 7° and the tube became shadowed at 115° by compartment A. A 72° sector, therefore, received no direct sunlight; the others received amounts ranging from 0 to the maximum value at the scribe line and with incidence angles ranging from grazing incidence to 75° (the maximum possible because of its spatial orientation) to the surface. The maximum exposure is 6784 hr for areas in the vicinity of the scribe line.

References

A-1. SCOTT, R. F.; LU, T.-D.; AND ZUCKERMAN, K. A.: "Movement of Surveyor 3 Spacecraft." *J. Geophysics Res.*, vol. 76, 1971, pp. 3414–3423.

A-2. STEINBACHER, R. H.: "Orientation of Camera With Lunar Surface and Sun." *Surveyor III Mission Report. Part III: Television Data*, TR 32-1177, Jet Propulsion Laboratory, Pasadena, Calif., 1967, pp. 15–19.

A-3. BLAIR, P. M.; AND BUETTNER, D. G.: "Description and Initial Examination of Returned Parts." *Surveyor III Parts and Materials; Evaluation of Lunar Effects*, P-70-54, Hughes Aircraft Co., Culver City, Calif., 1971, pp. 3–21.

Appendix B

Surveyor 3 Material Analysis Plan

N. L. Nickle and W. F. Carroll

The material analysis plan presented here is composed of a series of comprehensive charts (see figs. B-1 through B-19) of all science and engineering investigations that have been reported in this document. The plan, as shown, progresses from left to right on each figure. Each test was designed to ensure that the subsequent tests would not be significantly affected by those that preceded it.

Each box represents one or more tasks performed by the individual or firm named at the lower left. The upper-right corner shows the amount and type of material subjected to the test or tests listed within the box.

The materials that have been analyzed and listed in the following illustrations do not represent all the material returned from the Moon. Most parts disassembled from the television camera, for example, were not analyzed in detail if they performed according to specifications. An inventory of camera parts is presented in appendix C of this document.

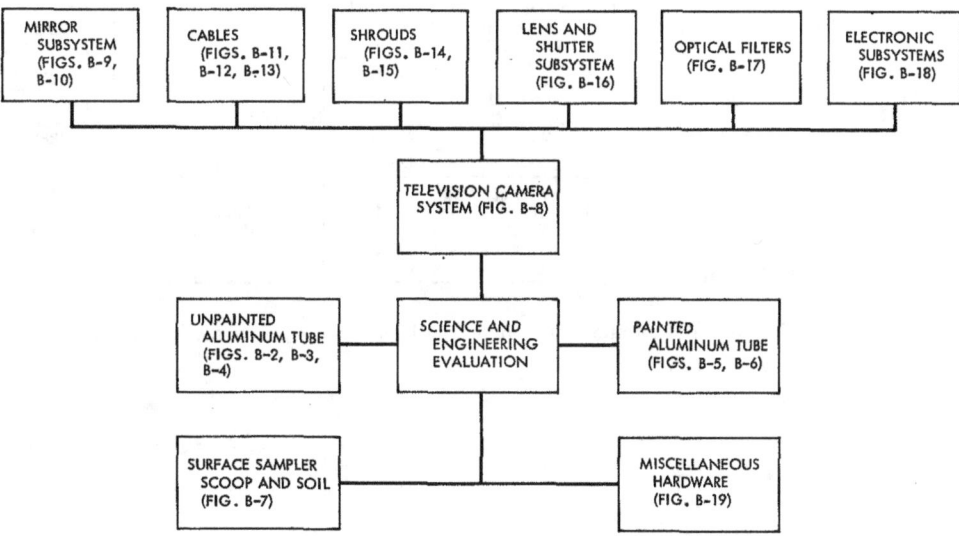

FIGURE B-1.—Categories of returned Surveyor 3 parts and references to subsequent figures describing the type of tests conducted, by whom, and the sequence used in the analysis.

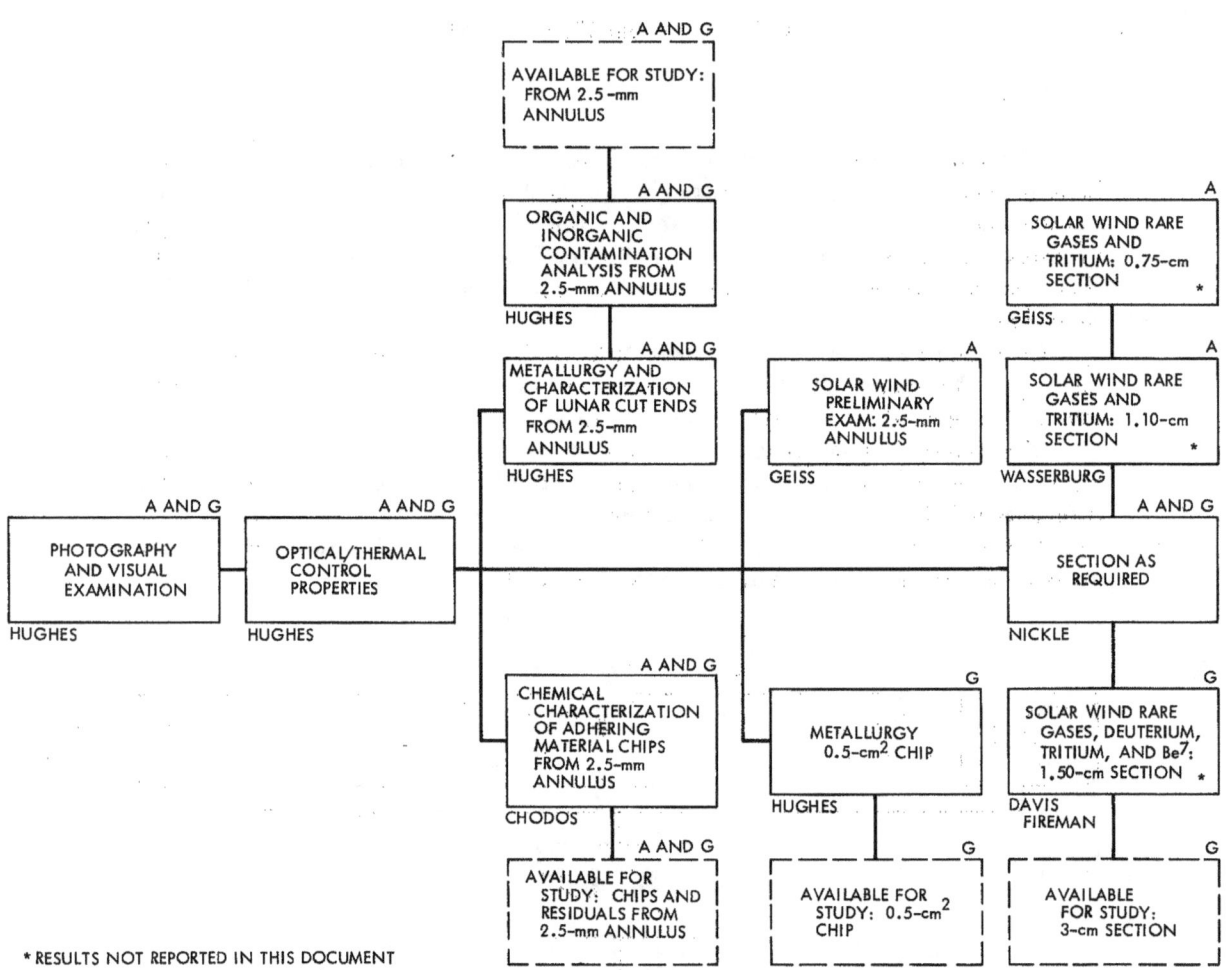

FIGURE B-2.—Unpainted aluminum tube: investigations conducted on sections A and G.

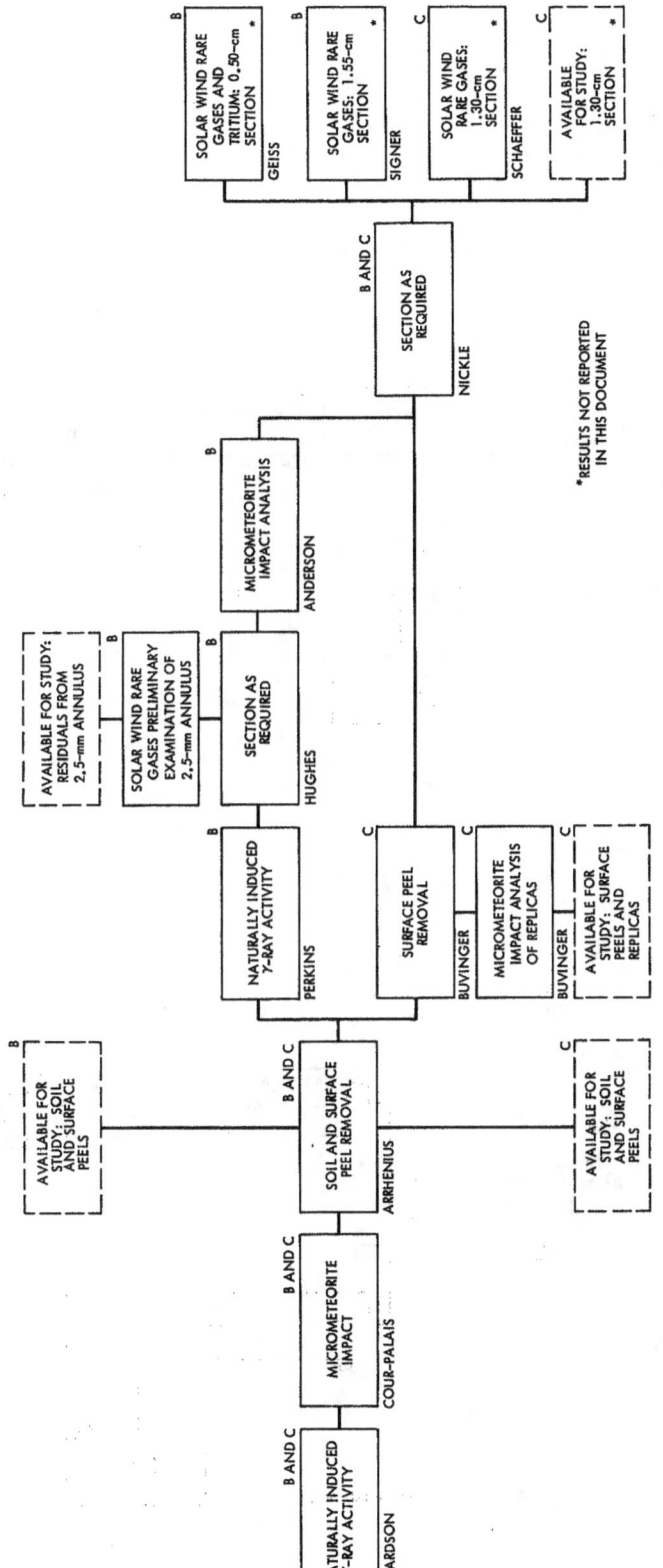

FIGURE B-3.—Unpainted aluminum tube: investigations conducted on sections B and C.

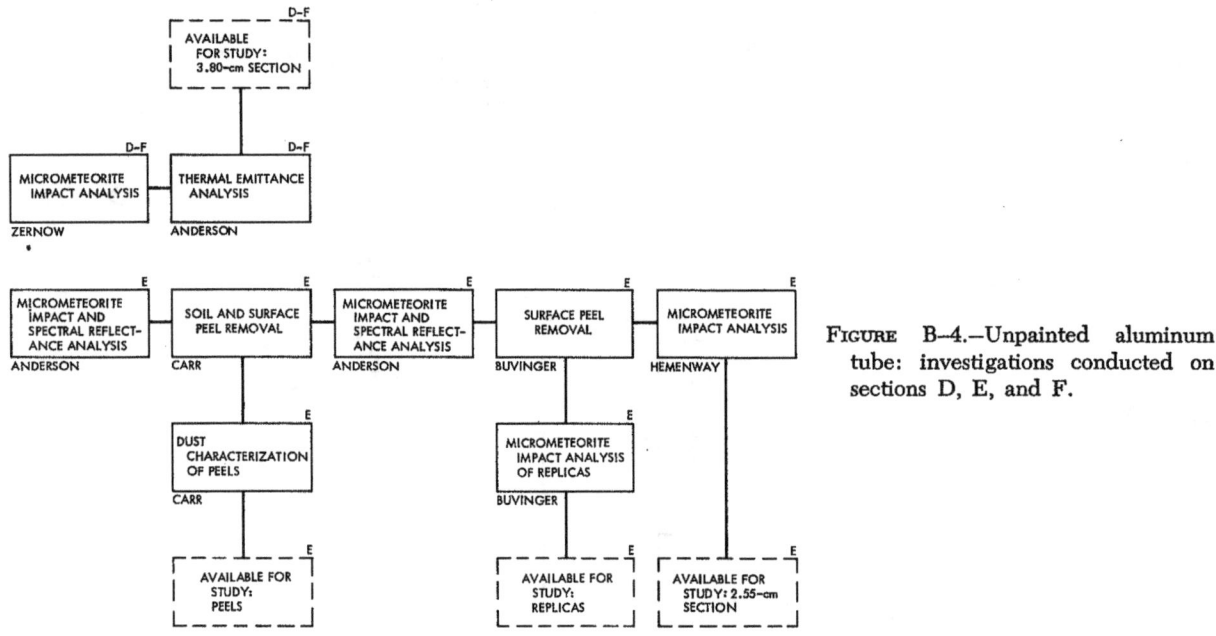

FIGURE B-4.—Unpainted aluminum tube: investigations conducted on sections D, E, and F.

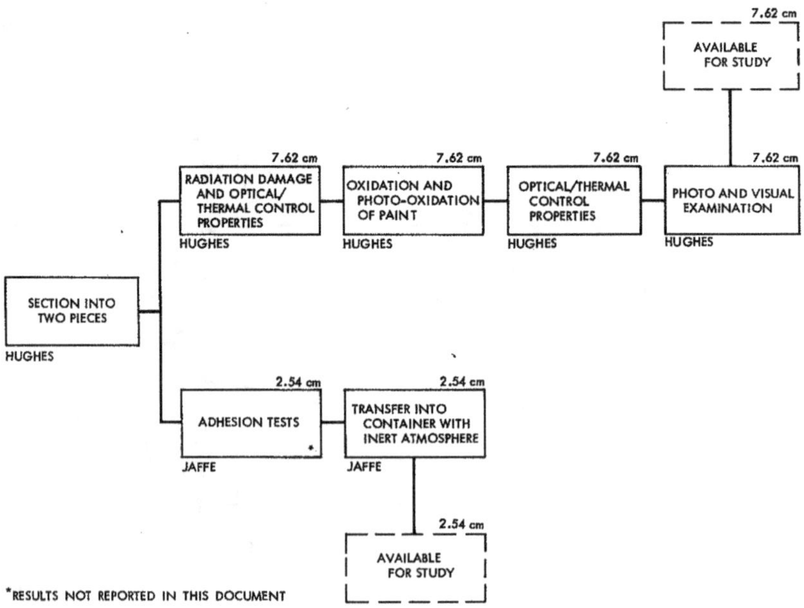

FIGURE B-5.—Painted aluminum tube: investigations conducted in a controlled environment on material returned in the sealed environmental sample container.

*RESULTS NOT REPORTED IN THIS DOCUMENT

APPENDIX B—SURVEYOR 3 MATERIAL ANALYSIS PLAN

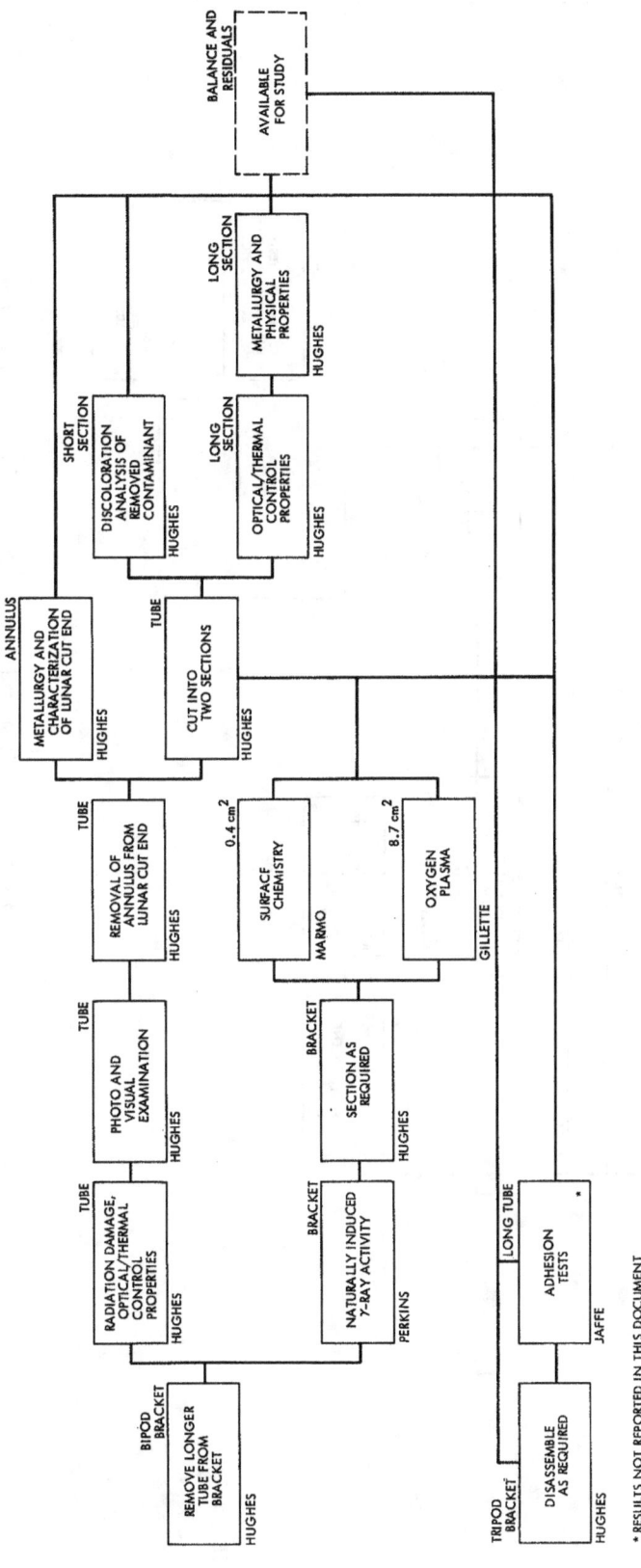

FIGURE B-6.—Painted aluminum components: investigations conducted on returned camera components.

* RESULTS NOT REPORTED IN THIS DOCUMENT

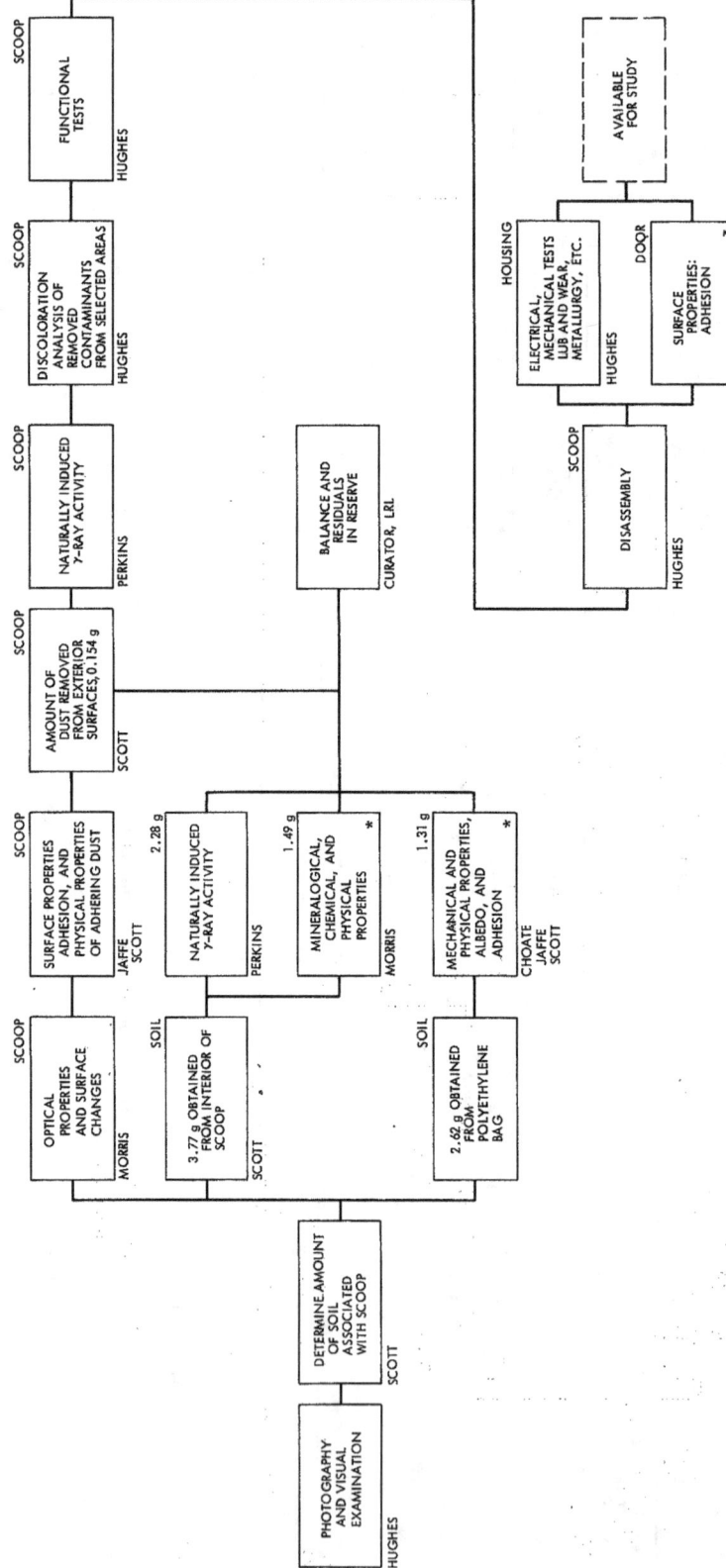

FIGURE B-7.—Surface sampler scoop and lunar soil: investigations conducted on the returned scoop and associated soil.

APPENDIX B—SURVEYOR 3 MATERIAL ANALYSIS PLAN

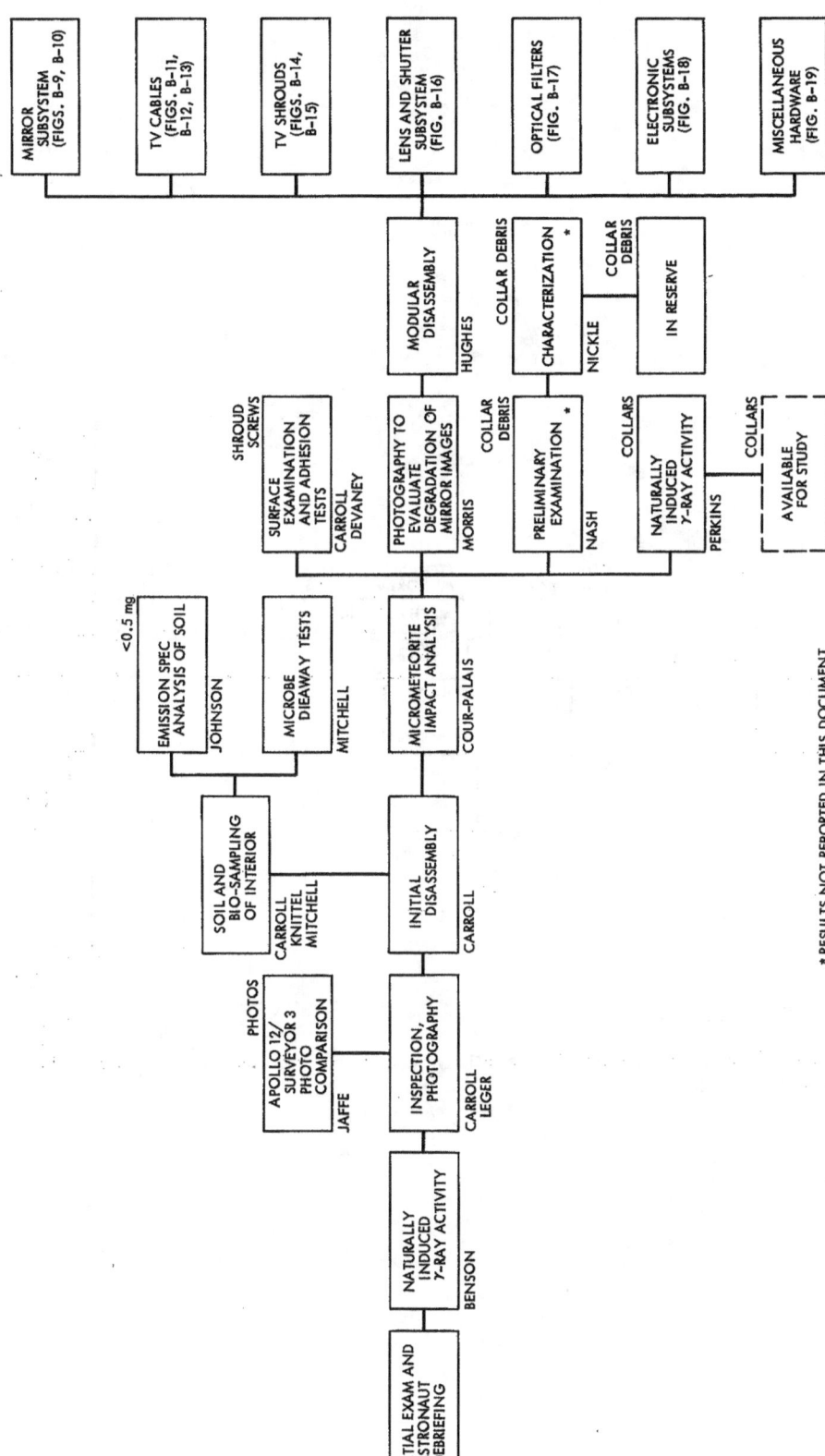

FIGURE B-8.—Television camera: investigations conducted on the complete camera and individual parts.

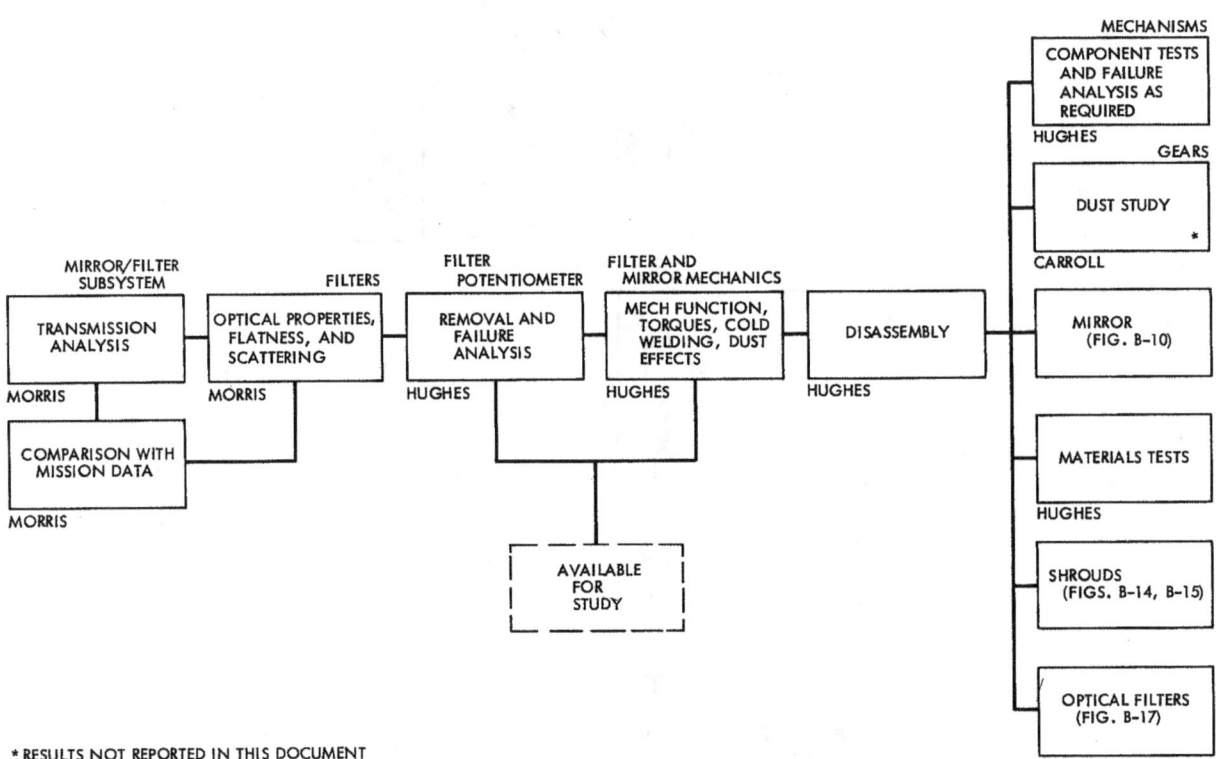

FIGURE B-9.—Mirror subsystem: investigations conducted on the mirror-optical filter-mechanical assembly.

APPENDIX B—SURVEYOR 3 MATERIAL ANALYSIS PLAN

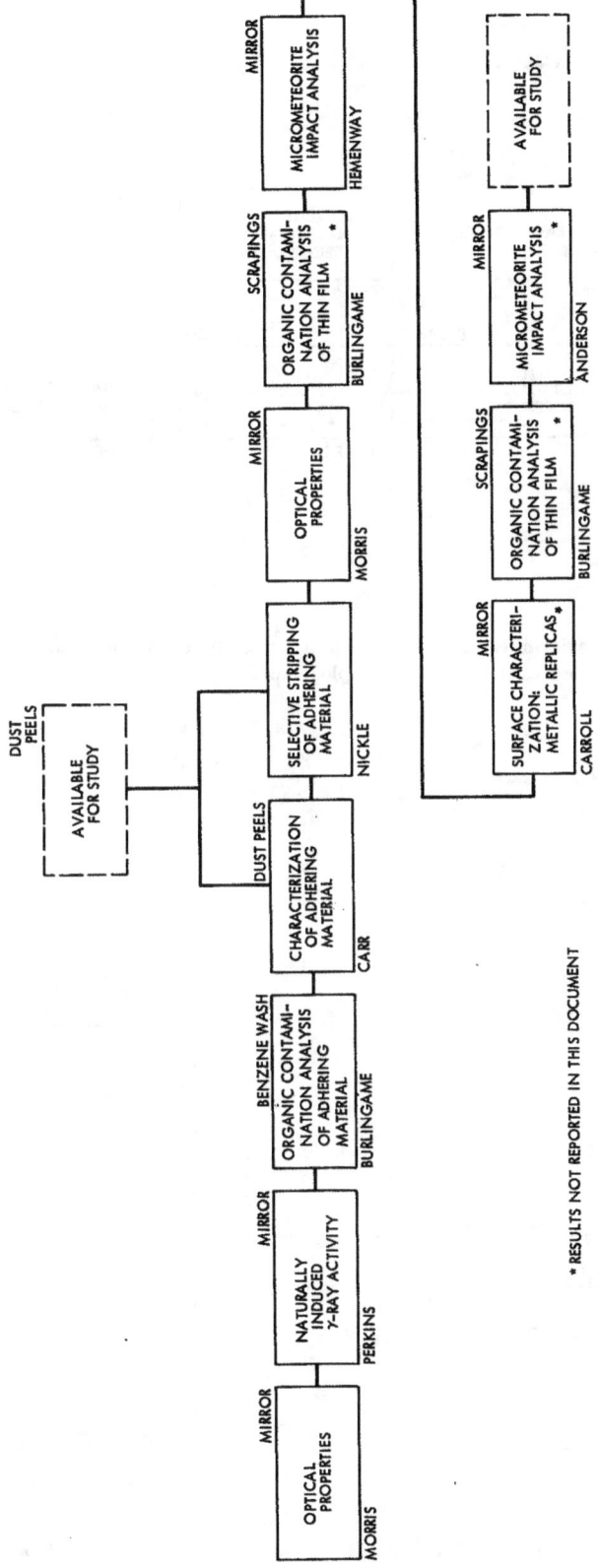

FIGURE B-10.—Mirror: investigations conducted on the mirror's surface.

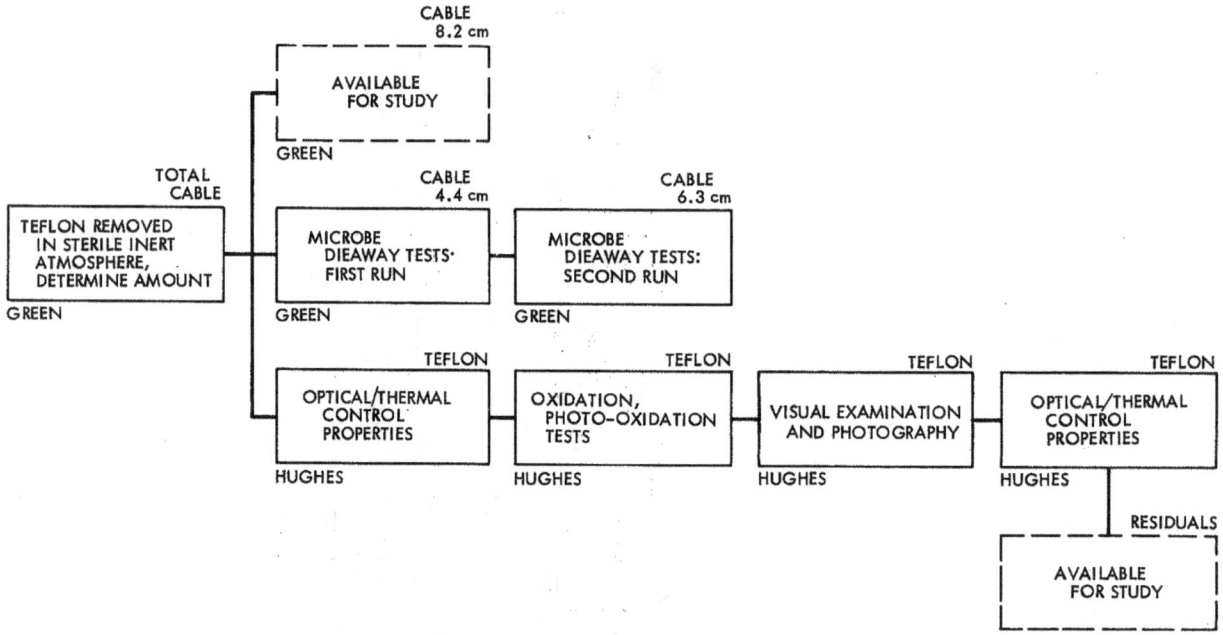

FIGURE B-11.—Television cable: investigations conducted on the Teflon-wrapped cable returned in the sealed environmental sample container.

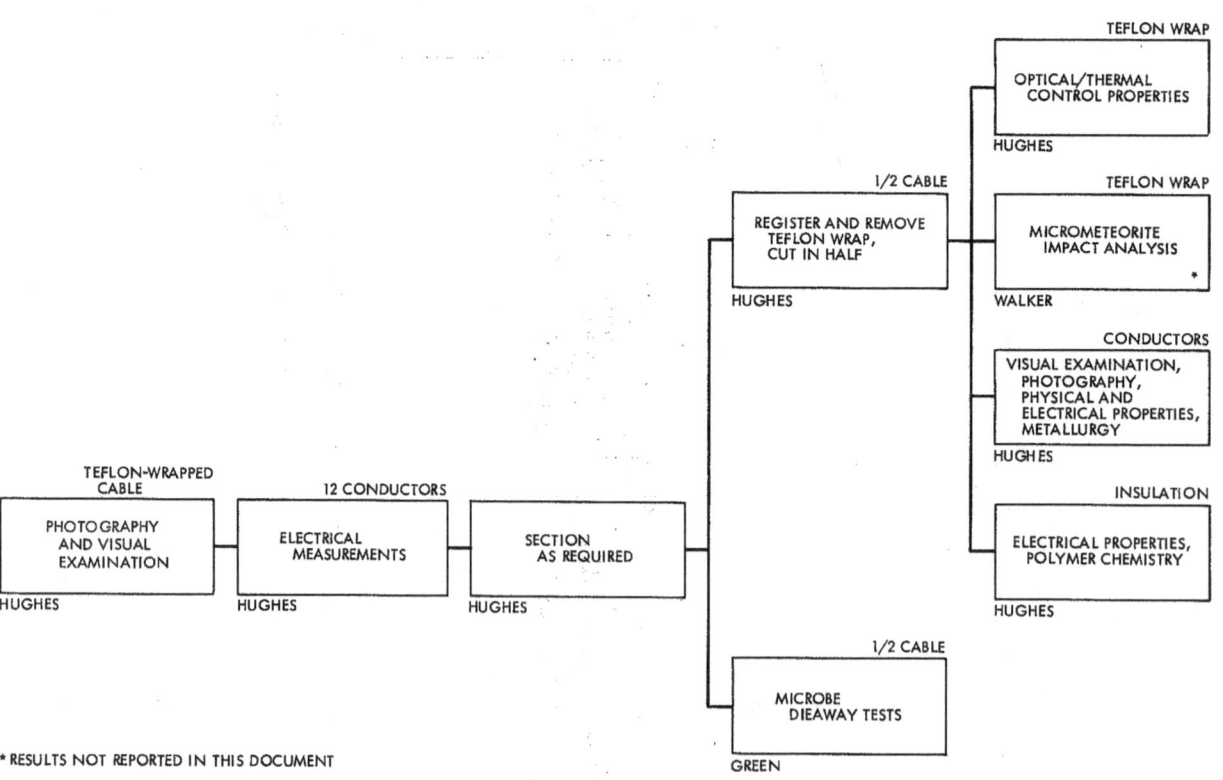

* RESULTS NOT REPORTED IN THIS DOCUMENT

FIGURE B-12.—Television cable: investigations conducted on the Teflon-wrapped cable.

APPENDIX B—SURVEYOR 3 MATERIAL ANALYSIS PLAN 271

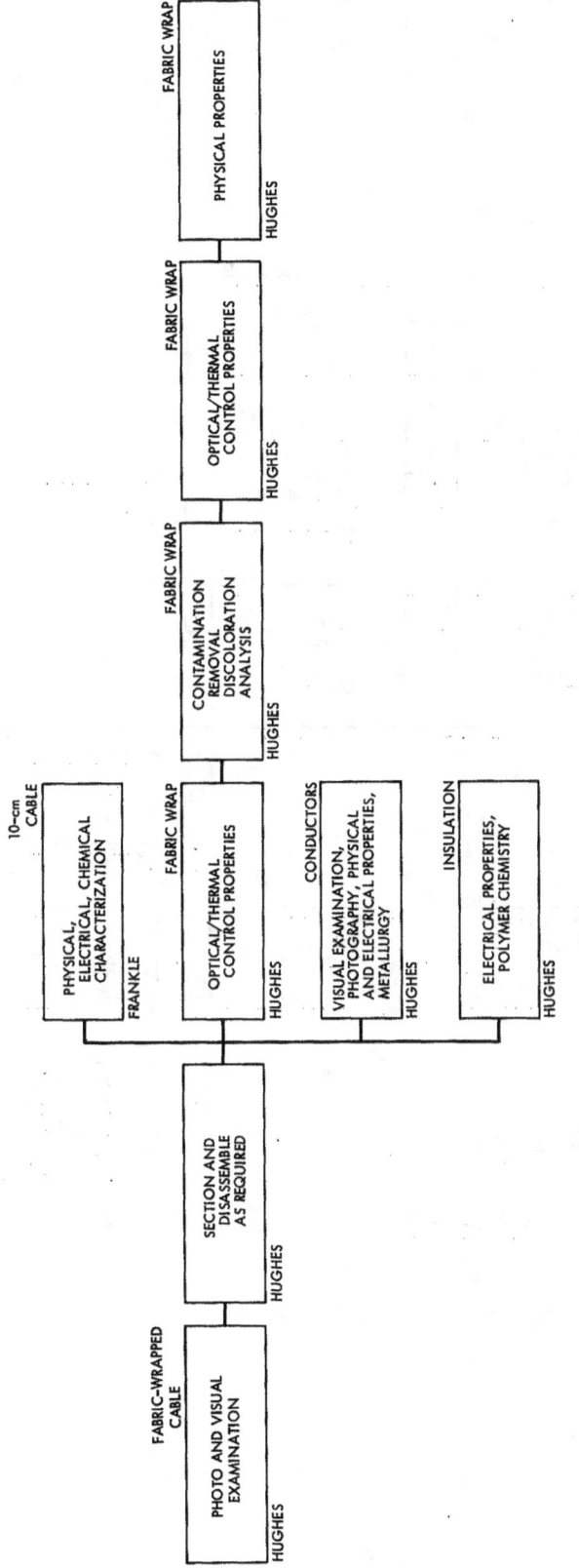

FIGURE B-13.—Television cable: investigations conducted on the fabric-wrapped cable.

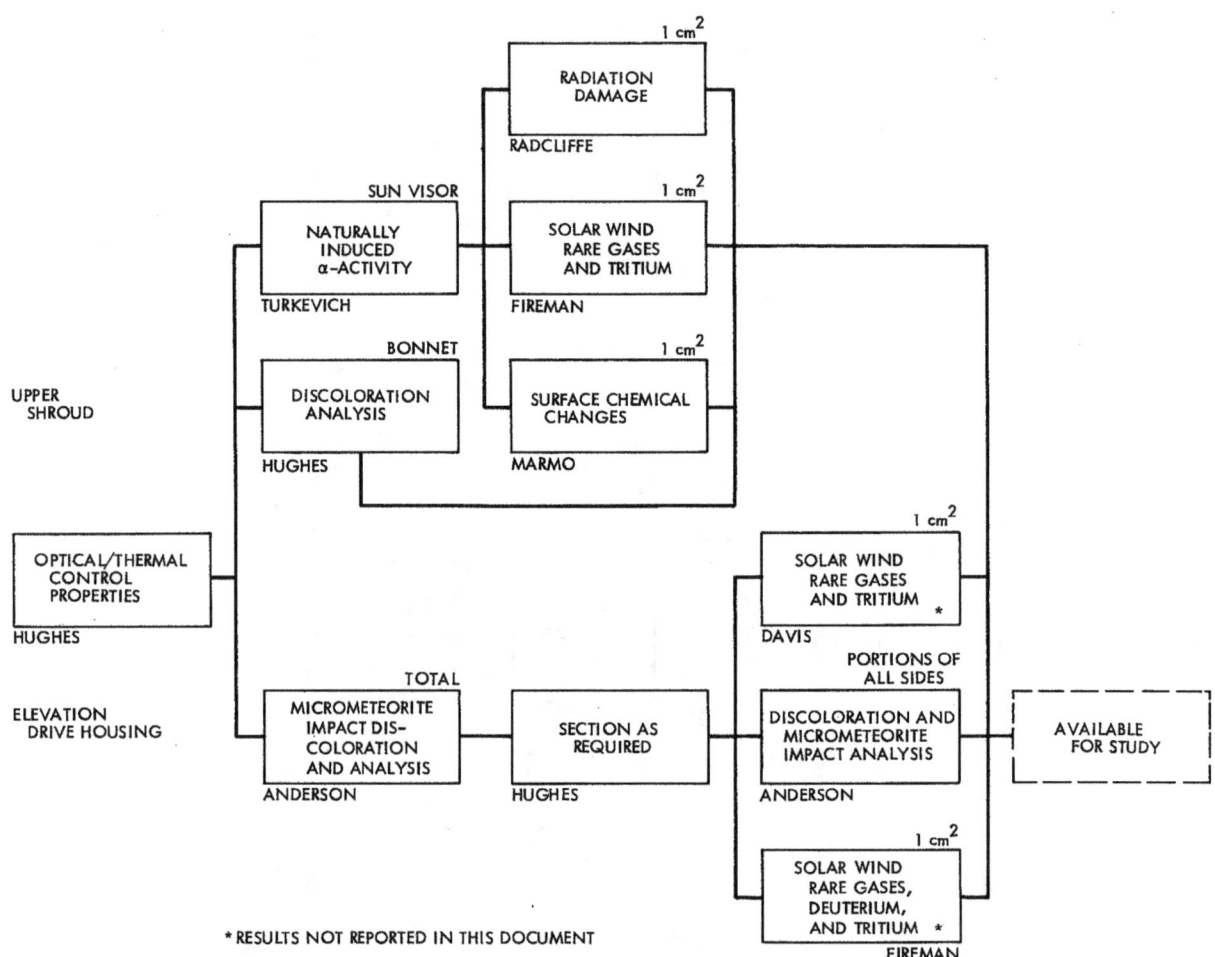

FIGURE B-14.—Television shrouds: investigations conducted on the upper shroud and the elevation drive housing.

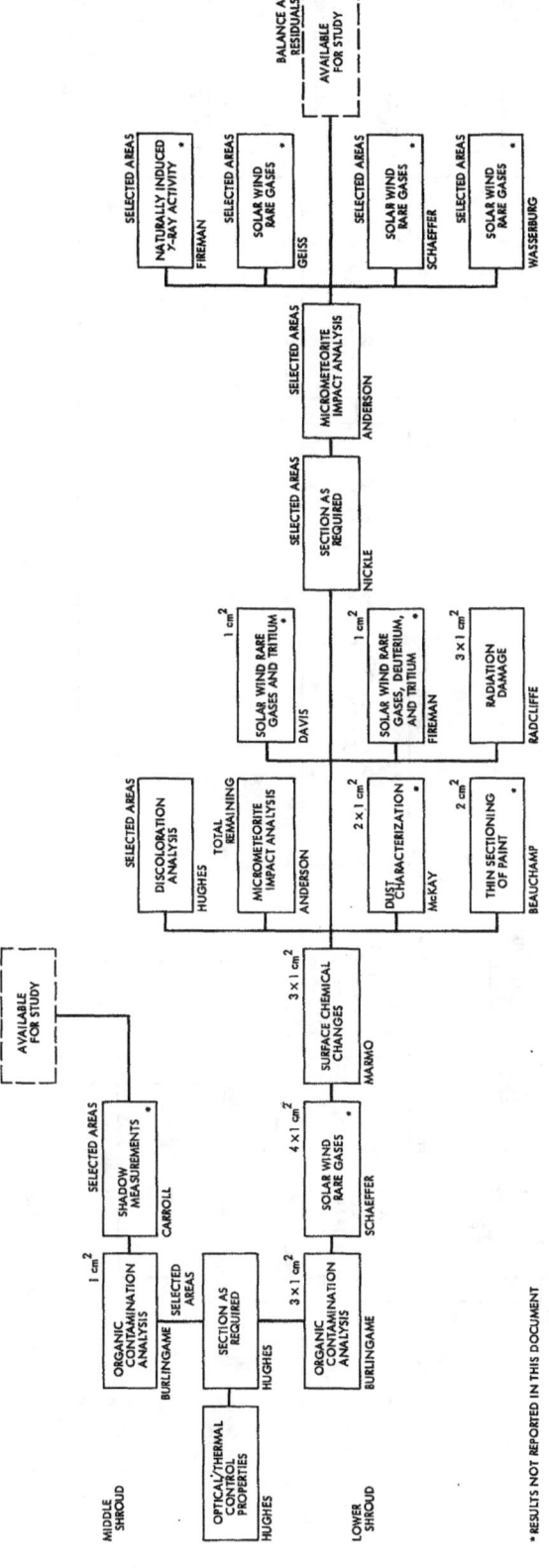

FIGURE B-15.—Television shrouds: investigations conducted on the middle and lower shrouds.

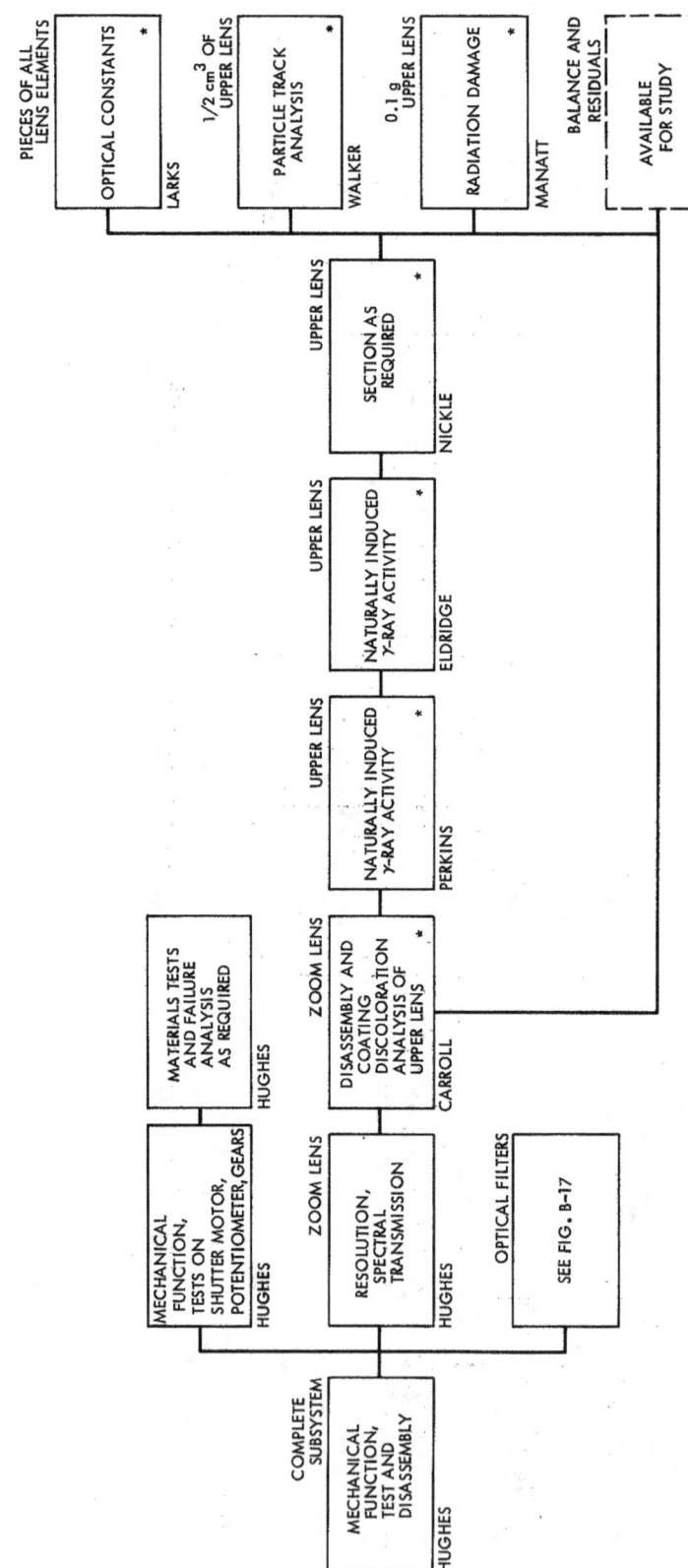

FIGURE B-16.—Television lens and shutter: investigations conducted on the zoom lens, shutter, and related components.

* RESULTS NOT REPORTED IN THIS DOCUMENT

APPENDIX B—SURVEYOR 3 MATERIAL ANALYSIS PLAN

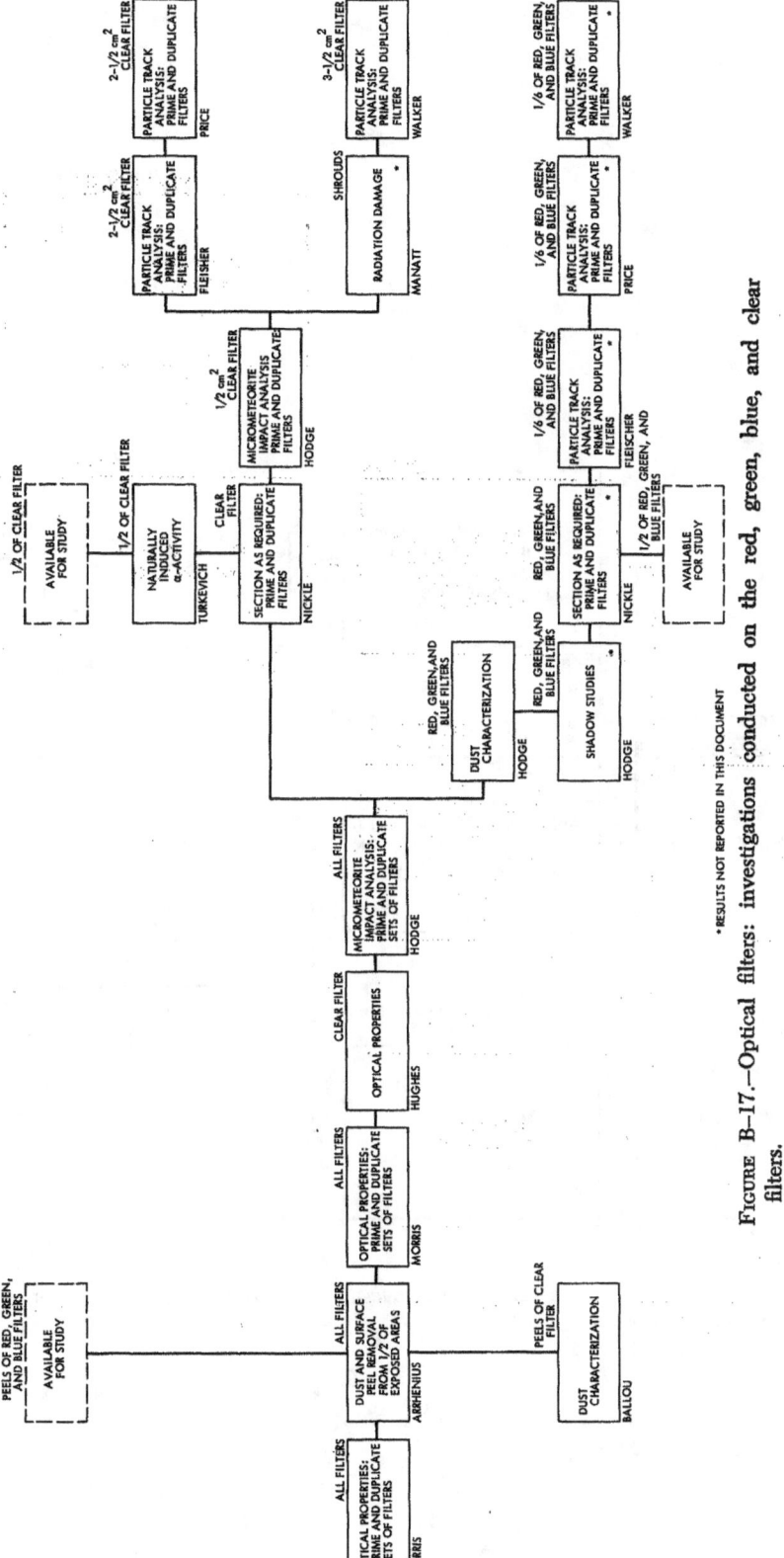

FIGURE B-17.—Optical filters: investigations conducted on the red, green, blue, and clear filters.

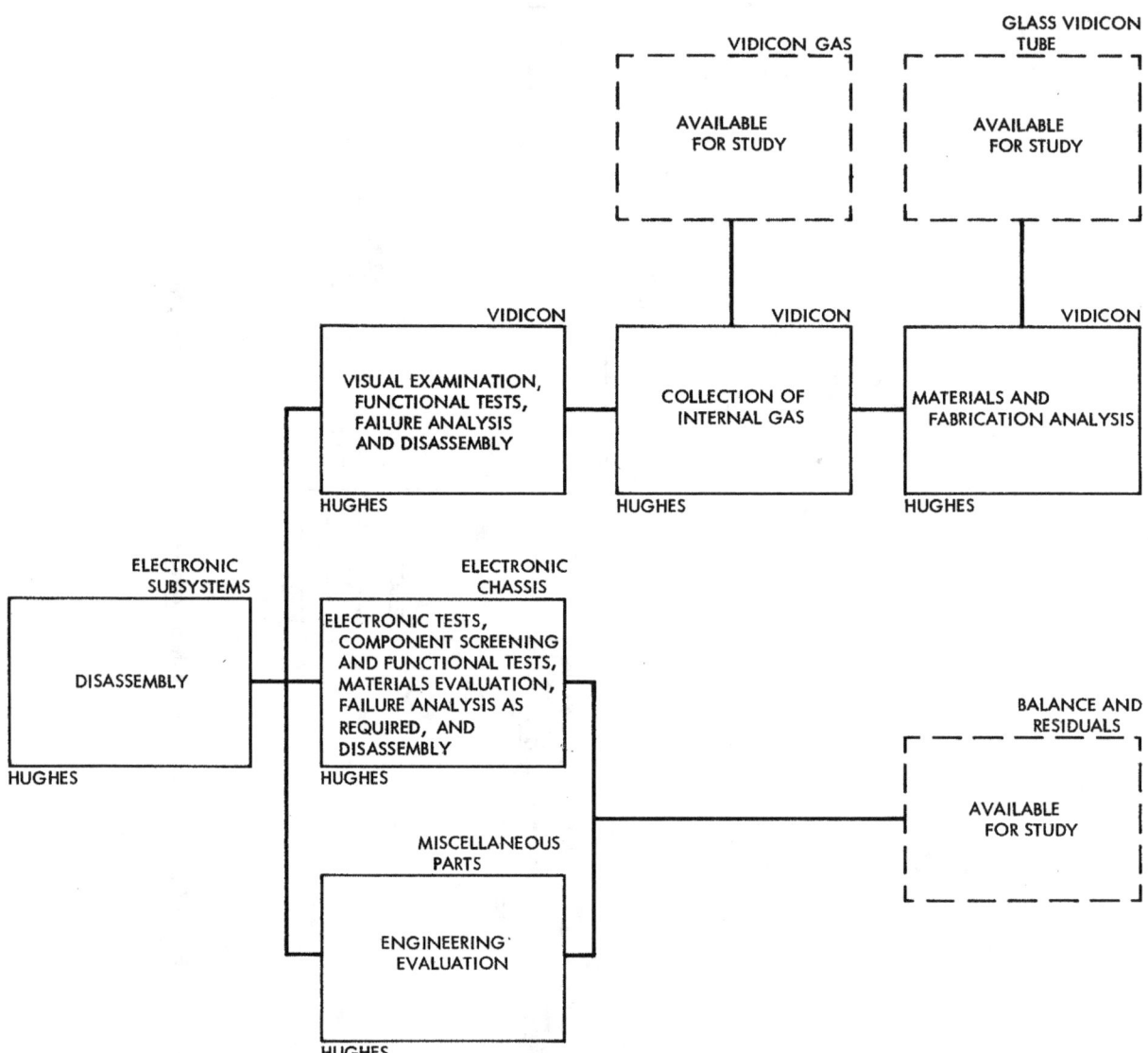

FIGURE B-18.—Electronics subsystems: investigations conducted on the electronic chassis and their components.

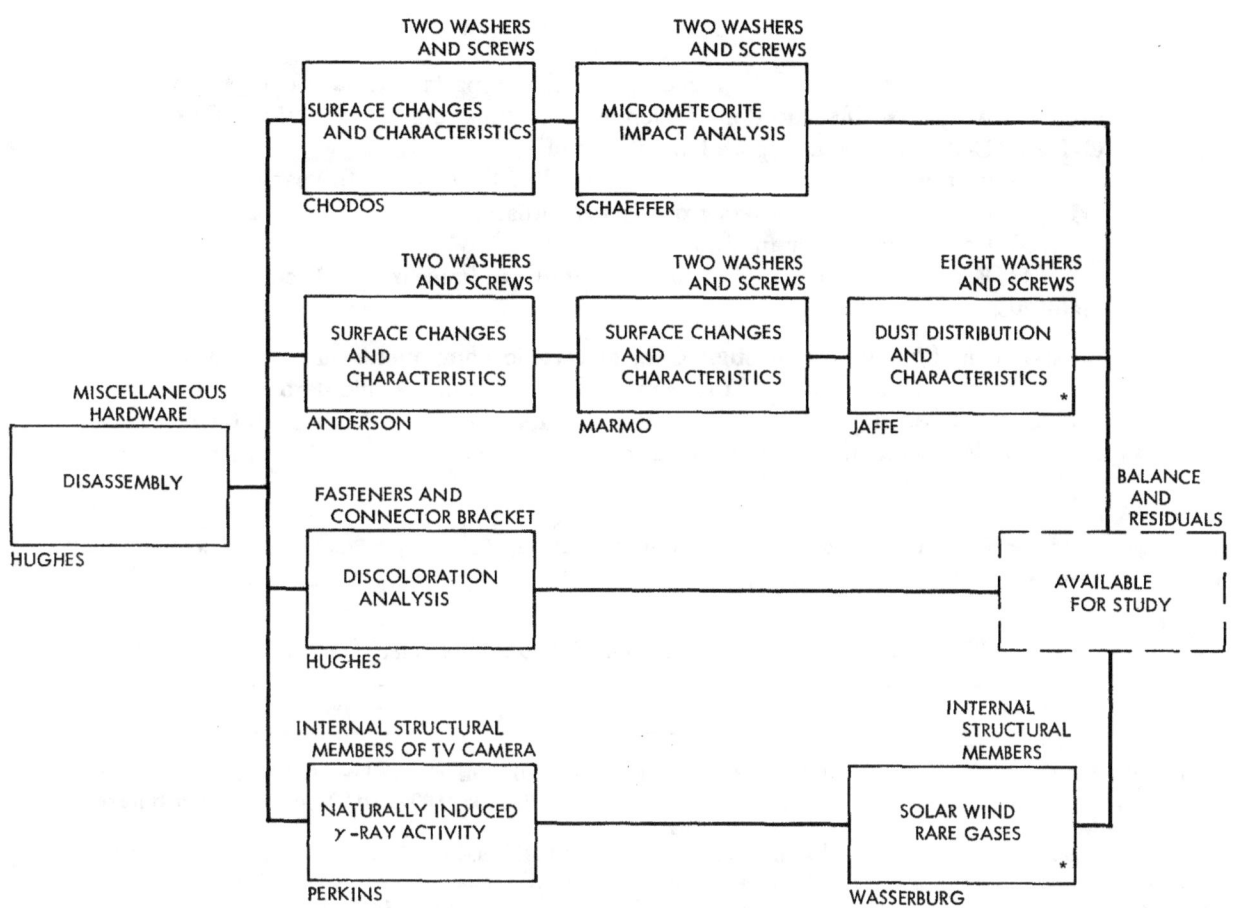

FIGURE B-19.—Miscellaneous hardware: investigations conducted on screws, washers, brackets, and supporting struts of the television camera.

Appendix C

Surveyor Television Camera—Selected Materials and Electronic Components

W. F. Carroll

This appendix presents the locations of various parts and has been prepared as a guide for possible additional science or engineering investigations. Tables C-1 and C-2 have been prepared with emphasis on—

(1) Exterior parts and surfaces that are directly exposed to space.
(2) Parts that shield others from space radiation.
(3) Representative or unique materials.
(4) Electronic devices that may contain unique or well characterized materials.

As shown in figures C-1 through C-5, electronic components are well distributed throughout the interior of the camera. Locations of specific items of interest can be identified on request. There may be additional materials of interest in this complex camera. On request, a review of documents can be made for availability, location, and quantity.[1]

[1] For further information, contact W. Carroll, Materials Section, Jet Propulsion Laboratory, Pasadena, Calif.

TABLE C-1.—*Location and identification of selected materials*

Item [a]	Material description [b]
1. Lower shroud	Aluminum alloy 6061 T-4, 0.031 in., polished per HP 9-29.
2. Paint	White inorganic paint per HP 4-135, 0.005 to 0.008 in., a baked mixture of $K_2Si_2O_3$ and $Al_2(Si_2O_3)_3$.
3. Lower shroud	Aluminum alloy 6061 T-4, 0.031-in.+0.000050-in. chromate coat per MIL-C-5541.
4. Screws	Torque-set screw A-286 CRES steel passivated (NAS 1631-C2).
5. Washer	Aluminum alloy 5052, 0.016 in. thick (NAS 620-A4L).
6. Rivet	Copper rod QQC-502, fully hard, 0.062-in. diameter, gold plated per MIL-G-45204, type II, class 2.
7. Radiator	Aluminum alloy 2024 T-4, 0.025 in. thick, polished bottom side per HP 9-29, painted top (item 2).
8. Support sleeve	Aluminum alloy 6061 T-6, 0.080-in. average+0.000050-in. chromate coat per MIL-C-5541.
9. Clamp ring	Aluminum alloy 7075 T-6, 0.060 to 0.125 in. thick.
10. Mirror blank	Beryllium plate, Brush Be Co. S-200C machined to 0.125- to 0.150-in. thickness.
11. Mirror base	Polished Kanigen electroless nickel, 0.003 to 0.005 in. thick.
12. Mirror coating	(a) Vacuum-deposited aluminum per MIL-M-13508, <2500 Å thick. (b) Silicon monoxide overcoating, <1500 Å thick.
13. Mirror blank	Aluminum foil, 0.005 in. thick (bonded to back).
14. Paint (back of mirror)	White organic paint (3M 202-A10) per HP 4-144, TiO_2 pigmented acrylic lacquer, 0.0035 to 0.0055 in. thick.

TABLE C-1.—*Location and identification of selected materials*—Continued

Item [a]	Material description [b]
15. Paint	Black organic paint (3M 101-C10) per HP 4-143, alkyd enamel with carbon containing proprietary pigment, 0.0025 to 0.0035 in. thick.
16. Mirror hood	Aluminum alloy 6061-0, various pieces 0.016 to 0.040 in. thick, heat treated to T-6 per MIL-H-6088, chromate coated.
17. Tie cord	Nylon, type P unwaxed class I, MIL-T-713.
18. Cable wrap	Aluminized FEP Teflon film, 0.002 or 0.005 in. thick, 1500-Å vacuum-deposited aluminum on inside surface.
19. Cable sleeve	Braided glass yarn per MIL-Y-1140 Form 1, E24, 0.020 in. wide (strand), 0.030-in. woven thickness.
20. Wire	Silver-coated copper (HMS 2-1306-22 type C) FEP/TFE insulation (0.006) +modified (0.010) polyimide, 19 wires in cable soldered to connection with ^{60}Sn ^{40}Pb (per QQ-S-571).
21. Coating	FEP Teflon 1/2-in.-wide semi-elliptical strip on mirror housing.
22. Pad (not shown)	Teflon 0.025 in. per MIL-P-22242 bonded between two parts of item 16 per HP 16-25 (epoxy adhesive, 0.008 in.).
23. Support arm	Aluminum alloy 7075 T-6 machined to 0.090-in. wall thickness (2¼ by 4¾ by ½ in. gross dimensions).
24. Motor support	Aluminum alloy 2024 T-4.
25. Motor	DC stepper motor, commercial.
26. Cover	Aluminum alloy sheet 6061-0, heat treated to T-6, 0.016 in. thick.
27. Housing	Aluminum alloy 2024 T-4, machined (gross dimensions 1.5 by 1.7 by 1.7 in.).
28. Connector bracket	Aluminum alloy 2024-0.
29. Pivot support	Aluminum alloy 2024 T-4, chromate coated (not painted).
30. Bolt	(a) Stainless bolt NAS 1101 C4 H-10. (b) Nut. (c) Washers (AN 960C 416L).
31. Lock-wire	MS 20995C80.
32. Fitting	Aluminum alloy.
33. Screw	Stainless steel.
34. Tube	Glass-fiber-reinforced epoxy.
35. Tube	Aluminum alloy.
36. Cable clamp	
37. Cover	Aluminum alloy, 6061-0, 0.016 in. thick.
38. Screws (not shown)	Stainless steel.
39. Tabs (part of 16)	Aluminum alloy, 6061 alloy, chromate coated, 0.00005 in. thick (not painted).
40. Spring (inside of 26)	Be-Cu strip per QQ-C-533, 0.016 by 0.250 by 0.10 in.
41. Bracket	Aluminum alloy 6061 T-4, 0.040 in. thick.
42. ECU cover	Conetic "AA": proprietary magnetic shielding, 0.014 in. thick.
43. Strap	Aluminum alloy 1100 H-14, 0.062 in. thick.
44. Lens	Optical glass.
45. Support structure	Aluminum alloy 6061 T-6, machined to 0.040 to 0.125 in.; black anodized 0.0004 in. thick, unknown dye.
46. Board	Epoxy glass, 0.031 in. thick, copper clad 0.0018, etched.
47. Spur gear (not shown)	17-4 PH stainless steel (Lubricant—see 64).
48. Board	(a) Aluminum alloy 2024 T-3, 0.016 in. thick. (b) Polyurethane foam core, 8 lb/ft³, 0.125 to 0.138 in. thick. (c) Epoxy glass board, 0.031 in. thick, copper clad 0.0018 in. etched.
49. Support	Aluminum alloy 6061 T-6, wall thickness 0.062 to 0.125 in. (approx ½ by 1 by 4½ in. gross dimensions)+0.00005-in. chromate coating.
50. Lock washer	Stainless steel.
51. Sleeving	TFE Teflon, extruded 0.009- to 0.040-in. wall thickness.
52. Solder	Tin 60, lead 40, per QQ-S-571.
53. Wire	Copper, silver clad 24 AWG; polyimide insulation 0.004 in. per HMS2-1293, type I.
54. Component bonding	(a) Adhesive, epoxy-polyamide+cabosil. (b) Adhesive, epoxy-polyamide+cabosil+Al_2O_3 per HP 16-99, type II.

TABLE C-1.—*Location and identification of selected materials*—Concluded

Item [a]	Material description [b]
55. Conformal coating	Epoxy-polyamide, 0.002 to 0.003 in., per HP 16-99, type I.
56. Conductive adhesive (not shown)	Silver-filled epoxy per HP 16-95, type II.
57. Clear filter	Dense flint glass, vacuum deposited Inconel back side.
58. Green filter	(a) Top, Schott, OG4. (b) Bottom, Chance, OGR3.
59. Blue filter	(a) Top, Schott, GG15. (b) Bottom, Schott, BG1.
60. Red filter	Corning 3-76.
61. Motor support	Aluminum alloy 2024 T-4, 0.075 to 0.150 in.
62. Potentiometer support	Aluminum alloy 2024 T-3, 0.090 in.
63. Gears	17-4 PH stainless steel per AMS 5643.
64. Lubricant	Proprietary, inorganic bonded MoS_2+additives, 0.0002 to 0.0005 in. thick.

[a] Item numbers are keyed to figs. C-1 through C-5. Circled numbers indicate "visible" materials or parts.

[b] Some measurements in this column are given in English rather than in metric units because they are taken directly from the engineering drawings.

NOTE: Filter thickness is 1.2 to 3.0 mm. Only No. 57 (clear) had any appreciable view of space; all others were shielded by hood and mechanisms.

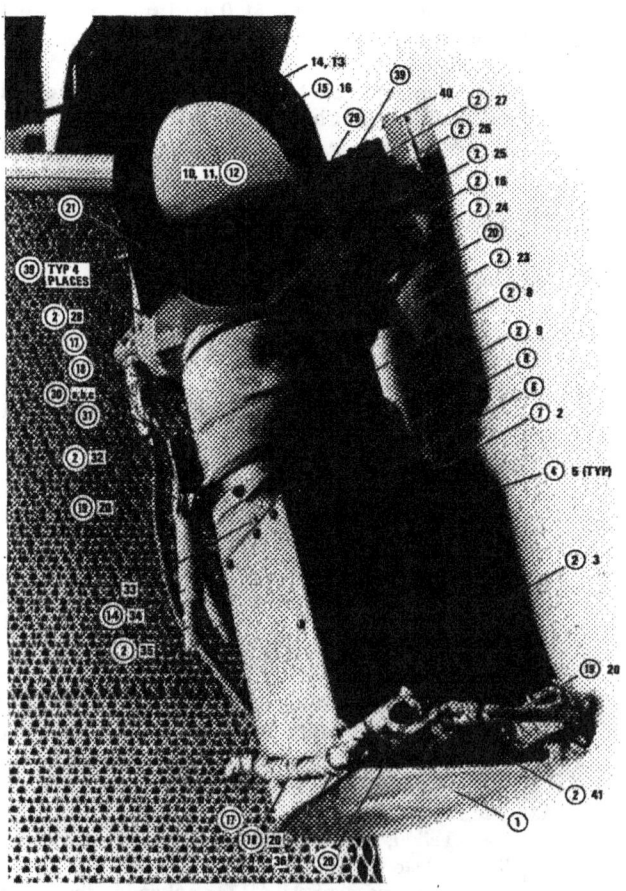

FIGURE C-1.—Returned Surveyor 3 television camera. The circled number is visible material.

TABLE C–2.—*Selected electronic components*

Generic type	Generic class	Manufacturer	Description	Camera quantity
Capacitor	KG	KEMET	Solid tantalum	76
Capacitor	118P	SPRAGUE	Metallized paper	26
Capacitor	HRDM	ELMOT	Silvered mica	5
Capacitor	CYFR	CORNING	Glass dielectric and case	9
Capacitor	VK	VIT	Ceramic dielectric	25
Resistor	CB-1/4	A/B	Carbon composition	379
Resistor	EB	A/B	Carbon composition	44
Resistor	GB	A/B	Carbon composition	2
Resistor	HB	A/B	Carbon composition	1
Resistor	CG-1/8	TI	Carbon film glass case, hermetic seal	230
Resistor	CG-1/4	TI	Carbon film glass case, hermetic seal	15
Resistor	NF85	MEPCO	Carbon film glass case, hermetic seal	16
Resistor	CAH	A/B	Metal film ceramic case, hermetic seal	33
Resistor	1120S	SAGE	Wirewound	2
Resistor	60001 (or R9900)	DAVEN (or ULTRNX)	Wirewound	2
Resistance temperature sensor	3XXX (or 1188)	RDF (or RSMT)	Platinum	2
Diode	1N3070	FA	Silicon, glass enclosure	261
Diode	1N969B	HU-HO-PS	Silicon, glass enclosure	4
Diode, zener	1N1313	HU-PS	Silicon	1
Diode, zener	1N1316	HU-PS	Silicon	1
Diode, zener	1N1317	HU-PS	Silicon	4
Diode, zener	1N827	MO-TR	Silicon, glass enclosure	1
Diode, zener	1N938B	MO-DI	Silicon, glass enclosure	1
Diode, high voltage	FA3075 SELECT	FA (FD200SERIES)	Silicon, glass enclosure	2
Diode	1N3206	MicroSem	Silicon, plastic package	1
Diode (RD750)	1N3730 SELECT	RA	Silicon, glass enclosure	37
Diode, zener	1N754A SELECT	CD	Silicon, glass enclosure	17
Diode	1N1317	PS/TRU	Silicon	1
Transistor	2N/18A	FA-PSI	Metal can, silicon	20
Transistor	2N722	FA	Metal can, silicon	31
Transistor	2N871	FA	Metal can, silicon	36
Transistor	2N859	SPR	Metal can, silicon	49
Transistor	2N930	TI	Metal can, silicon	45
Transistor	2N2586	TI	Metal can, silicon	5
Transistor	2N871	FA	Metal can, silicon	10
Transistor	2N2891	FA	Metal can, silicon	2
Transistor	2N2150	TI	Metal can, silicon	2
Transistor	2N2151	TI	Metal can, silicon	1
Transistor	2N1132 MATCHED	FA	Metal can, silicon	3
Transistor	2N2192A	FA	Metal can, silicon	5
Transistor	2N2193A	FA	Metal can, silicon	2
Transistor	2N1936	TI	Metal can, silicon	1
Transistor	2N2193 SELECT	FA	Metal can, silicon	1
Transistor	2N871 SELECT	FA	Metal can, silicon	6
Transistor	2N3891	FA-MH		18
Transistor	2N2880	MH		5
Transistor	2N2707A	FA-MO-TI		2
Diode, zener	FSP358-1	FAS		14
Diode (SCR)	2N1930 SELECT	TI	Silicon, metal can	4
Diode (SCR)	2N2323	GE-TR	Silicon	1

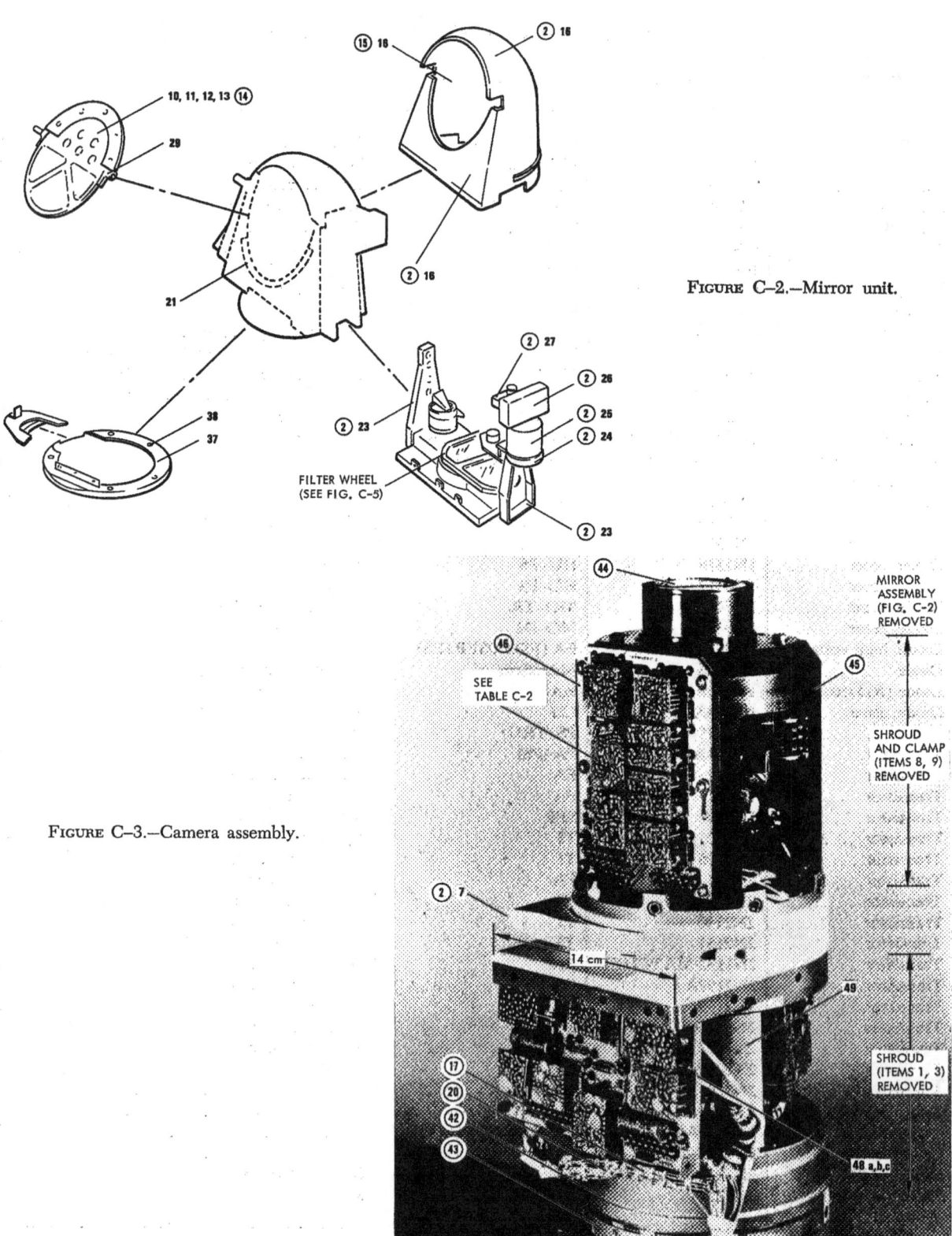

Figure C-2.—Mirror unit.

Figure C-3.—Camera assembly.

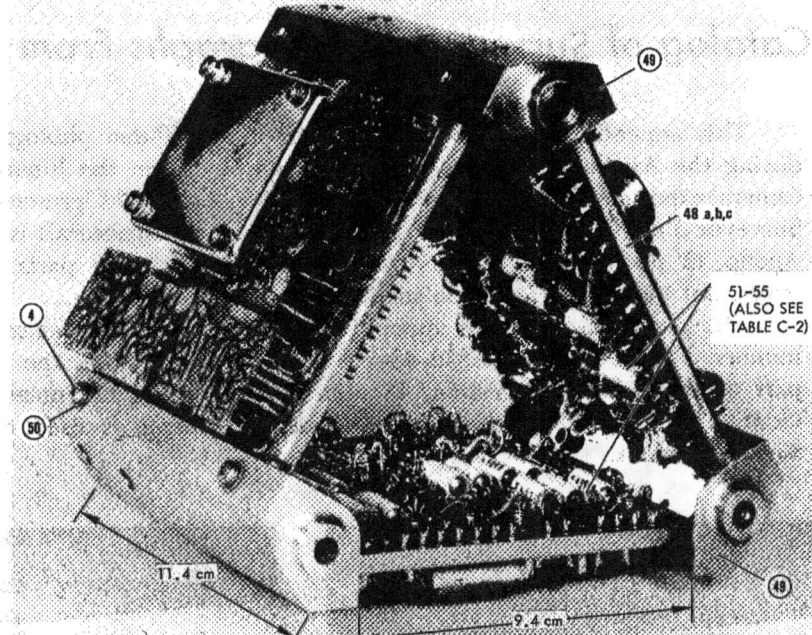

FIGURE C-4.—Multichassis assembly.

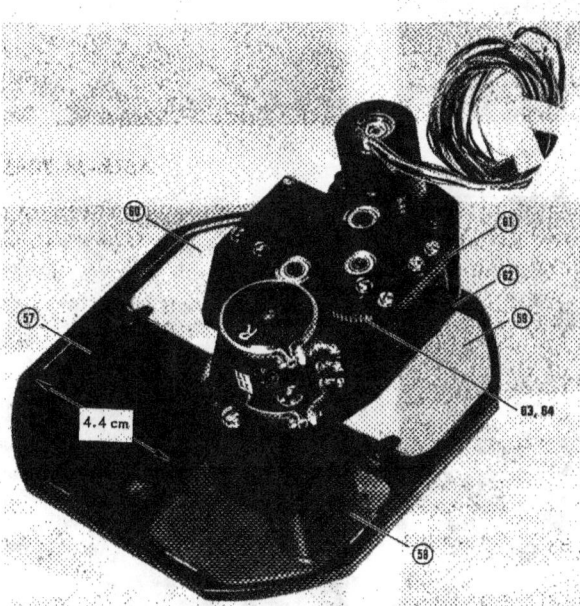

FIGURE C-5.—Filter-wheel assembly. The filters are identified by relative position on the returned Surveyor 3 camera. The clear filter (No. 57) was over the lens and thus was exposed to space.

Appendix D

Catalog of Surveyor 3 Photographs From Apollo 12

This appendix consists of proof prints of the 70-mm photography exposed during the Apollo 12 mission. (The photography from the lunar multispectral camera experiment is not included.) The 56 photographs presented pertain to Surveyor 3 activities or are photographs in which the spacecraft is visible. Other Apollo 12 photographs of Surveyor 3 exist, primarily as parts of panoramic series, but the spacecraft is too small to be recognized easily.

The photography in this appendix has been sorted by magazine and frame number. For example, in AS12–48–7084, the AS12 identifies the photograph as part of the Apollo 12 mission, 48 identifies the magazine number, and 7084 identifies the frame number. This numbering scheme is used throughout all Apollo missions.

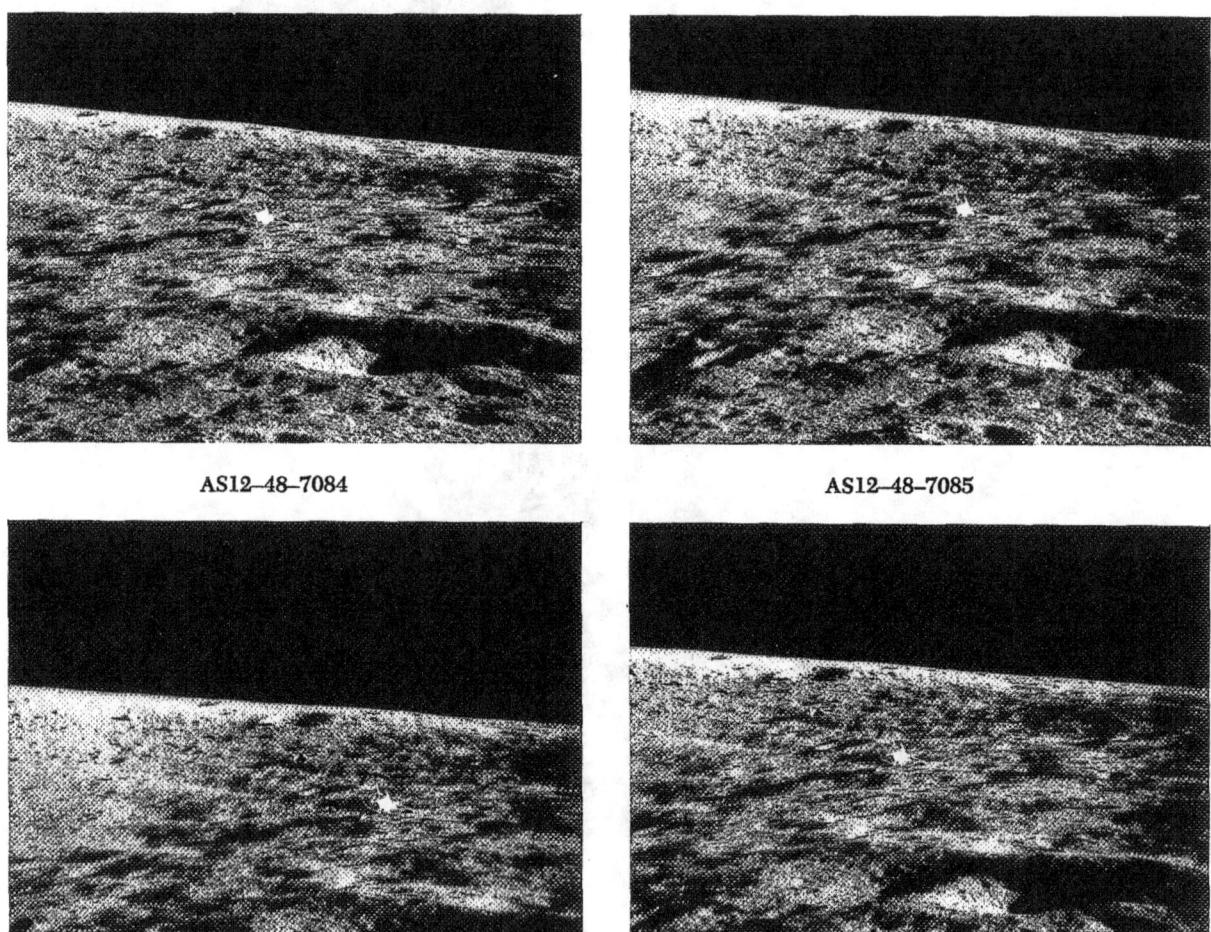

AS12–48–7084

AS12–48–7085

AS12–48–7086

AS12–48–7087

APPENDIX D—CATALOG OF SURVEYOR 3 PHOTOGRAPHS 285

AS12-48-7088

AS12-48-7089

AS12-48-7090

AS12-48-7091

AS12-48-7092

AS12-48-7093

AS12-48-7094

AS12-48-7095

AS12-48-7096

AS12-48-7097

AS12-48-7098

AS12-48-7099

APPENDIX D—CATALOG OF SURVEYOR 3 PHOTOGRAPHS 287

AS12-48-7100

AS12-48-7101

AS12-48-7102

AS12-48-7103

AS12-48-7104

AS12-48-7105

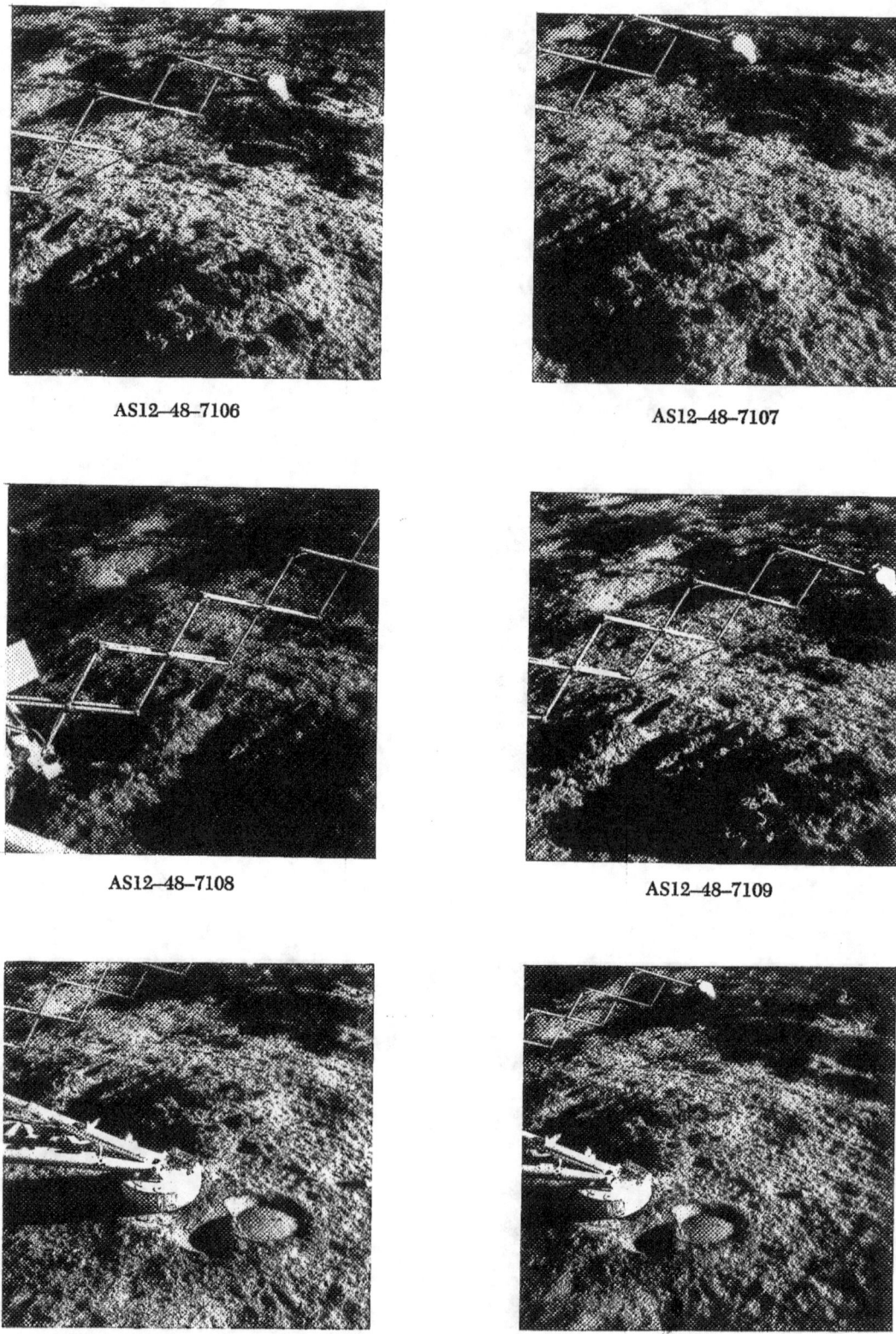

AS12-48-7106

AS12-48-7107

AS12-48-7108

AS12-48-7109

AS12-48-7110

AS12-48-7111

AS12-48-7112

AS12-48-7113

AS12-48-7114

AS12-48-7115

AS12-48-7116

AS12-48-7117

AS12-48-7118

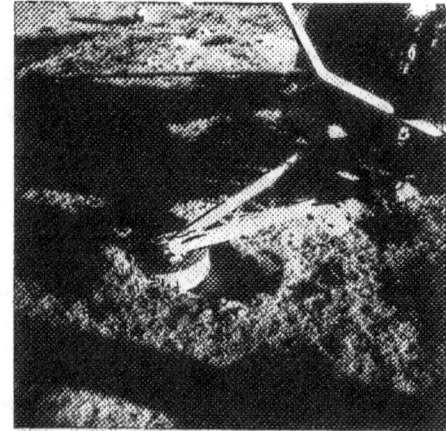
AS12-48-7119

AS12-48-7120

AS12-48-7121

AS12-48-7122

AS12-48-7123

APPENDIX D—CATALOG OF SURVEYOR 3 PHOTOGRAPHS

AS12-48-7124

AS12-48-7125

AS12-48-7126

AS12-48-7127

AS12-48-7128

AS12-48-7129

AS12-48-7130

AS12-48-7131

AS12-48-7132

AS12-48-7133

AS12-48-7134

AS12-48-7135

AS12-48-7136

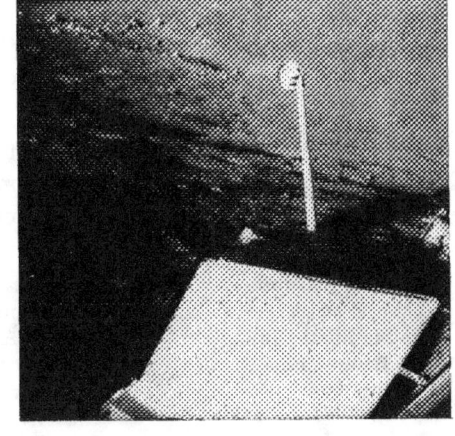

AS12-48-7137

AS12-48-7138

AS12-48-7139

Appendix E
Index of Contributing Authors

The Surveyor 3 components returned to Earth by the Apollo 12 astronauts were studied by 40 teams of investigators. The results of some of these studies are presented in this document. Contributing authors and their affiliations are presented below.

D. L. ANDERSON	Ames Research Center, Moffett Field, Calif.
D. J. BARBER	Essex University, Colchester, United Kingdom
P. M. BLAIR, JR.	Hughes Aircraft Co., Culver City, Calif.
D. BROWNLEE	University of Washington, Seattle
W. BUCHER	University of Washington, Seattle
F. BÜHLER	University of Bern, Switzerland
A. L. BURLINGAME	University of California, Berkeley
E. A. BUVINGER	Wright-Patterson Air Force Base, Ohio
M. H. CARR	U.S. Geological Survey, Menlo Park, Calif.
W. F. CARROLL	Jet Propulsion Laboratory, Pasadena, Calif.
A. A. CHODOS	California Institute of Technology, Pasadena
G. M. COMSTOCK	General Electric Co., Schenectady, N.Y.
B. G. COUR-PALAIS	Manned Spacecraft Center, Houston, Tex.
R. COWSIK	TATA Institute of Fundamental Research, Bombay, India
G. CROZAZ	National Belge de la Recherche Scientifique, France
B. E. CUNNINGHAM	Ames Research Center, Moffett Field, Calif.
R. G. DAHMS	Ames Research Center, Moffett Field, Calif.
R. DAVIS	Jet Propulsion Laboratory, Pasadena, Calif.
J. R. DEVANEY	Jet Propulsion Laboratory, Pasadena, Calif.
W. R. DUFF	U.S. Steel Corp. Research Center, Monroeville, Pa.
P. EBERHARDT	University of Bern, Switzerland
T. E. ECONOMOU	University of Chicago, Illinois
W. L. ELLIS	Manned Spacecraft Center, Houston, Tex.
K. C. EVANS	Jet Propulsion Laboratory, Pasadena, Calif.
M. S. FAVERO	U.S. Public Health Service, Phoenix, Ariz.
H. FECHTIG	Max-Planck-Institut fur Kernphysik, Heidelberg, Germany
W. D. FELIX	Battelle Northwest Laboratories, Richland, Wash.
E. L. FIREMAN	Smithsonian Astrophysical Observatory, Cambridge, Mass.
R. M. FISHER	U.S. Steel Corp. Research Center, Monroeville, Pa.
R. E. FLAHERTY	Manned Spacecraft Center, Houston, Tex.
R. L. FLEISCHER	General Electric Co., Schenectady, N.Y.
E. L. GAFFORD	Battelle Northwest Laboratories, Richland, Wash.
J. GEISS	University of Bern, Switzerland
R. B. GILLETTE	The Boeing Co., Seattle, Wash.
M. GOLDFINE	Jet Propulsion Laboratory, Pasadena, Calif.
R. H. GREEN	Jet Propulsion Laboratory, Pasadena, Calif.
F. C. GROSS	Goddard Space Flight Center, Greenbelt, Md.
D. S. HALLGREN	The Dudley Observatory, Albany, N.Y.
H. R. HART, JR.	General Electric Co., Schenectady, N.Y.
E. I. HAWTHORNE	Hughes Aircraft Co., Culver City, Calif.
C. L. HEMENWAY	The Dudley Observatory, Albany, N.Y.
R. W. HIGH	Manned Spacecraft Center, Houston, Tex.
P. HODGE	University of Washington, Seattle

APPENDIX E—INDEX OF CONTRIBUTING AUTHORS

H. Holt	U.S. Geological Survey, Flagstaff, Ariz.
I. D. Hutcheon	University of California, Berkeley
S. Jacobs	Manned Spacecraft Center, Houston, Tex.
L. D. Jaffe	Jet Propulsion Laboratory, Pasadena, Calif.
D. J. Kessler	Manned Spacecraft Center, Houston, Tex.
M. D. Knittel	Jet Propulsion Laboratory, Pasadena, Calif.
A. T. Laudate	The Dudley Observatory, Albany, N.Y.
L. Leger	Manned Spacecraft Center, Houston, Tex.
T.-D. Lu	California Institute of Technology, Pasadena
F. F. Marmo	GCA Corp., Bedford, Mass.
D. S. McKay	Manned Spacecraft Center, Houston, Tex.
A. Mehl	Max-Planck-Institut fur Kernphysik, Heidelberg, Germany
B. Milwitzky	NASA Headquarters, Washington, D.C.
F. J. Mitchell	Texas A&M University, College Station
K. Moll	U.S. Geological Survey, Flagstaff, Ariz.
R. G. Morgan	Ames Research Center, Moffett Field, Calif.
G. Neukum	Max-Planck-Institut fur Kernphysik, Heidelberg, Germany
N. L. Nickle	Jet Propulsion Laboratory, Pasadena, Calif.
J. J. Park	Goddard Space Flight Center, Greenbelt, Md.
R. W. Perkins	Battelle Northwest Laboratories, Richland, Wash.
P. B. Price	University of California, Berkeley
S. J. Proudfoot	U.S. Geological Survey, Menlo Park, Calif.
S. V. Radcliffe	Case Western Reserve University, Cleveland, Ohio
W. D. Radigan	The Dudley Observatory, Albany, N.Y.
R. S. Rajan	University of California, Berkeley
L. A. Rancitelli	Battelle Northwest Laboratories, Richland, Wash.
J. Rennilson	California Institute of Technology, Pasadena
D. M. Robertson	Battelle Northwest Laboratories, Richland, Wash.
F. G. Satkiewicz	GCA Corp., Bedford, Mass.
E. Schneider	Max-Planck-Institut fur Kernphysik, Heidelberg, Germany
R. P. Schwarz	The Dudley Observatory, Albany, N.Y.
J. Schwarzmüller	University of Bern, Switzerland
R. F. Scott	California Institute of Technology, Pasadena
B. R. Simoneit	University of California, Berkeley
R. S. Strebin, Jr.	Battelle Northwest Laboratories, Richland, Wash.
H. Tenny	Battelle Northwest Laboratories, Richland, Wash.
L. E. Thomas	U.S. Steel Corp. Research Center, Monroeville, Pa.
A. L. Turkevich	University of Chicago, Illinois
R. M. Walker	Washington University, St. Louis, Mo.
N. A. Wogman	Battelle Northwest Laboratories, Richland, Wash.
L. Zernow	Shock Hydrodynamics Inc., Sherman Oaks, Calif.
H. A. Zook	Manned Spacecraft Center, Houston, Tex.
K. A. Zuckerman	California Institute of Technology, Pasadena

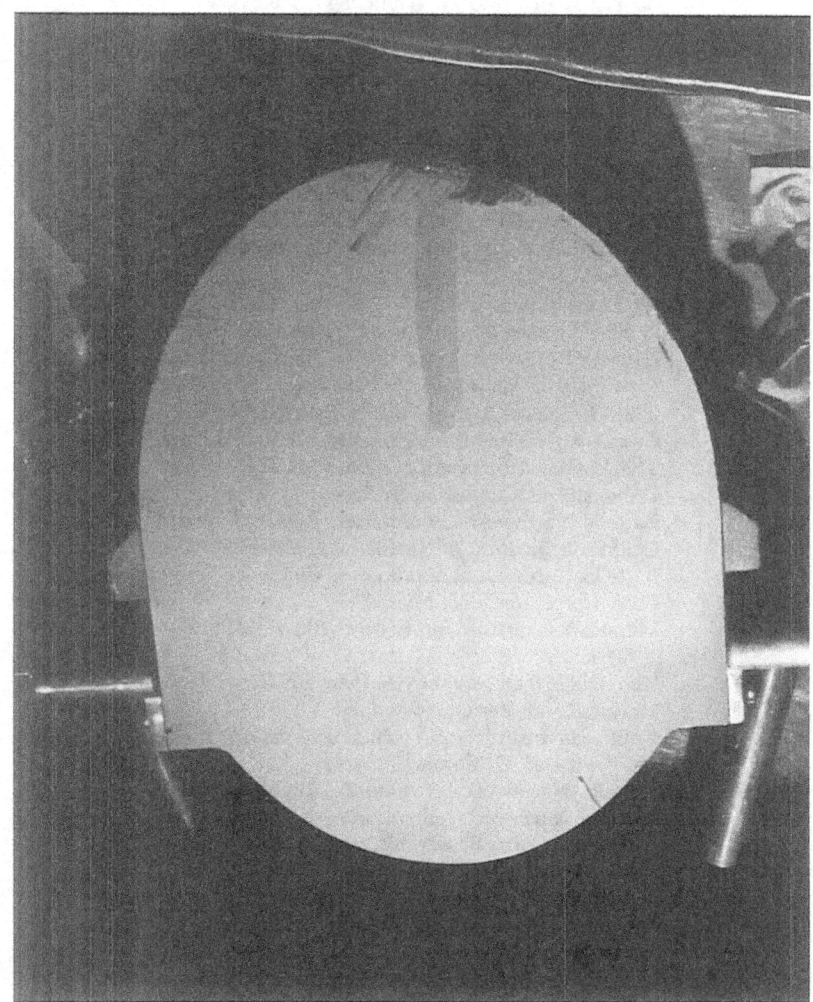

FIGURE 2.—The mirror illuminated by white light shows a spectral band running between the trunnions, and a "shadow line" running diagonally below the spectral band. The shadow line was produced by the masking effect of the front opening of the camera protecting the lower part of the mirror from impinging lunar soil particles.

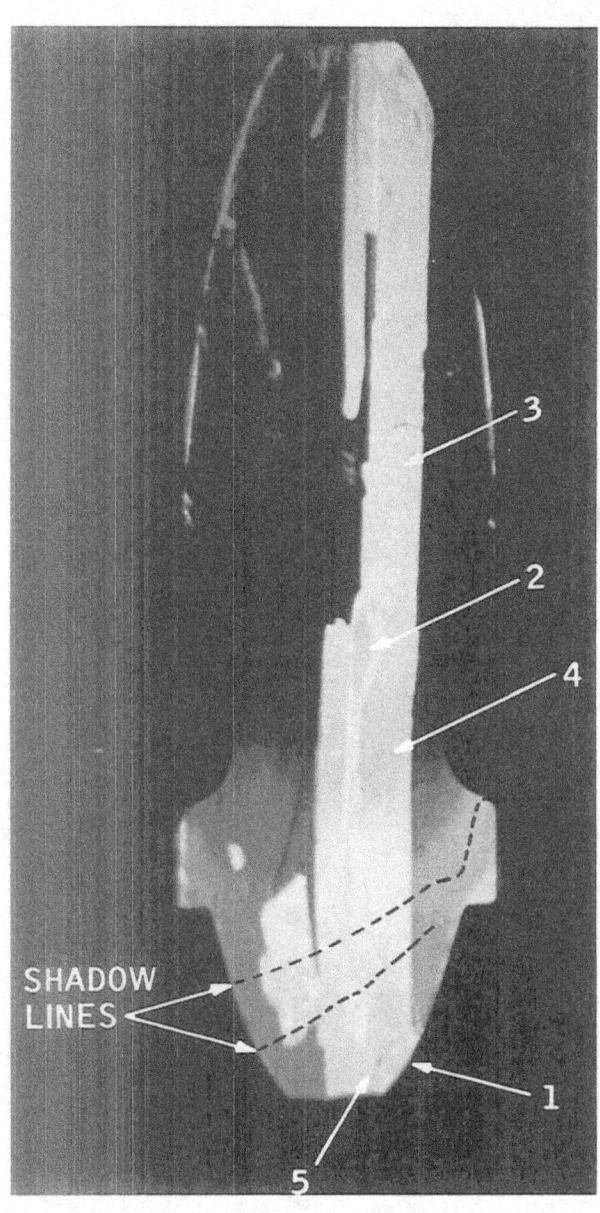
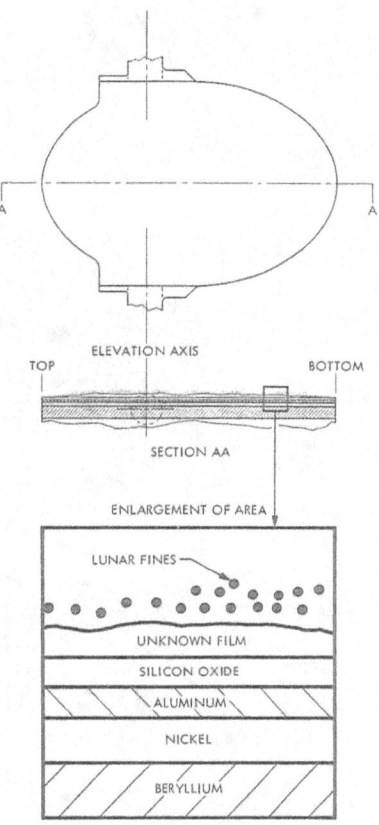

FIGURE 19.—The photograph at left shows the colors observed by the white light ellipsometry experiment. The numbered areas correspond to the curves in figure 20. Two faint boundaries of lunar fines are marked and discussed in the test. Sketch above is assumed cross section of the mirror. Not shown to scale.

FIGURE 11.—Comparison of right-hand side of Surveyor 3 surface sampler with left-hand side of laboratory surface sampler. Ultraviolet stimulation, visible recorded on film.